Bayesian Econometrics

Bayesian Econometrics

Gary Koop

Department of Economics
University of Glasgow

WILEY

Telephone (+44) 1243 779777

Email (for orders and customer service enquiries): cs-books@wiley.co.uk
Visit our Home Page on www.wileyeurope.com or www.wiley.com

Reprinted with corrections February 2006

Other Wiley Editorial Offices

John Wiley & Sons Inc., 111 River Street, Hoboken, NJ 07030, USA

Jossey-Bass, 989 Market Street, San Francisco, CA 94103-1741, USA

Wiley-VCH Verlag GmbH, Boschstr. 12, D-69469 Weinheim, Germany

John Wiley & Sons Australia Ltd, 33 Park Road, Milton, Queensland 4064, Australia

John Wiley & Sons (Asia) Pte Ltd, 2 Clementi Loop #02-01, Jin Xing Distripark, Singapore 129809

John Wiley & Sons Canada Ltd, 22 Worcester Road, Etobicoke, Ontario, Canada M9W 1L1

Wiley also publishes its books in a variety of electronic formats. Some content that appears in print may not be available in electronic books.

British Library Cataloguing in Publication Data

A catalogue record for this book is available from the British Library

ISBN 0-470-84567-8

Typeset in 10/12pt Times by Laserwords Private Limited, Chennai, India

This book is printed on acid-free paper responsibly manufactured from sustainable forestry in which at least two trees are planted for each one used for paper production.

To Lise

Contents

Preface

Bayesian methods are increasingly becoming attractive to researchers in many fields. Econometrics, however, is a field in which Bayesian methods have had relatively less influence. A key reason for this absence is the lack of a suitable advanced undergraduate or graduate level textbook. Existing Bayesian books are either out-dated, and hence do not cover the computational advances that have revolutionized the field of Bayesian econometrics since the late 1980s, or do not provide the broad coverage necessary for the student interested in empirical work applying Bayesian methods. For instance, Arnold Zellner's seminal Bayesian econometrics book (Zellner, 1971) was published in 1971. Dale Poirier's influential book (Poirier, 1995) focuses on the methodology and statistical theory underlying Bayesian and frequentist methods, but does not discuss models used by applied economists beyond regression. Other important Bayesian books, such as Bauwens, Lubrano and Richard (1999), deal only with particular areas of econometrics (e.g. time series models). In writing this book, my aim has been to fill the gap in the existing set of Bayesian textbooks, and create a Bayesian counterpart to the many popular non-Bayesian econometric textbooks now available (e.g. Greene, 1995). That is, my aim has been to write a book that covers a wide range of models and prepares the student to undertake applied work using Bayesian methods.

This book is intended to be accessible to students with no prior training in econometrics, and only a single course in mathematics (e.g. basic calculus). Students will find a previous undergraduate course in probability and statistics useful; however Appendix B offers a brief introduction to these topics for those without the prerequisite background. Throughout the book, I have tried to keep the level of mathematical sophistication reasonably low. In contrast to other Bayesian and comparable frequentist textbooks, I have included more computer-related material. Modern Bayesian econometrics relies heavily on the computer, and developing some basic programming skills is essential for the applied Bayesian. The required level of computer programming skills is not that high, but I expect that this aspect of Bayesian econometrics might be most unfamiliar to the student brought up in the world of spreadsheets and click-and-press computer packages. Accordingly, in addition to discussing computation in detail in the book itself, the website associated with the book contains MATLAB programs for performing Bayesian analysis in a wide variety of models. In general, the focus of the book is on application rather than theory. Hence, I expect that the applied economist interested in using Bayesian methods will find it more useful than the theoretical econometrician.

I would like to thank the numerous people (some anonymous) who gave me helpful comments at various stages in the writing of this book, including: Luc Bauwens, Jeff Dorfman, David Edgerton, John Geweke, Bill Griffiths, Frank Kleibergen, Tony Lancaster, Jim LeSage, Michel Lubrano, Brendan McCabe, Bill McCausland, Richard Paap, Rodney Strachan, and Arnold Zellner. In addition, I would like to thank Steve Hardman for his expert editorial advice. All I know about Bayesian econometrics comes through my work with a series of exceptional co-authors: Carmen Fernandez, Henk Hoek, Eduardo Ley, Kai Li, Jacek Osiewalski, Dale Poirier, Simon Potter, Mark Steel, Justin Tobias, and Herman van Dijk. Of these, I would like to thank Mark Steel, in particular, for patiently responding to my numerous questions about Bayesian methodology and requests for citations of relevant papers. Finally, I wish to express my sincere gratitude to Dale Poirier, for his constant support throughout my professional life, from teacher and PhD supervisor, to valued co-author and friend.

1
An Overview of Bayesian Econometrics

1.1 BAYESIAN THEORY

Bayesian econometrics is based on a few simple rules of probability. This is one of the chief advantages of the Bayesian approach. All of the things that an econometrician would wish to do, such as estimate the parameters of a model, compare different models or obtain predictions from a model, involve the same rules of probability. Bayesian methods are, thus, universal and can be used any time a researcher is interested in using data to learn about a phenomenon.

To motivate the simplicity of the Bayesian approach, let us consider two random variables, A and B.[1] The rules of probability imply:

$$p(A, B) = p(A|B)p(B)$$

where $p(A, B)$ is the *joint probability*[2] of A and B occurring, $p(A|B)$ is the probability of A occurring conditional on B having occurred (i.e. the *conditional probability* of A given B), and $p(B)$ is the *marginal probability* of B. Alternatively, we can reverse the roles of A and B and find an expression for the joint probability of A and B:

$$p(A, B) = p(B|A)p(A).$$

Equating these two expressions for $p(A, B)$ and rearranging provides us with *Bayes' rule*, which lies at the heart of Bayesian econometrics:

$$p(B|A) = \frac{p(A|B)p(B)}{p(A)} \tag{1.1}$$

[1]This chapter assumes the reader knows the basic rules of probability. Appendix B provides a brief introduction to probability for the reader who does not have such a background or would like a reminder of this material.

[2]We are being slightly sloppy with terminology here and in the following material in that we should always say 'probability density' if the random variable is continuous and 'probability function' if the random variable is discrete (see Appendix B). For simplicity, we simply drop the word 'density' or 'function'.

Econometrics is concerned with using data to learn about something the researcher is interested in. Just what the 'something' is depends upon the context. However, in economics we typically work with models which depend upon parameters. For the reader with some previous training in econometrics, it might be useful to have in mind the regression model. In this model interest often centers on the coefficients in the regression, and the researcher is interested in estimating these coefficients. In this case, the coefficients are the parameters under study. Let y be a vector or matrix of data and θ be a vector or matrix which contains the parameters for a model which seeks to explain y.[3] We are interested in learning about θ based on the data, y. Bayesian econometrics uses Bayes' rule to do so. In other words, the Bayesian would replace B by θ and A by y in (1.1) to obtain:

$$p(\theta|y) = \frac{p(y|\theta)p(\theta)}{p(y)} \tag{1.2}$$

Bayesians treat $p(\theta|y)$ as being of fundamental interest. That is, it directly addresses the question "Given the data, what do we know about θ?". The treatment of θ as a random variable is controversial among some econometricians. The chief competitor to Bayesian econometrics, often called *frequentist econometrics*, says that θ is not a random variable. However, Bayesian econometrics is based on a subjective view of probability, which argues that our uncertainty about anything unknown can be expressed using the rules of probability. In this book, we will not discuss such methodological issues (see Poirier (1995) for more detail). Rather, we will take it as given that econometrics involves learning about something unknown (e.g. coefficients in a regression) given something known (e.g. data) and the conditional probability of the unknown given the known is the best way of summarizing what we have learned.

Having established that $p(\theta|y)$ is of fundamental interest for the econometrician interested in using data to learn about parameters in a model, let us now return to (1.2). Insofar as we are only interested in learning about θ, we can ignore the term $p(y)$, since it does not involve θ. We can then write:

$$p(\theta|y) \propto p(y|\theta)p(\theta) \tag{1.3}$$

The term $p(\theta|y)$ is referred to as the *posterior density*, the p.d.f. for the data given the parameters of the model, $p(y|\theta)$, as the *likelihood function* and $p(\theta)$ as the *prior density*. You often hear this relationship referred to as "posterior is proportional to likelihood times prior". At this stage, this may seem a little abstract, and the manner in which priors and likelihoods are developed to allow for the calculation of the posterior may be unclear. Things should become clearer to you in the following chapters, where we will develop likelihood functions and priors in specific contexts. Here we provide only a brief general discussion of what these are.

[3]Appendix A contains a brief introduction to matrix algebra.

The prior, $p(\theta)$, does not depend upon the data. Accordingly, it contains any non-data information available about θ. In other words, it summarizes what you know about θ prior to seeing the data. As an example, suppose θ is a parameter which reflects returns to scale in a production process. In many cases, it is reasonable to assume that returns to scale are roughly constant. Thus, before you look at the data, you have prior information about θ, in that you would expect it to be approximately one. Prior information is a controversial aspect of Bayesian methods. In this book, we will discuss both informative and noninformative priors for various models. In addition, in later chapters, we will discuss empirical Bayes methods. These use data-based information to choose the prior and, hence, violate a basic premise of Bayesian methods. Nevertheless, empirical Bayes methods are becoming increasingly popular for the researcher who is interested in practical, objective, tools that seem to work well in practice.[4]

The likelihood function, $p(y|\theta)$, is the density of the data conditional on the parameters of the model. It is often referred to as the data generating process. For instance, in the linear regression model (which will be discussed in the next chapter), it is common to assume that the errors have a Normal distribution. This implies that $p(y|\theta)$ is a Normal density, which depends upon parameters (i.e. the regression coefficients and the error variance).

The posterior, $p(\theta|y)$, is the density which is of fundamental interest. It summarizes all we know about θ after (i.e. posterior to) seeing the data. Equation (1.3) can be thought of as an updating rule, where the data allows us to update our prior views about θ. The result is the posterior which combines both data and non-data information.

In addition to learning about parameters of a model, an econometrician might be interested in comparing different models. A model is formally defined by a likelihood function and a prior. Suppose we have m different models, M_i for $i = 1, \ldots, m$, which all seek to explain y. M_i depends upon parameters θ^i. In cases where many models are being entertained, it is important to be explicit about which model is under consideration. Hence, the posterior for the parameters calculated using M_i is written as

$$p(\theta^i|y, M_i) = \frac{p(y|\theta^i, M_i)p(\theta^i|M_i)}{p(y|M_i)} \tag{1.4}$$

and the notation makes clear that we now have a posterior, likelihood, and prior for each model.

The logic of Bayesian econometrics suggests that we use Bayes' rule to derive a probability statement about what we do not know (i.e. whether a model is a correct one or not) conditional on what we do know (i.e. the data). This means the *posterior model probability* can be used to assess the degree of support for

[4]Carlin and Louis (2000) is a good reference for the reader interested in developing a deeper understanding of empirical Bayes methods.

M_i. Using (1.1) with $B = M_i$ and $A = y$, we obtain

$$p(M_i|y) = \frac{p(y|M_i)p(M_i)}{p(y)} \tag{1.5}$$

Of the terms in (1.5), $p(M_i)$ is referred to as the *prior model probability*. Since it does not involve the data, it measures how likely we believe M_i to be the correct one before seeing the data. $p(y|M_i)$ is called the *marginal likelihood*, and is calculated using (1.4) and a few simple manipulations. In particular, if we integrate both sides of (1.4) with respect to θ^i, use the fact that $\int p(\theta^i|y, M_i)d\theta^i = 1$ (since probability density functions integrate to one), and rearrange, we obtain:

$$p(y|M_i) = \int p(y|\theta^i, M_i)p(\theta^i|M_i)d\theta^i \tag{1.6}$$

Note that the marginal likelihood depends only upon the prior and the likelihood. In subsequent chapters, we discuss how (1.6) can be calculated in practice.

Since the denominator in (1.5) is often hard to calculate directly, it is common to compare two models, i and j, using the *posterior odds ratio*, which is simply the ratio of their posterior model probabilities:

$$PO_{ij} = \frac{p(M_i|y)}{p(M_j|y)} = \frac{p(y|M_i)p(M_i)}{p(y|M_j)p(M_j)} \tag{1.7}$$

Note that, since $p(y)$ is common to both models, it cancels out when we take the ratio. As we will discuss in subsequent chapters, there are special techniques in many cases for calculating the posterior odds ratio directly. If we calculate the posterior odds ratio comparing every pair of models, and we assume that our set of models is exhaustive (in that $p(M_1|y) + p(M_2|y) + \cdots + p(M_m|y) = 1$), then we can use posterior odds ratios to calculate the posterior model probabilities given in (1.5). For instance, if we have $m = 2$ models then we can use the two equations

$$p(M_1|y) + p(M_2|y) = 1$$

and

$$PO_{12} = \frac{p(M_1|y)}{p(M_2|y)}$$

to work out

$$p(M_1|y) = \frac{PO_{12}}{1 + PO_{12}}$$

and

$$p(M_2|y) = 1 - p(M_1|y)$$

Thus, knowledge of the posterior odds ratio allows us to figure out the posterior model probabilities.

To introduce some more jargon, econometricians may be interested in model comparison when equal prior weight is attached to each model. That is, $p(M_i) = p(M_j)$ or, equivalently, the *prior odds ratio* which is $\frac{p(M_i)}{p(M_j)}$ is set to one. In this case, the posterior odds ratio becomes simply the ratio of marginal likelihoods, and is given a special name, the *Bayes Factor,* defined as:

$$BF_{ij} = \frac{p(y|M_i)}{p(y|M_j)} \tag{1.8}$$

Finally, econometricians are often interested in prediction. That is, given the observed data, y, the econometrician may be interested in predicting some future unobserved data y^*. Our Bayesian reasoning says that we should summarize our uncertainty about what we do not know (i.e. y^*) through a conditional probability statement. That is, prediction should be based on the *predictive density* $p(y^*|y)$ (or, if we have many models, we would want to make explicit the dependence of a prediction on a particular model, and write $p(y^*|y, M_i)$). Using a few simple rules of probability, we can write $p(y^*|y)$ in a convenient form. In particular, since a marginal density can be obtained from a joint density through integration (see Appendix B), we can write:

$$p(y^*|y) = \int p(y^*, \theta|y)d\theta$$

However, the term inside the integral can be rewritten using another simple rule of probability:

$$p(y^*|y) = \int p(y^*|y, \theta)p(\theta|y)d\theta \tag{1.9}$$

As we shall see in future chapters, the form for the predictive in (1.9) is quite convenient, since it involves the posterior.

On one level, this book could end right here. These few pages have outlined all the basic theoretical concepts required for the Bayesian to learn about parameters, compare models and predict. We stress what an enormous advantage this is. Once you accept that unknown things (i.e. θ, M_i and y^*) are random variables, the rest of Bayesian approach is non-controversial. It simply uses the rules of probability, which are mathematically true, to carry out statistical inference. A benefit of this is that, if you keep these simple rules in mind, it is hard to lose sight of the big picture. When facing a new model (or reading a new chapter in the book), just remember that Bayesian econometrics requires selection of a prior and a likelihood. These can then be used to form the posterior, (1.3), which forms the basis for all inference about unknown parameters in a model. If you have many models and are interested in comparing them, you can use posterior model probabilities (1.5), posterior odds ratios (1.7), or Bayes Factors (1.8). To obtain any of these, we usually have to calculate the marginal likelihood (1.6). Prediction is done through the predictive density, $p(y^*|y)$, which is usually calculated using (1.9). These few equations can be used to carry out statistical inference in *any* application you may wish to consider.

The rest of this book can be thought of as simply examples of how (1.5)–(1.9) can be used to carry out Bayesian inference for various models which have been commonly-used by others. Nevertheless, we stress that Bayesian inference can be done with any model using the techniques outlined above and, when confronting an empirical problem, you should not necessarily feel constrained to work with one of the off-the-shelf models described in this book.

1.2 BAYESIAN COMPUTATION

The theoretical and conceptual elegance of the Bayesian approach has made it an attractive one for many decades. However, until recently, Bayesians have been in a distinct minority in the field of econometrics, which has been dominated by the frequentist approach. There are two main reasons for this: prior information and computation. With regards to the former, many researchers object to the use of 'subjective' prior information in the supposedly 'objective' science of economics. There is a long, at times philosophical, debate about the role of prior information in statistical science, and the present book is not the place to attempt to summarize this debate. The interested reader is referred to Poirier (1995), which provides a deeper discussion of this issue and includes an extensive bibliography. Briefly, most Bayesians would argue that the entire model building process can involve an enormous amount of non-data information (e.g. econometricians must decide which models to work with, which variables to include, what criteria to use to compare models or estimate parameters, which empirical results to report, etc.). The Bayesian approach is honest and rigorous about precisely how such non-data information is used. Furthermore, if prior information is available, it should be used on the grounds that more information is preferred to less. As a final line of defense, Bayesians have developed *noninformative priors* for many classes of model. That is, the Bayesian approach allows for the use of prior information if you wish to use it. However, if you do not wish to use it, you do not have to do so. Regardless of how a researcher feels about prior information, it should in no way be an obstacle to the adoption of Bayesian methods.

Computation is the second, and historically more substantive, reason for the minority status of Bayesian econometrics. That is, Bayesian econometrics has historically been computationally difficult or impossible to do for all but a few specific classes of model. The computing revolution of the last 20 years has overcome this hurdle and has led to a blossoming of Bayesian methods in many fields. However, this has made Bayesian econometrics a field which makes heavy use of the computer, and a great deal of this book is devoted to a discussion of computation. In essence, the ideas of Bayesian econometrics are simple, since they only involve the rules of probability. However, to use Bayesian econometrics in practice often requires a lot of number crunching.

To see why computational issues are so important, let us return to the basic equations which underpin Bayesian econometrics. The equations relating to model comparison and prediction either directly or indirectly involve integrals (i.e. (1.6) and (1.9) involve integrals, and (1.6) is a building block for (1.7) and (1.8)). In some (rare) cases, analytical solutions for these integrals are available. That is, you can sit down with pen and paper and work out the integrals. However, we usually need the computer to evaluate the integrals for us, and many algorithms for doing so have been developed.

The equation defining the posterior does not involve any integrals, but presentation of information about the parameters can often involve substantial computation. This arises since, although $p(\theta|y)$ summarizes all we know about θ after seeing the data, it is rarely possible to present all the information about $p(\theta|y)$ when writing up a paper. In cases where $p(\theta|y)$ has a simple form or θ is one-dimensional, it is possible to do so, for instance, by graphing the posterior density. However, in general, econometricians choose to present various numerical summaries of the information contained in the posterior, and these can involve integration. For instance, it is common to present a point estimate, or best guess, of what θ is. Bayesians typically use decision theory to justify a particular choice of a point estimate. In this book, we will not discuss decision theory. The reader is referred to Poirier (1995) or Berger (1985) for excellent discussions of this topic (see also Exercise 1 below). Suffice it to note here that various intuitively plausible point estimates such as the mean, median, and mode of the posterior can be justified in a decision theoretical framework.

Let us suppose you want to use the mean of the posterior density (or *posterior mean*) as a point estimate, and suppose θ is a vector with k elements, $\theta = (\theta_1, \dots, \theta_k)'$. The posterior mean of any element of θ is calculated as (see Appendix B)

$$E(\theta_i|y) = \int \theta_i p(\theta|y) d\theta \tag{1.10}$$

Apart from a few simple cases, it is not possible to evaluate this integral analytically, and once again we must turn to the computer.

In addition to a point estimate, it is usually desirable to present a measure of the degree of uncertainty associated with the point estimate. The most common such measure is the *posterior standard deviation*, which is the square root of the *posterior variance*. The latter is calculated as

$$var(\theta_i|y) = E(\theta_i^2|y) - \{E(\theta_i|y)\}^2$$

which requires evaluation of the integral in (1.10), as well as

$$E(\theta_i^2|y) = \int \theta_i^2 p(\theta|y) d\theta$$

Depending on the context, the econometrician may wish to present many other features of the posterior. For instance, interest may center on whether a particular parameter is positive. In this case, the econometrician would calculate

$$p(\theta_i \geq 0|y) = \int_0^\infty p(\theta|y) d\theta$$

and, once again, an integral is involved.

All of these posterior features which the Bayesian may wish to calculate have the form:

$$E[g(\theta)|y] = \int g(\theta)p(\theta|y)d\theta \tag{1.11}$$

where $g(\theta)$ is a *function of interest*. For instance, $g(\theta) = \theta_i$ when calculating the posterior mean of θ_i and $g(\theta) = 1(\theta_i \geq 0)$ when calculating the probability that θ_i is positive, where $1(A)$ is the indicator function which equals 1 if condition A holds and equals zero otherwise. Even the predictive density in (1.9) falls in this framework if we set $g(\theta) = p(y^*|y, \theta)$. Thus, most things a Bayesian would want to calculate can be put in the form (1.11). The chief exceptions which do not have this form are the marginal likelihood and quantiles of the posterior density (e.g. in some cases, one may wish to calculate the posterior median and posterior interquartile range, and these cannot be put in the form of (1.11)). These exceptions will be discussed in the context of particular models in subsequent chapters.

At this point, a word of warning is called for. Throughout this book, we focus on evaluating $E[g(\theta)|y]$ for various choices of $g(.)$. Unless otherwise noted, for every model and $g(.)$ discussed in this book, $E[g(\theta)|y]$ exists. However, for some models it is possible that $E[g(\theta)|y]$ does not exist. For instance, for the Cauchy distribution, which is the t distribution with one degree of freedom (see Appendix B, Definition B.26), the mean does not exist. Hence, if we had a model which had a Cauchy posterior distribution, $E[\theta|y]$ would not exist. When developing methods for Bayesian inference in a new model, it is thus important to prove that $E[g(\theta)|y]$ does exist. Provided that $p(\theta|y)$ is a valid probability density function, quantiles will exist. So, if you are unsure that $E[g(\theta)|y]$ exists, you can always present quantile-based information (e.g. the median and interquartile range).

In rare cases, (1.11) can be worked out analytically. However, in general, we must use the computer to calculate (1.11). There are many methods for doing this, but the predominant approach in modern Bayesian econometrics is *posterior simulation*. There are a myriad of posterior simulators which are commonly used in Bayesian econometrics, and many of these will be discussed in future chapters in the context of particular models. However, all these are applications or extensions of *laws of large numbers* or *central limit theorems*. In this book, we do not discuss these concepts of *asymptotic distribution theory* in any detail. The interested reader is referred to Poirier (1995) or Greene (2000). Appendix B provides some simple cases, and these can serve to illustrate the basic ideas of posterior simulation.

A straightforward implication of the law of large numbers given in Appendix B (see Definition B.31 and Theorem B.19) is:

Theorem 1.1: Monte Carlo integration

Let $\theta^{(s)}$ for $s = 1, \ldots, S$ be a random sample from $p(\theta|y)$, and define

$$\widehat{g}_S = \frac{1}{S}\sum_{s=1}^{S} g(\theta^{(s)}) \tag{1.12}$$

then $\widehat{g}_S$ converges to $E[g(\theta)|y]$ as S goes to infinity.

In practice, this means that, if we can get the computer to take a random sample from the posterior, (1.12) allows us to approximate $E[g(\theta)|y]$ by simply averaging the function of interest evaluated at the random sample. To introduce some jargon, this sampling from the posterior is referred to as *posterior simulation*, and $\theta^{(s)}$ is referred to as a *draw* or *replication*. Theorem 1.1 describes the simplest posterior simulator, and use of this theorem to approximate $E[g(\theta)|y]$ is referred to as *Monte Carlo integration*.

Monte Carlo integration can be used to approximate $E[g(\theta)|y]$, but only if S were infinite would the approximation error go to zero. The econometrician can, of course, choose any value for S (although larger values of S will increase the computational burden). There are many ways of gauging the approximation error associated with a particular value of S. Some of these will be discussed in subsequent chapters. However, many are based on extensions of the central limit theorem given in Appendix B, Definition B.33 and Theorem B.20. For the case of Monte Carlo integration, this central limit theorem implies:

Theorem 1.2: A numerical standard error

Using the setup and definitions of Theorem 1.1,

$$\sqrt{S}\{\widehat{g}_S - E[g(\theta)|y]\} \rightarrow N(0, \sigma_g^2) \tag{1.13}$$

as S goes to infinity, where $\sigma_g^2 = var[g(\theta)|y]$.

Theorem 1.2 can be used to obtain an estimate of the approximation error in a Monte Carlo integration exercise by using properties of the Normal distribution. For instance, using the fact that the standard Normal has 95% of its probability located within 1.96 standard deviations from its mean yields the approximate result that:

$$\Pr\left[-1.96\frac{\sigma_g}{\sqrt{S}} \leq \widehat{g}_S - E[g(\theta)|y] \leq 1.96\frac{\sigma_g}{\sqrt{S}}\right] = 0.95$$

By controlling S, the econometrician can ensure that $\widehat{g}_S - E[g(\theta)|y]$ is sufficiently small with a high degree of probability. In practice, σ_g is unknown, but the Monte Carlo integration procedure allows us to approximate it. The term $\frac{\sigma_g}{\sqrt{S}}$ is known as the *numerical standard error*, and the econometrician can simply report it as a measure of approximation error. Theorem 1.2 also implies, for example, that if $S = 10\,000$ then the numerical standard error is 1%, as big as the posterior standard deviation. In many empirical contexts, this may be a nice way of expressing the approximation error implicit in Monte Carlo integration.

Unfortunately, it is not always possible to do Monte Carlo integration. Algorithms exist for taking random draws from many common densities (e.g. the

Normal, the Chi-squared).[5] However, for many models, the posteriors do not have one of these common forms. In such cases, development of posterior simulators is a more challenging task. In subsequent chapters, we describe many types of posterior simulators. However, we introduce Monte Carlo integration here so as to present the basic ideas behind posterior simulation in a simple case.

1.3 BAYESIAN COMPUTER SOFTWARE

There are several computer software packages that are useful for doing Bayesian analysis in certain classes of model. However, Bayesian econometrics still tends to require a bit more computing effort than frequentist econometrics. For the latter, there are many canned packages that allow the user to simply click on an icon in order to carry out a particular econometric procedure. Many would argue that this apparent advantage is actually a disadvantage, in that it encourages the econometrician to simply use whatever set of techniques is available in the computer package. This can lead to the researcher simply presenting whatever estimates, test statistics, and diagnostics that are produced, regardless of whether they are appropriate for the application at hand. Bayesian inference forces the researcher to think in terms of the models (i.e. likelihoods and priors), which are appropriate for the empirical question under consideration. The myriad of possible priors and likelihoods make it difficult to construct a Bayesian computer package that can be used widely. For this reason, many Bayesian econometricians create their own programs in matrix programming languages such as MATLAB, Gauss, or Ox. This is not that difficult to do. It is also well worth the effort, since writing a program is a very good way of forcing yourself to fully understand an econometric procedure. In this book, the empirical illustrations are carried out using MATLAB, which is probably the most commonly-used computer language for Bayesian econometrics and statistics. The website associated with this book contains copies of the programs used in the empirical illustrations, and the reader is encouraged to experiment with these programs as a way of learning Bayesian programming. Furthermore, some of the questions at the end of each chapter require the use of the computer, and provide another route for the reader to develop some basic programming skills.

For readers who do not wish to develop programming skills, there are some Bayesian computer packages that allow for simple analysis of standard classes of models. BUGS, an acronym for Bayesian Inference Using Gibbs Sampling (see Best *et al.*, 1995), handles a fairly wide class of models using a common posterior simulation technique called Gibbs sampling. More directly relevant for econometricians is Bayesian Analysis, Computation and Communication (BACC), which handles a wide range of common models (see McCausland and Stevens, 2001).

[5] Draws made by the computer follow a particular algorithm and, hence, are not formally random. It is more technically correct to call draws generated by the computer *pseudo-random*. Devroye (1986) provides a detailed discussion of pseudo-random number generation.

The easiest way to use BACC is as a dynamically linked library to another popular language such as MATLAB. In other words, BACC can be treated as a set of MATLAB commands. For instance, instead of programming up a posterior simulator for analysis of the regression model discussed in Chapter 4, BACC allows for Bayesian inference to be done using one simple MATLAB command. Jim LeSage's Econometrics Toolbox (see LeSage, 1999) also contains many MATLAB functions that can be used for aspects of Bayesian inference. The empirical illustrations in this book which involve posterior simulation use his random number generators. At the time of writing, BUGS, BACC, and the Econometrics Toolbox were available on the web for free for educational purposes. Many other Bayesian software packages exist, although most are more oriented towards the statistician than the econometrician. Appendix C of Carlin and Louis (2000) provides much more information about relevant software.

1.4 SUMMARY

In this chapter, we have covered all the basic issues in Bayesian econometrics at a high level of abstraction. We have stressed that the ability to put all the general theory in one chapter, involving only basic concepts in probability, is an enormous advantage of the Bayesian approach. The basic building blocks of the Bayesian approach are the likelihood function and the prior, the product of these defines the posterior (see (1.3)), which forms the basis for inference about the unknown parameters in a model. Different models can be compared using *posterior model probabilities* (see (1.5)), which require the calculation of *marginal likelihoods* (1.6). Prediction is based on the *predictive density* (1.9). In most cases, it is not possible to work with all these building blocks analytically. Hence, Bayesian computation is an important topic. *Posterior simulation* is the predominant method of Bayesian computation.

Future chapters go through particular models, and show precisely how these abstract concepts become concrete in practical contexts. The logic of Bayesian econometrics set out in this chapter provides a template for the organization of following chapters. Chapters will usually begin with a likelihood function and a prior. Then a posterior is derived along with computational methods for posterior inference and model comparison. The reader is encouraged to think in terms of this likelihood/prior/posterior/computation organizational structure both when reading this book and when beginning a new empirical project.

1.5 EXERCISES

1.5.1 Theoretical Exercises

Remember that Appendix B describes basic concepts in probability, including definitions of common probability distributions.

1. *Decision Theory.* In this book, we usually use the posterior mean as a point estimate. However, in a formal decision theoretic context, the choice of a point estimate of θ is made by defining a loss function and choosing the point estimate which minimizes expected loss. Thus, if $C(\widetilde{\theta}, \theta)$ is the loss (or cost) associated with choosing $\widetilde{\theta}$ as a point estimate of θ, then we would choose $\widetilde{\theta}$ which minimizes $E[C(\widetilde{\theta}, \theta)|y]$ (where the expectation is taken with respect to the posterior of θ). For the case where θ is a scalar, show the following:
   (a) *Squared error loss function.* If $C(\widetilde{\theta}, \theta) = (\widetilde{\theta} - \theta)^2$ then $\widetilde{\theta} = E(\theta|y)$.
   (b) *Asymmetric linear loss function.* If

$$C(\widetilde{\theta}, \theta) = \begin{cases} c_1|\widetilde{\theta} - \theta| \text{ if } \widetilde{\theta} \leq \theta \\ c_2|\widetilde{\theta} - \theta| \text{ if } \widetilde{\theta} > \theta \end{cases}$$

   where $c_1 > 0$ and $c_2 > 0$ are constants, then $\widetilde{\theta}$ is the $\frac{c_1}{c_1+c_2}$th quantile of $p(\theta|y)$.
   (c) *All-or-nothing loss function.* If

$$C(\widetilde{\theta}, \theta) = \begin{cases} c \text{ if } \widetilde{\theta} \neq \theta \\ 0 \text{ if } \widetilde{\theta} = \theta \end{cases}$$

   where $c > 0$ is a constant, then $\widetilde{\theta}$ is the mode of $p(\theta|y)$.
2. Let $y = (y_1, \ldots, y_N)'$ be a random sample where $p(y_i|\theta) = f_G(y_i|\theta, 2)$. Assume a Gamma prior for θ: $p(\theta) = f_G(\theta|\underline{\theta}, \underline{\nu})$:
   (a) Derive $p(\theta|y)$ and $E(\theta|y)$.
   (b) What happens to $E(\theta|y)$ as $\underline{\nu} \to 0$? In what sense is such a prior 'noninformative'?
3. Let $y = (y_1, \ldots, y_N)'$ be a random sample, where

$$p(y_i|\theta) = \begin{cases} \theta^{y_i}(1-\theta)^{y_i} \text{ if } 0 \leq y_i \leq 1 \\ 0 \quad \text{otherwise} \end{cases}$$

   (a) Derive the posterior for θ assuming a prior $\theta \sim U(0, 1)$. Derive $E(\theta|y)$.
   (b) Repeat part (a) assuming a prior of the form:

$$p(\theta) = \begin{cases} \dfrac{\Gamma(\underline{\alpha} + \underline{\beta})}{\Gamma(\underline{\alpha})\Gamma(\underline{\beta})}\theta^{\underline{\alpha}-1}(1-\theta)^{\underline{\beta}-1} \text{ if } 0 < \theta < 1 \\ 0 \quad \text{otherwise} \end{cases}$$

   where $\underline{\alpha}$ and $\underline{\beta}$ are prior hyperparameters.

1.5.2 Computer-Based Exercises

4. Suppose that the posterior for a parameter, θ, is $N(0, 1)$:
   (a) Create a computer program which carries out Monte Carlo integration (see (1.12)) to estimate the posterior mean and variance of θ. (Note: Virtually any relevant computer package such as MATLAB or Gauss will have a function which takes random draws from the standard Normal.)
   (b) How may replications are necessary to ensure that the Monte Carlo estimates of the posterior mean and variance are equal to their true values of 0 and 1 to three decimal places?
   (c) To your computer program, add code which calculates numerical standard errors (see (1.13)). Experiment with calculating posterior means, standard deviations, and numerical standard errors for various values of S. Do the numerical standard errors give a reliable indication of the accuracy of approximation in the Monte Carlo integration estimates?

2
The Normal Linear Regression Model with Natural Conjugate Prior and a Single Explanatory Variable

2.1 INTRODUCTION

The regression model is the workhorse of econometrics. A detailed motivation and discussion of the regression model can be found in any standard econometrics text (e.g. Greene (2000), Gujarati (1995), Hill, Griffiths and Judge (1997), or Koop (2000). Briefly, the linear regression model posits a relationship between a *dependent variable*, y, and k *explanatory variables*, $x_1, \ldots, x_k$, of the form:

$$y = \beta_1 + \beta_2 x_2 + \cdots + \beta_k x_k + \varepsilon$$

where ε is the regression error, and x_1 is implicitly set to 1 to allow for an intercept.

It is not hard for any economist to think of many examples where a particular variable depends upon others. For instance, an individual's wage depends upon her education, experience, and other characteristics. The level of GDP in a country depends upon the size and quality of its workforce, its capital stock, and many other characteristics. The production costs of a firm depends upon the amount of outputs produced, as well as input prices, etc. The empirical example used in the next chapter involves data on houses in Windsor, Canada. Interest centers on the factors which influence house prices, the dependent variable. The explanatory variables are the lot size of the property, the number of bedrooms, number of bathrooms, and number of storeys in the house. Note that this example (like most in economics) involves many explanatory variables and, hence, we have many parameters. With many parameters, the notation becomes very complicated unless matrix algebra is used. To introduce the basic concepts and motivation for the linear regression model with minimal matrix algebra, we begin with a simple case where there is only one explanatory variable. Subsequently, in Chapter 3, we move to the general case involving many explanatory variables.

2.2 THE LIKELIHOOD FUNCTION

Let y_i and x_i denote the observed data on the dependent and explanatory variables, respectively, for individual i for $i = 1, \ldots, N$. We use the term 'individual' to denote the unit of observation, but we could have data on firms, products, time periods, etc. To simplify the mathematics, we do not allow for an intercept and, hence, the linear regression model becomes:

$$y_i = \beta x_i + \varepsilon_i \tag{2.1}$$

where ε_i is an error term. There are many justifications for inclusion of an error term. It can reflect measurement error, or the fact that the linear relationship between x and y is only an approximation of the true relationship. More simply, you can imagine the linear regression model as fitting a straight line with slope β through an XY-plot of the data. In all but the most trivial cases, it is not possible to fit a straight line through all N data points. Hence, it is inevitable that error will result.

Assumptions about ε_i and x_i determine the form of the likelihood function. The standard assumptions (which we will free up in later chapters) are:

1. ε_i is Normally distributed with mean 0, variance σ^2, and ε_i and ε_j are independent of one another for $i \neq j$. Shorthand notation for this is: ε_i is i.i.d. $N(0, \sigma^2)$, where i.i.d. stands for 'independent and identically distributed'.
2. The x_i are either fixed (i.e. not random variables) or, if they are random variables, they are independent of ε_i with a probability density function, $p(x_i|\lambda)$ where λ is a vector of parameters that does not include β and σ^2.

The assumption that the explanatory variables are not random is a standard one in the physical sciences, where experimental methods are common. That is, as part of the experimental setup, the researcher chooses particular values for x and they are not random. In most economic applications, such an assumption is not reasonable. However, the assumption that the distribution of x is independent of the error and with a distribution which does not depend upon the parameters of interest is often a reasonable one. In the language of economics, you can think of it as implying that x is an exogenous variable.

The likelihood function is defined as the joint probability density function for all the data conditional on the unknown parameters (see (1.3)). As shorthand notation, we can stack all our observations of the dependent variable into a vector of length N:

$$y = \begin{bmatrix} y_1 \\ y_2 \\ \cdot \\ \cdot \\ y_N \end{bmatrix}$$

or, equivalently (and more compactly), $y = (y_1, y_2, \ldots, y_N)'$. Similarly, for the explanatory variable, we define $x = (x_1, x_2, \ldots, x_N)'$. The likelihood function then becomes $p(y, x|\beta, \sigma^2, \lambda)$. The second assumption above implies that we can write the likelihood function as:

$$p(y, x|\beta, \sigma^2, \lambda) = p(y|x, \beta, \sigma^2)p(x|\lambda)$$

Insofar as the distribution of x is not of interest, we can then work with the likelihood function conditional on x, $p(y|x, \beta, \sigma^2)$. For simplicity of notation, we will not explicitly include x in our conditioning set for the regression model. It should be remembered that the regression model (whether handled using Bayesian or frequentist methods) implicitly involves working with the conditional distribution of y given x, and not the joint distribution of these two random vectors.

The assumptions about the errors can be used to work out the precise form of the likelihood function. In particular, using some basic rules of probability and (2.1), we find:

- $p(y_i|\beta, \sigma^2)$ is Normal (see Appendix B, Theorem B.10).
- $E(y_i|\beta, \sigma^2) = \beta x_i$ (see Appendix B, Theorem B.2).
- $var(y_i|\beta, \sigma^2) = \sigma^2$ (see Appendix B, Theorem B.2).

Using the definition of the Normal density (Appendix B, Definition B.24) we obtain

$$p(y_i|\beta, \sigma^2) = \frac{1}{\sqrt{2\pi\sigma^2}} \exp\left[-\frac{(y_i - \beta x_i)^2}{2\sigma^2}\right]$$

Finally, since, for $i \neq j$, ε_i and ε_j are independent of one another, it follows that y_i and y_j are also independent of one another and, thus, $p(y|\beta, \sigma^2) = \prod_{i=1}^{N} p(y_i|\beta, \sigma^2)$ and, hence, the likelihood function is given by:

$$p(y|\beta, \sigma^2) = \frac{1}{(2\pi)^{\frac{N}{2}}\sigma^N} \exp\left[-\frac{1}{2\sigma^2}\sum_{i=1}^{N}(y_i - \beta x_i)^2\right] \tag{2.2}$$

For future derivations, it proves convenient to rewrite the likelihood in a slightly different way. It can be shown that:[1]

$$\sum_{i=1}^{N}(y_i - \beta x_i)^2 = \nu s^2 + (\beta - \widehat{\beta})^2 \sum_{i=1}^{N} x_i^2$$

where

$$\nu = N - 1 \tag{2.3}$$

$$\widehat{\beta} = \frac{\sum x_i y_i}{\sum x_i^2} \tag{2.4}$$

[1]To prove this, write $\sum(y_i - \beta x_i)^2 = \sum\{(y_i - \widehat{\beta} x_i) - (\beta - \widehat{\beta})x_i\}^2$ and then expand out the right-hand side.

and

$$s^2 = \frac{\sum(y_i - \widehat{\beta}x_i)^2}{\nu} \tag{2.5}$$

For the reader with a knowledge of frequentist econometrics, note that $\widehat{\beta}$, s^2 and ν are the Ordinary Least Squares (OLS) estimator for β, standard error and degrees of freedom, respectively. They are also sufficient statistics (see Poirier, 1995, p. 222) for (2.2). Furthermore, for many technical derivations, it is easier to work with the error precision rather than the variance. The error precision is defined as: $h = \frac{1}{\sigma^2}$.

Using these results, we can write the likelihood function as:

$$p(y|\beta, h) = \frac{1}{(2\pi)^{\frac{N}{2}}} \left\{ h^{\frac{1}{2}} \exp\left[-\frac{h}{2}(\beta - \widehat{\beta})^2 \sum_{i=1}^{N} x_i^2 \right] \right\} \left\{ h^{\frac{\nu}{2}} \exp\left[-\frac{h\nu}{2s^{-2}} \right] \right\} \tag{2.6}$$

For future reference, note that the first term in curly brackets looks like the kernel of a Normal density for β, and the second term looks almost like a Gamma density for h (see Appendix B, Definitions B.24 and B.22).

2.3 THE PRIOR

Priors are meant to reflect any information the researcher has before seeing the data which she wishes to include. Hence, priors can take any form. However, it is common to choose particular classes of priors that are easy to interpret and/or make computation easier. *Natural conjugate priors* typically have both such advantages. A conjugate prior distribution is one which, when combined with the likelihood, yields a posterior that falls in the same class of distributions. A natural conjugate prior has the additional property that it has the same functional form as the likelihood function. These properties mean that the prior information can be interpreted in the same way as likelihood function information. In other words, the prior can be interpreted as arising from a fictitious data set from the same process that generated the actual data.

In the simple linear regression model, we must elicit a prior for β and h, which we denote by $p(\beta, h)$. The fact that we are not conditioning on the data means that $p(\beta, h)$ is a prior density, the posterior density will be denoted by $p(\beta, h|y)$. It proves convenient to write $p(\beta, h) = p(\beta|h)p(h)$ and think in terms of a prior for $\beta|h$ and one for h. The form of the likelihood function in (2.6) suggests that the natural conjugate prior will involve a Normal distribution for $\beta|h$ and a Gamma distribution for h. This is indeed the case. The name given to a distribution such as this which is a product of a Gamma and a (conditional) Normal is the *Normal-Gamma*. Appendix B, Definition B.26 provides further details on this distribution. Using notation introduced in Appendix B, if

$$\beta|h \sim N(\underline{\beta}, h^{-1}\underline{V})$$

and

$$h \sim G(\underline{s}^{-2}, \underline{\nu})$$

then the natural conjugate prior for β and h is denoted by:

$$\beta, h \sim NG(\underline{\beta}, \underline{V}, \underline{s}^{-2}, \underline{\nu}) \tag{2.7}$$

The researcher would then choose particular values of the so-called *prior hyperparameters* $\underline{\beta}, \underline{V}, \underline{s}^{-2}$ and $\underline{\nu}$ to reflect her prior information. The exact interpretation of these hyperparameters becomes clearer once you have seen their role in the posterior and, hence, we defer a deeper discussion of prior elicitation until the next section.

Throughout this book, we use bars under parameters (e.g. $\underline{\beta}$) to denote parameters of a prior density, and bars over parameters (e.g. $\overline{\beta}$) to denote parameters of a posterior density.

2.4 THE POSTERIOR

The posterior density summarizes all the information, both prior and data-based, that we have about the unknown parameters, β and h. It is proportional to (see (1.3)) the likelihood (2.2) times the prior density (see (2.7)). For the sake of brevity, we do not provide all the algebraic details here. Poirier (1995, p. 527) or Zellner (1971, pp. 60–61) provide closely related derivations. Messy, but conceptually straightforward, manipulations can be used to show that the posterior density is also of Normal-Gamma form, confirming that the prior of the previous section is indeed a natural conjugate one.

Formally, we have a posterior of the form

$$\beta, h|y \sim NG(\overline{\beta}, \overline{V}, \overline{s}^{-2}, \overline{\nu}) \tag{2.8}$$

where

$$\overline{V} = \frac{1}{\underline{V}^{-1} + \sum x_i^2} \tag{2.9}$$

$$\overline{\beta} = \overline{V}(\underline{V}^{-1}\underline{\beta} + \widehat{\beta}\sum x_i^2) \tag{2.10}$$

$$\overline{\nu} = \underline{\nu} + N \tag{2.11}$$

and $\overline{s}^{-2}$ is defined implicitly through

$$\overline{\nu}\overline{s}^2 = \underline{\nu}\underline{s}^2 + \nu s^2 + \frac{(\widehat{\beta} - \underline{\beta})^2}{\underline{V} + \left(\frac{1}{\sum x_i^2}\right)} \tag{2.12}$$

In regression modeling, the coefficient on the explanatory variable, β, is usually the primary focus, since it is a measure of the marginal effect of the explanatory variable on the dependent variable. The posterior mean, $E(\beta|y)$, is a commonly-used point estimate, and $var(\beta|y)$ is a commonly-used metric for the uncertainty associated with the point estimate. Using the basic rules of probability, the posterior mean can be calculated as:

$$E(\beta|y) = \iint \beta p(\beta, h|y) dh d\beta = \int \beta p(\beta|y) d\beta$$

This equation motivates interest in the marginal posterior density, $p(\beta|y)$. Fortunately, this can be calculated analytically using the properties of the Normal-Gamma distribution (see Appendix B, Theorem B.15). In particular, these imply that, if we integrate out h (i.e. use the fact that $p(\beta|y) = \int p(\beta, h|y)dh$), the marginal posterior distribution for β is a t distribution. In terms of the notation of Appendix B, Definition B.25:

$$\beta|y \sim t(\overline{\beta}, \overline{s}^2\overline{V}, \overline{\nu}) \tag{2.13}$$

and it follows from the definition of the t distribution that

$$E(\beta|y) = \overline{\beta} \tag{2.14}$$

and

$$var(\beta|y) = \frac{\overline{\nu}\overline{s}^2}{\overline{\nu} - 2}\overline{V} \tag{2.15}$$

The error precision, h, is usually of less interest than β, but the properties of the Normal-Gamma imply immediately that:

$$h|y \sim G(\overline{s}^{-2}, \overline{\nu}) \tag{2.16}$$

and hence that

$$E(h|y) = \overline{s}^{-2} \tag{2.17}$$

and

$$var(h|y) = \frac{2\overline{s}^{-4}}{\overline{\nu}}. \tag{2.18}$$

Equations (2.9)–(2.18) provide insight into how Bayesian methods combine prior and data information in a very simple model and, hence, it is worth discussing them in some detail. Note, first, that the results the Bayesian econometrician would wish to report all can be written out analytically, and do not involve integration. In Chapter 1, we stressed that Bayesian inference often required posterior simulation. The linear regression model with Normal-Gamma natural conjugate prior is one case where posterior simulation is not required.

The frequentist econometrician would often use $\widehat{\beta}$, the ordinary least squares estimate of β. The common Bayesian point estimate, $\overline{\beta}$, is a weighted average of the OLS estimate and the prior mean, $\underline{\beta}$. The weights are proportional to $\sum x_i^2$ and $\underline{V}^{-1}$, respectively. The latter of these reflects the confidence in the prior. For instance, if the prior variance you select is high, you are saying you are very uncertain about what likely values of β are. As a result, $\underline{V}^{-1}$will be small and little weight will be attached to $\underline{\beta}$, your best prior guess at what β is. The term $\sum x_i^2$ plays a similar role with respect to data-based information. Loosely speaking, it reflects the degree of confidence that the data have in its best guess for β, the OLS estimate $\widehat{\beta}$. Readers knowledgeable of frequentist econometrics will recognize $(\sum x_i^2)^{-1}$ as being proportional to the variance of $\widehat{\beta}$. Alternative intuition can be obtained by considering the simplest case, where $x_i = 1$ for $i = 1, \ldots, N$. Then $\sum x_i^2 = N$, and the weight attached to $\widehat{\beta}$ will simply be the sample size, a reasonable measure for the amount of information in the data. Note that, for both prior mean and the OLS estimate, the posterior mean attaches weight proportional to their precisions (i.e. the inverse of their variances). Hence, Bayesian methods combine data and prior information in a sensible way.

In frequentist econometrics, the variance of the OLS estimator for the regression model given in (2.1) is $s^2(\sum x_i^2)^{-1}$. This variance would be used to obtain frequentist standard errors and carry out various hypothesis tests (e.g. the frequentist t-statistic for testing $\beta = 0$ is $\frac{\widehat{\beta}}{\sqrt{s^2(\sum x_i^2)^{-1}}}$). The Bayesian analogue is the posterior variance of β given in (2.15), which has a very similar form, but incorporates both prior and data information. For instance, (2.9) can be informally interpreted as saying "posterior precision is an average of prior precision ($\underline{V}^{-1}$) and data precision ($\sum x_i^2$)". Similarly, (2.12) has an intuitive interpretation of "posterior sum of squared errors ($\overline{\nu s}^2$) is the sum of prior sum of squared errors ($\underline{\nu s}^2$), OLS sum of squared errors (νs^2), and a term which measures the conflict between prior and data information".

The other equations above also emphasize the intuition that the Bayesian posterior combines data and prior information. Furthermore, the natural conjugate prior implies that the prior can be interpreted as arising from a fictitious data set (e.g. $\underline{\nu}$ and N play the same role in (2.11) and (2.12) and, hence, $\underline{\nu}$ can be interpreted as a prior sample size).

For the reader trained in frequentist econometrics, it is useful to draw out the similarities and differences between what a Bayesian would do and what a frequentist would do. The latter might calculate $\widehat{\beta}$ and its variance, $s^2(\sum x_i^2)^{-1}$, and estimate σ^2 by s^2. The former might calculate the posterior mean and variance of β (i.e. $\overline{\beta}$ and $\frac{\overline{\nu s}^2}{\overline{\nu}-2}\overline{V}$) and estimate $h = \sigma^{-2}$ by its posterior mean, $\overline{s}^{-2}$. These are very similar strategies, except for two important differences. First, the Bayesian formulae all combine prior and data information. Secondly, the Bayesian interprets β as a random variable, whereas the frequentist interprets $\widehat{\beta}$ as a random variable.

The fact that the natural conjugate prior implies prior information enters in the same manner as data information helps with prior elicitation. For instance, when choosing particular values for $\underline{\beta}, \underline{V}, \underline{s}^{-2}$ and $\underline{\nu}$ it helps to know that $\underline{\beta}$ is equivalent to the OLS estimate from an imaginary data set of $\underline{\nu}$ observations with an imaginary $\sum x_i^2$ equal to $\underline{V}^{-1}$ and an imaginary s^2 given by $\underline{s}^2$. However, econometrics is a public science where empirical results are presented to a wide variety of readers. In many cases, most readers may be able to agree on what a sensible prior might be (e.g. economic theory often specifies what reasonable parameter values might be). However, in cases where different researchers can approach a problem with very different priors, a Bayesian analysis with only a single prior can be criticized. There are two main Bayesian strategies for surmounting such a criticism. First, a *prior sensitivity analysis* can be carried out. This means that empirical results can be presented using various priors. If empirical results are basically the same for various sensible priors, then the reader is reassured that researchers with different beliefs can, after looking at the data, come to agreement. If results are sensitive to choice of prior, then the data is not enough to force agreement on researchers with different prior views. The Bayesian approach allows for the scientifically honest finding of such a state of affairs. There is a substantive literature which finds bounds on, for example, the posterior mean of a parameter. We do not discuss this so-called *extreme bounds analysis* literature in any detail. A typical result in this literature is of the form: "for any possible choice of $\underline{V}$, $\overline{\beta}$ must lie between specified upper and lower bounds". Poirier (1995, pp. 532–536) provides an introduction to this literature, and further references (see also Exercise 6 in Chapter 3).

A second strategy for prior elicitation in cases where wide disagreement about prior choice could arise is to use a *noninformative prior*. The Bayesian literature on noninformative priors is too voluminous to survey here. Poirier (1995, pp. 318–331) and Zellner (1971, pp. 41–53) provide detailed discussion about this issue (see also Chapter 12, Section 12.3). Suffice it to note here that, in many cases, it is desirable for data information to be predominant over prior information. In the context of the natural conjugate prior above, it is clear how one can do this. Given the 'fictitious prior sample' interpretation of the natural conjugate prior, it can be seen that setting $\underline{\nu}$ small relative to N and $\underline{V}$ to a large value will ensure that prior information plays little role in the posterior formula (see (2.9)–(2.12)). We refer to such a prior as a *relatively noninformative prior*.

Taking the argument in the previous paragraph to the limit suggests that we can create a purely noninformative prior by setting $\underline{\nu} = 0$ and $\underline{V}^{-1} = 0$ (i.e. $\underline{V} \to \infty$). Such choices are indeed commonly made, and they imply $\beta, h|y \sim NG(\overline{\beta}, \overline{V}, \overline{s}^{-2}, \overline{\nu})$, where

$$\overline{V} = \frac{1}{\sum x_i^2} \tag{2.19}$$

$$\overline{\beta} = \widehat{\beta} \tag{2.20}$$

$$\overline{\nu} = N \tag{2.21}$$

and

$$\overline{\nu}\overline{s}^2 = \nu s^2 \tag{2.22}$$

With this noninformative prior, all of these formulae involve only data information and, in fact, are equal to ordinary least squares results.

In one sense, this noninformative prior has very attractive properties and, given the close relationship with OLS results, provides a bridge between the Bayesian and frequentist approaches. However, it has one undesirable property: this prior 'density' is not, in fact, a valid density, in that it does not integrate to one. Such priors are referred to as *improper*. The Bayesian literature has many examples of problems caused by the use of improper priors. We will see below problems which occur in a model comparison exercise when improper priors are used.

To see the impropriety of this noninformative prior, note that the posterior results (2.19)–(2.22) can be justified by as combining the likelihood function with the following 'prior density':

$$p(\beta, h) = \frac{1}{h}$$

where h is defined over the interval $(0, \infty)$. If you try integrating this 'prior density' over $(0, \infty)$, you will find that the result is ∞, not one as would occur for a valid p.d.f. Bayesians often write this prior as:

$$p(\beta, h) \propto \frac{1}{h} \tag{2.23}$$

but it should be stressed that this notation is not formally correct, since $p(\beta, h)$ is not a valid density function.

It is worth digressing and noting that noninformative priors tend to be improper in most models. To see why this is, consider a continuous scalar parameter θ, which is defined on an interval $[a, b]$. A researcher who wishes to be noninformative about θ would allocate equal prior weight to each equally sized sub-interval (e.g. each interval of width 0.01 should be equally likely). This implies that a Uniform prior over the interval $[a, b]$ is a sensible noninformative prior for θ. However, in most models we do not know a and b, so they should properly be set to $-\infty$ and ∞, respectively. Unfortunately, any Uniform density which yields non-zero probability to a finite bounded interval will integrate to infinity over $(-\infty, \infty)$. Formally, we should not even really speak of the Uniform density in this case, since it is only defined for finite values of a and b. Thus, any Uniform 'noninformative' prior will be improper.

2.5 MODEL COMPARISON

Suppose we have two simple regression models, M_1 and M_2, which purport to explain y. These models differ in their explanatory variables. We distinguish the two models by adding subscripts to the variables and parameters. That is, M_j for $j = 1, 2$ is based on the simple linear regression model:

$$y_i = \beta_j x_{ji} + \varepsilon_{ji} \tag{2.24}$$

for $i = 1, \ldots, N$. Assumptions about ε_{ji} and x_{ji} are the same as those about ε_i and x_i in the previous section (i.e. ε_{ji} is i.i.d. $N(0, h_j^{-1})$, and x_{ji} is either not random or exogenous for $j = 1, 2$).

For the two models, we write the Normal-Gamma natural conjugate priors as:

$$\beta_j, h_j|M_j \sim NG(\underline{\beta}_j, \underline{V}_j, \underline{s}_j^{-2}, \underline{\nu}_j) \tag{2.25}$$

which implies posteriors of the form:

$$\beta_j, h_j|y, M_j \sim NG(\overline{\beta}_j, \overline{V}_j, \overline{s}_j^{-2}, \overline{\nu}_j) \tag{2.26}$$

where

$$\overline{V}_j = \frac{1}{\underline{V}_j^{-1} + \sum x_{ji}^2} \tag{2.27}$$

$$\overline{\beta}_j = \overline{V}_j(\underline{V}_j^{-1}\underline{\beta}_j + \widehat{\beta}_j \sum x_{ji}^2) \tag{2.28}$$

$$\overline{\nu}_j = \underline{\nu}_j + N \tag{2.29}$$

and $\overline{s}_j^{-2}$ is defined implicitly through

$$\overline{\nu}_j \overline{s}_j^2 = \underline{\nu}_j \underline{s}_j^2 + \nu_j s_j^2 + \frac{(\widehat{\beta}_j - \underline{\beta}_j)^2}{\underline{V}_j + \left(\frac{1}{\sum x_{ji}^2}\right)} \tag{2.30}$$

$\widehat{\beta}_j$, s_j^2 and ν_j are OLS quantities analogous to those defined in (2.3)–(2.5). In other words, everything is as in (2.7)–(2.12), except that we have added j subscripts to distinguish between the two models.

Equations (2.26)–(2.30) can be used to carry out posterior inference in either of the two models. However, our purpose here is to discuss model comparison. As described in Chapter 1, a chief tool of Bayesian model comparison is the posterior odds ratio:

$$PO_{12} = \frac{p(y|M_1)p(M_1)}{p(y|M_2)p(M_2)}$$

The prior model probabilities, $p(M_i)$ for $i = 1, 2$, must be selected before seeing the data. The noninformative choice, $p(M_1) = p(M_2) = \frac{1}{2}$, is commonly made. The marginal likelihood, $p(y|M_j)$, is calculated as:

$$p(y|M_j) = \iint p(y|\beta_j, h_j)p(\beta_j, h_j)d\beta_j dh_j \tag{2.31}$$

Unlike with many models, in the Normal linear regression model with natural conjugate prior, the integrals in (2.31) can be calculated analytically. Poirier (1995, pp. 542–543) or Zellner (1971, pp. 72–75) provide details of this calculation, which allows us to write:

$$p(y|M_j) = c_j \left(\frac{\overline{V}_j}{\underline{V}_j}\right)^{\frac{1}{2}} (\overline{\nu}_j \overline{s}_j^2)^{-\frac{\overline{\nu}_j}{2}} \tag{2.32}$$

for $j = 1, 2$, where

$$c_j = \frac{\Gamma\left(\frac{\overline{\nu}_j}{2}\right)\left(\underline{\nu}_j \underline{s}_j^2\right)^{\frac{\underline{\nu}_j}{2}}}{\Gamma\left(\frac{\underline{\nu}_j}{2}\right)\pi^{\frac{N}{2}}} \tag{2.33}$$

and $\Gamma()$ is the Gamma function.[2] The posterior odds ratio comparing M_1 to M_2 becomes:

$$PO_{12} = \frac{c_1\left(\frac{\overline{V}_1}{\underline{V}_1}\right)^{\frac{1}{2}}\left(\overline{\nu}_1 \overline{s}_1^2\right)^{-\frac{\overline{\nu}_1}{2}} p(M_1)}{c_2\left(\frac{\overline{V}_2}{\underline{V}_2}\right)^{\frac{1}{2}}\left(\overline{\nu}_2 \overline{s}_2^2\right)^{-\frac{\overline{\nu}_2}{2}} p(M_2)} \tag{2.34}$$

The posterior odds ratio can be used to calculate the posterior model probabilities, $p(M_j|y)$, using the relationships:

$$p(M_1|y) = \frac{PO_{12}}{1 + PO_{12}}$$

and

$$p(M_2|y) = \frac{1}{1 + PO_{12}}$$

A discussion of (2.34) offers insight into the factors which enter a Bayesian comparison of models. First, the greater is the prior odds ratio, $\frac{p(M_1)}{p(M_2)}$, the higher the support for M_1. Note, secondly, that $\overline{\nu}_j \overline{s}_j^2$ contains the term $\nu_j s_j^2$ which is the sum of squared errors (see (2.3) and (2.5)). The sum of squared errors is a common measure of the model fit, with lower values indicating a better model fit. Hence, the posterior odds ratio rewards models which fit the data better. Thirdly, other things being equal, the posterior odds ratio will indicate support for the model where there is the greatest coherency between prior and data information (i.e. $(\widehat{\beta}_j - \underline{\beta}_j)^2$ enters $\overline{\nu}_j \overline{s}_j^2$). Finally, $\left(\frac{\overline{V}_1}{\underline{V}_1}\right)$ is the ratio of posterior to prior variances. This term can be interpreted as saying, all else being equal, the model with more prior information (i.e. smaller prior variance) relative to posterior information receives most support.

As we shall see in the next chapter, posterior odds ratios also contain a reward for parsimony in that, all else being equal, posterior odds favor the model with fewer parameters. The two models compared here have the same number of parameters (i.e. β_j and h_j) and, hence, this reward for parsimony is not evident. However, in general, this is an important feature of posterior odds ratios.

[2]See Poirier (1995, p. 98) for a definition of the Gamma function. All that you need to know here is that the Gamma function is calculated by the type of software used for Bayesian analysis (e.g. MATLAB or Gauss).

Under the noninformative variant of the natural conjugate prior (i.e. $\underline{\nu}_j = 0$, $\underline{V}_j^{-1} = 0$), the marginal likelihood is not defined and, hence, the posterior odds ratio is undefined. This is one problem with the use of noninformative priors for model comparison (we will see another problem in the next chapter). However, in the present context, a common solution to this problem is to set $\underline{\nu}_1 = \underline{\nu}_2$ equal to an arbitrarily small number and do the same with $\underline{V}_1^{-1}$ and $\underline{V}_2^{-1}$. Also, set $\underline{s}_1^2 = \underline{s}_2^2$. Under these assumptions, the posterior odds ratio is defined and simplifies and becomes arbitrarily close to:

$$PO_{12} = \frac{\left(\dfrac{1}{\sum x_{1i}^2}\right)^{\frac{1}{2}} (\nu_1 s_1^2)^{-\frac{N}{2}} p(M_1)}{\left(\dfrac{1}{\sum x_{2i}^2}\right)^{\frac{1}{2}} (\nu_2 s_2^2)^{-\frac{N}{2}} p(M_2)} \tag{2.35}$$

In this case, the posterior odds ratio reflects only the prior odds ratio, the relative goodness of fit of the two models, and the ratio of terms involving $\frac{1}{\sum x_{ji}^2}$, which reflect the precision of the posterior for M_j. However, as we shall see in the next chapter, this solution to the problem which arises from the use of the noninformative prior will not work when the number of parameters is different in the two models being compared.

In this section, we have shown how a Bayesian would compare two models. If you have many models, you can compare any or all pairs of them or calculate posterior model probabilities for each model (see the discussion after (1.7) in Chapter 1).

2.6 PREDICTION

Now let us drop the j subscript and return to the single model with likelihood and prior defined by (2.6) and (2.7). Equations (2.8)–(2.12) describe Bayesian methods for learning about the parameters β and h, based on a data set with N observations. Suppose interest centers on predicting an unobserved data point generated from the same model. Formally, assume we have the equation:

$$y^* = \beta x^* + \varepsilon^* \tag{2.36}$$

where y^* is not observed. Other than this, all the assumptions of this model are the same as for the simple regression model discussed previously (i.e. ε^* is independent of ε_i for $i = 1, \ldots, N$ and is $N(0, h^{-1})$, and the β in (2.36) is the same as the β in (2.1)). It is also necessary to assume x^* is observed. To understand why the latter assumption is necessary, consider an application where the dependent variable is a worker's salary, and the explanatory variable is some characteristic of the worker (e.g. years of education). If interest focuses on predicting the wage of a new worker, we would have to know her years of education in order to form a meaningful prediction.

As described in Chapter 1, Bayesian prediction is based on calculating:

$$p(y^*|y) = \iint p(y^*|y, \beta, h)p(\beta, h|y)d\beta\, dh \tag{2.37}$$

The fact that ε^* is independent of ε_i implies that y and y^* are independent of one another and, hence, $p(y^*|y, \beta, h) = p(y^*|\beta, h)$. The terms inside the integral in (2.37) are thus the posterior, $p(\beta, h|y)$, and $p(y^*|\beta, h)$. Using a similar reasoning to that used for deriving the likelihood function, we find that

$$p(y^*|\beta, h) = \frac{h^{\frac{1}{2}}}{(2\pi)^{\frac{1}{2}}} \exp\left[-\frac{h}{2}(y^* - \beta x^*)^2\right] \tag{2.38}$$

Multiplying (2.38) by the posterior given in (2.8) and integrating as described in (2.37) yields (Zellner, 1971, pp. 72–75):

$$p(y^*|y) \propto [\overline{\nu} + (y^* - \overline{\beta}x^*)^2\overline{s}^{-2}(1 + \overline{V}x^{*2})^{-1}]^{-\frac{\overline{\nu}+1}{2}} \tag{2.39}$$

It can be verified (see Appendix B, Definition B.25) that this is a univariate t-density with mean $\overline{\beta}x^*$, variance $\frac{\overline{\nu}\overline{s}^2}{\overline{\nu}-2}(1 + \overline{V}x^{*2})$, and degrees of freedom $\overline{\nu}$. In other words,

$$y^*|y \sim t(\overline{\beta}x^*, \overline{s}^2\{1 + \overline{V}x^{*2}\}, \overline{\nu}) \tag{2.40}$$

These results can be used to provide point predictions and measures of uncertainty associated with the point prediction (e.g. the predictive standard deviation).

Our discussion of prediction is a logical place to introduce an important Bayesian concept: model averaging. In the previous section, we have shown how to calculate posterior model probabilities, $p(M_j|y)$, for $j = 1, 2$. These can be used to select one of the two models to work with. However, it is not always desirable to simply choose the one model with highest posterior model probability and throw away the other (or others). *Bayesian model averaging* involves keeping all models, but presenting results averaged over all models. In terms of the rules of probability, it is simple to derive:

$$p(y^*|y) = p(y^*|y, M_1)p(M_1|y) + p(y^*|y, M_2)p(M_2|y) \tag{2.41}$$

In words, insofar as a interest centers on $p(y^*|y)$, you should not simply choose one model and work with, e.g., $p(y^*|y, M_1)$, but rather average results over the two models with weights given by the posterior model probabilities. Using the properties of the expected value operator (see Appendix B, Definition B.8), it follows immediately that:

$$E(y^*|y) = E(y^*|y, M_1)p(M_1|y) + E(y^*|y, M_2)p(M_2|y)$$

which can be used to calculate point predictions averaged over the two models. If $g(.)$ is any function of interest (see (1.11)), then the preceding result generalizes to

$$E[g(y^*)|y] = E[g(y^*)|y, M_1]p(M_1|y) + E[g(y^*)|y, M_2]p(M_2|y) \quad (2.42)$$

which can be used to calculate other functions of the predictive such as the predictive variance.

These results can be generalized to the case of many models and to the case where the function of interest involves parameters instead of y^*. Bayesian model averaging is discussed in much greater detail in Chapter 11.

2.7 EMPIRICAL ILLUSTRATION

The regression model outlined in this chapter is probably too simple to be used for any serious empirical work. For one thing, to simplify the algebra, we have not included an intercept in the model. Furthermore, virtually any serious application will involve several explanatory variables. Hence, to illustrate the basic concepts discussed in this chapter, we will work with a data set artificially generated by the computer. That is, we set $N = 50$. We begin by generating values of the explanatory variable, x_i, which are i.i.d. draws from the $N(0, 1)$ distribution for $i = 1, \ldots, 50$. We then generate values for the errors, ε_i, which are $N(0, h^{-1})$. Finally, we use the explanatory variables and errors to generate the dependent variable $y_i = \beta x_i + \varepsilon_i$. We set $\beta = 2$ and $h = 1$. We use two priors, the noninformative one given in (2.23) and the informative natural conjugate prior given in (2.7) with $\underline{\beta} = 1.5$, $\underline{V} = 0.25$, $\underline{\nu} = 10$ and $\underline{s}^{-2} = 1$. The choices of data generating process and prior hyperparameter values are purely illustrative.

Tables 2.1 and 2.2 present prior and posterior properties of the model parameters, β and h, respectively, using (2.7)–(2.22). Figure 2.1 plots posteriors for β under the informative and noninformative priors as well as the informative prior itself (the noninformative prior for β is simply a flat line). From (2.13) it follows that the plotted p.d.f.s are all t-densities. Posterior properties based on the noninformative prior reflect only likelihood function information and are equivalent to frequentist OLS quantities (see (2.19)–(2.22)). For this reason, the

Table 2.1 Prior and Posterior Properties of β

	Prior	Posterior	
	Informative	Using Noninformative Prior	Using Informative Prior
Mean	1.50	2.06	1.96
St. Deviation	0.56	0.24	0.22

Table 2.2 Prior and Posterior Properties of h

	Prior	Posterior	
	Informative	Using Noninformative Prior	Using Informative Prior
Mean	1.00	1.07	1.04
St. Deviation	0.45	0.21	0.19

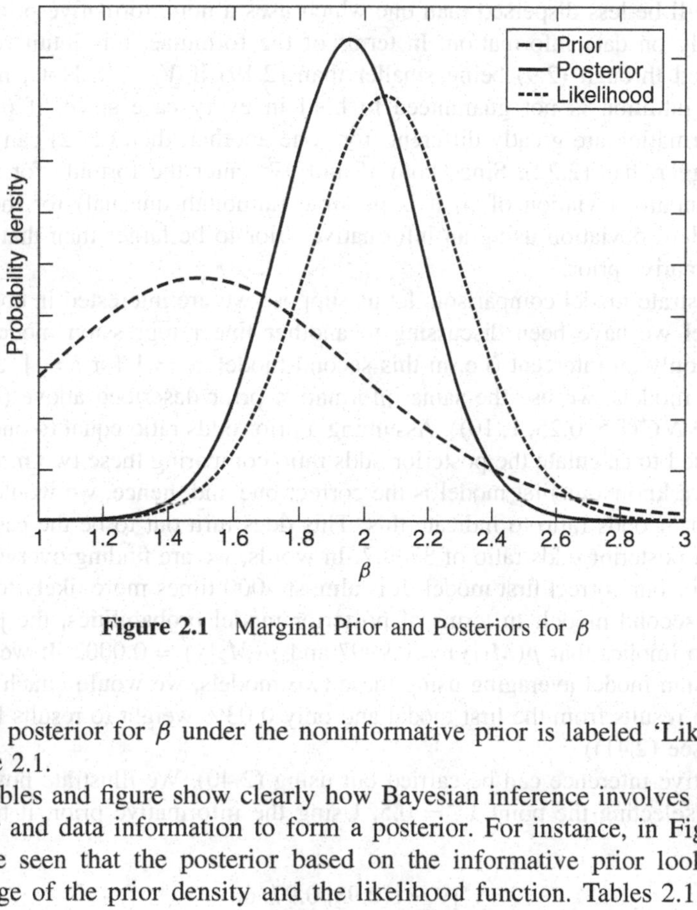

Figure 2.1 Marginal Prior and Posteriors for β

marginal posterior for β under the noninformative prior is labeled 'Likelihood' in Figure 2.1.

The tables and figure show clearly how Bayesian inference involves combining prior and data information to form a posterior. For instance, in Figure 2.1, it can be seen that the posterior based on the informative prior looks to be an average of the prior density and the likelihood function. Tables 2.1 and 2.2 show that the posterior means of both parameters, $E(\beta|y)$ and $E(h|y)$, using the informative prior lie between the relevant prior mean and the likelihood-based quantity (i.e. the posterior mean using the noninformative prior). The prior we have selected contains less information than the data. This can be seen either in the figure (i.e. the prior p.d.f. is more dispersed than the likelihood) or in the tables (i.e. the prior standard deviations are larger than the likelihood-based quantities).

Remember that, since the data set has been artificially created, we know that the true parameter values are $\beta = 2$ and $h = 1$. You would, of course, never expect a point estimate like a posterior mean or an OLS quantity to be precisely equal to the true value. However, the posterior means are quite close to their true values relative to their posterior standard deviations. Note also that the posterior standard deviations using the informative prior are slightly smaller than those using the noninformative prior. This reflects the intuitive notion that, in general, more information allows for more precise estimation. That is, it is intuitively sensible that a posterior which combines both prior and data information will be less dispersed than one which uses a noninformative prior and is based only on data information. In terms of the formulae, this intuitive notion is captured through (2.9) being smaller than (2.19) if $\underline{V} > 0$. Note, however, that this intuition is not guaranteed to hold in every case since, if prior and data information are greatly different from one another, then (2.12) can become much bigger than (2.22). Since both $\overline{V}$ and $\overline{\nu s}^2$ enter the formula for the posterior standard deviation of β, it is possible (although unusual) for the posterior standard deviation using an informative prior to be larger than that using a noninformative prior.

To illustrate model comparison, let us suppose we are interested in comparing the model we have been discussing to another linear regression model which contains only an intercept (i.e. in this second model $x_i = 1$ for $i = 1, \ldots, 50$). For both models, we use the same informative prior described above (i.e. both priors are $NG(1.5, 0.25, 1, 10)$). Assuming a prior odds ratio equal to one, (2.34) can be used to calculate the posterior odds ratio comparing these two models. Of course, we know our first model is the correct one and, hence, we would expect the posterior odds ratio to indicate this. This does turn out to be the case, since we find a posterior odds ratio of 3749.7. In words, we are finding overwhelming support for our correct first model. It is almost 4000 times more likely to be true than the second model. In terms of posterior model probabilities, the posterior odds ratio implies that $p(M_1|y) = 0.9997$ and $p(M_2|y) = 0.0003$. If we were to do Bayesian model averaging using these two models, we would attach 99.97% weight to results from the first model and only 0.03% weight to results from the second (see (2.41)).

Predictive inference can be carried out using (2.40). We illustrate how this is done by selecting the point $x^* = 0.5$. Using the informative prior, it turns out that

$$y^*|y \sim t(0.98, 0.97, 60)$$

Using the noninformative prior, it turns out that

$$y^*|y \sim t(1.03, 0.95, 50)$$

Either of these probability statements can be used to present point predictions, predictive standard deviations, or any other predictive function of interest you may wish to calculate.

2.8 SUMMARY

In this chapter, we have gone through a complete Bayesian analysis (i.e. likelihood, prior, posterior, model comparison and prediction), for the Normal linear regression model with a single explanatory variable and a so-called natural conjugate prior. For the parameters of this model, β and h, this prior has a Normal-Gamma distribution. The natural conjugate nature of the prior means that the posterior also has a Normal-Gamma distribution. For this prior, posterior and predictive inference and model comparison can be done analytically and no posterior simulation is required. Other themes introduced in this chapter include the concept of a noninformative prior and Bayesian model averaging.

2.9 EXERCISES

2.9.1 Theoretical Exercises

1. Prove the result in (2.8). Hint: This is a standard derivation proved in many other textbooks such as Poirier (1995, p. 527) or Zellner (1971, pp. 60–61), and you may wish to examine some of these references if you are having trouble.
2. For this question, assume the likelihood function is as described in Section 2.2 with known error precision, $h = 1$, and $x_i = 1$ for $i = 1, \ldots, N$.
   (a) Assume a Uniform prior for β such that $\beta \sim U(\underline{\alpha}, \underline{\gamma})$. Derive the posterior $p(\beta|y)$.
   (b) What happens to $p(\beta|y)$ as $\underline{\alpha} \rightarrow -\infty$ and $\underline{\gamma} \rightarrow \infty$?
   (c) Use the change-of-variable theorem (Appendix B, Theorem B.21) to derive the prior for a one-to-one function of the regression coefficient, $g(\beta)$, assuming that β has the Uniform prior given in (a). Sketch the implied prior for several choices of $g()$ (e.g. $g(\beta) = \log(\beta), g(\beta) = \frac{\exp(\beta)}{1+\exp(\beta)}, g(\beta) = \exp(\beta)$, etc.).
   (d) Consider what happens to the priors in part (c) as $\underline{\alpha} \rightarrow -\infty$ and $\underline{\gamma} \rightarrow \infty$.
   (e) Given your answers to part (d), discuss whether a prior which is 'noninformative' when the model is parameterized in one way is also 'noninformative' when the model is parameterized in a different way.

2.9.2 Computer-Based Exercises

Remember that some data sets and MATLAB programs are available on the website associated with this book.

3. *Generating artificial data sets.* This is an important skill since they can be used to understand the properties of models and investigate the performance of a particular computer algorithm. Since you have chosen the values for

parameters, you know roughly what answer you would hope your econometric methods should give.

(a) Generate several artificial data sets from the Normal linear regression model by using the following steps: (i) Choose values for β, h and N (e.g. $\beta = 2$, $h = 1$ and $N = 100$); (ii) Generate N values for the explanatory variable from a distribution of your choice (e.g. take $N = 100$ draws from the $U(0, 1)$ distribution); (iii) Generate N values of the errors by taking N i.i.d. draws from the $N(0, h^{-1})$; and (iv) Construct data on the dependent variables using your chosen value for β and the data generated in steps (ii) and (iii) (i.e. use $y_i = \beta x_i + \varepsilon_i$ for $i = 1, \ldots, N$).

(b) Make XY-plots of each data set to see how your choices of β, h, and N are reflected in the data.

4. *Bayesian inference in the Normal linear regression model: prior sensitivity.*

(a) Generate an artificial data set with $\beta = 2$, $h = 1$ and $N = 100$ using the $U(0, 1)$ distribution to generate the explanatory variable.

(b) Assume a prior of the form $\beta, h \sim NG(\underline{\beta}, \underline{V}, \underline{s}^{-2}, \underline{\nu})$ with $\underline{\beta} = 2$, $\underline{V} = 1$, $\underline{s}^{-2} = 1$, $\underline{\nu} = 1$, and calculate the posterior means and standard deviations of β and h. Calculate the Bayes factor comparing the model with $\beta = 0$ to that with $\beta \neq 0$. Calculate the predictive mean and standard deviation for an individual with $x = 0.5$.

(c) How does your answer to part (b) change if $\underline{V} = 0.01$? What if $\underline{V} = 0.1$? What if $\underline{V} = 10$? What if $\underline{V} = 100$? What if $\underline{V} = 1\,000\,000$?

(d) How does your answer to part (b) change if $\underline{\nu} = 0.01$? What if $\underline{\nu} = 0.1$? What if $\underline{\nu} = 10$? What if $\underline{\nu} = 100$? What if $\underline{\nu} = 1\,000\,000$?

(e) Set the prior mean of β different from the value used to generate the data (e.g. $\underline{\beta} = 0$) and repeat part (c).

(f) Set the prior mean of h far from its true value (e.g. $\underline{s}^{-2} = 100$) and repeat part (d).

(g) In light of your findings in parts (b) through (f) discuss the sensitivity of posterior means, standard deviations and Bayes factors to changes in the prior.

(h) Repeat parts (a) through (g) using more informative (e.g. $N = 1000$) and less informative (e.g. $N = 10$) data sets.

(i) Repeat parts (a) through (h) using different values for β and h to generate artificial data.

3
The Normal Linear Regression Model with Natural Conjugate Prior and Many Explanatory Variables

3.1 INTRODUCTION

In this chapter, we extend the results of the previous chapter to the more reasonable case where the linear regression model has several explanatory variables. The structure of this chapter is very similar to the previous one. The primary difference is that this chapter uses matrix algebra. Despite what many students beginning their study of econometrics might think, matrix algebra is a great simplifier. It offers a useful compact notation for writing out and manipulating formulae and simplifies many derivations. Appendix A offers a very brief introduction to the parts of matrix algebra which will be used in this book. The reader who is unfamiliar with matrix algebra should read this appendix before reading this chapter. Poirier (1995), Greene (2000), or Judge *et al.* (1985) all have good chapters on matrix algebra (and additional references), and the reader is referred to these for further detail.

The steps and derivations in this chapter are, apart from the introduction of matrix algebra, virtually identical to those in the previous chapter. Hence, some readers may find it useful to flip back and forth between this chapter and the previous one. That is, it is easier to understand or motivate derivations or results in matrix form if you first understand them without matrix algebra. Throughout this chapter, we point out similarities between the matrix formulae and their counterparts in the previous chapter as a way of easing the transition to matrix algebra.

3.2 THE LINEAR REGRESSION MODEL IN MATRIX NOTATION

Suppose we have data on a dependent variable, y_i, and k explanatory variables, $x_{i1}, \ldots, x_{ik}$ for $i = 1, \ldots, N$. The linear regression model is given by

$$y_i = \beta_1 + \beta_2 x_{i2} + \cdots + \beta_k x_{ik} + \varepsilon_i \tag{3.1}$$

Our notation is such that x_{i1} is implicitly set to 1 to allow for an intercept. This model can be written more compactly in matrix notation by defining the $N \times 1$ vectors:

$$y = \begin{bmatrix} y_1 \\ y_2 \\ . \\ . \\ y_N \end{bmatrix}$$

and

$$\varepsilon = \begin{bmatrix} \varepsilon_1 \\ \varepsilon_2 \\ . \\ . \\ \varepsilon_N \end{bmatrix}$$

the $k \times 1$ vector

$$\beta = \begin{bmatrix} \beta_1 \\ \beta_2 \\ . \\ . \\ \beta_k \end{bmatrix}$$

and the $N \times k$ matrix

$$X = \begin{bmatrix} 1 & x_{12} & . & . & x_{1k} \\ 1 & x_{22} & . & . & x_{2k} \\ . & . & . & . & . \\ . & . & . & . & . \\ 1 & x_{N2} & . & . & x_{Nk} \end{bmatrix}$$

and writing

$$y = X\beta + \varepsilon \tag{3.2}$$

Using the definition of matrix multiplication (see Appendix A, Definition A.4), it can be verified that (3.2) is equivalent to the N equations defined by (3.1).

3.3 THE LIKELIHOOD FUNCTION

The likelihood can be derived in the same manner as in the previous chapter, with the exception that we use matrix notation. Assumptions about ε and X determine the form of the likelihood function. The matrix generalizations of the assumptions in the previous chapter are:

1. ε has a multivariate Normal distribution with mean 0_N and covariance matrix $\sigma^2 I_N$, where 0_N is an N-vector with all elements equal to 0, and I_N is the $N \times N$ identity matrix. Notation for this is: ε is $N(0_N, h^{-1} I_N)$ where $h = \sigma^{-2}$.
2. All elements of X are either fixed (i.e. not random variables) or, if they are random variables, they are independent of all elements of ε with a probability density function, $p(X|\lambda)$, where λ is a vector of parameters that does not include β and h.

The *covariance matrix* of a vector is a matrix that contains the variances of all the elements of the vector on the diagonal and the covariances between different elements filling out the rest of the matrix. In the present context, this means:

$$var(\varepsilon) \equiv \begin{bmatrix} var(\varepsilon_1) & cov(\varepsilon_1, \varepsilon_2) & \ldots & cov(\varepsilon_1, \varepsilon_N) \\ cov(\varepsilon_1, \varepsilon_2) & var(\varepsilon_2) & \ldots & . \\ . & cov(\varepsilon_2, \varepsilon_3) & \ldots & . \\ . & . & \ldots & cov(\varepsilon_{N-1}, \varepsilon_N) \\ cov(\varepsilon_1, \varepsilon_N) & . & \ldots & var(\varepsilon_N) \end{bmatrix}$$

$$= \begin{bmatrix} h^{-1} & 0 & . & . & 0 \\ 0 & h^{-1} & . & . & . \\ . & . & . & . & . \\ . & . & . & . & 0 \\ 0 & . & . & 0 & h^{-1} \end{bmatrix}$$

In other words, the statement that $var(\varepsilon) = h^{-1} I_N$ is a compact notation for $var(\varepsilon_i) = h^{-1}$ and $cov(\varepsilon_i, \varepsilon_j) = 0$ for $i, j = 1, \ldots, N$ and $i \neq j$.

The second assumption implies that we can proceed conditionally on X and treat $p(y|X, \beta, h)$ as the likelihood function. As in the previous chapter, we drop the X from the conditioning set to simplify the notation.

Using the definition of the multivariate Normal density, we can write the likelihood function as:

$$p(y|\beta, h) = \frac{h^{\frac{N}{2}}}{(2\pi)^{\frac{N}{2}}} \left\{ \exp\left[-\frac{h}{2} (y - X\beta)'(y - X\beta) \right] \right\} \tag{3.3}$$

Comparing this equation to (2.2), it can be seen that $(y - X\beta)'(y - X\beta)$ enters in the same manner as $\sum (y_i - \beta x_i)^2$, and it can be confirmed that matrix constructs of the form $a'a$, where a is a vector, are sums of squares.

It proves convenient to write the likelihood function in terms of OLS quantities comparable to (2.3)–(2.5). These are (see Greene (2000), or any other frequentist econometrics textbook which uses matrix algebra):

$$\nu = N - k \tag{3.4}$$

$$\widehat{\beta} = (X'X)^{-1}X'y \tag{3.5}$$

and

$$s^2 = \frac{(y - X\widehat{\beta})'(y - X\widehat{\beta})}{\nu} \tag{3.6}$$

Using a matrix generalization of the derivation in Chapter 2 (see the material between (2.2) and (2.6)), it can be shown that the likelihood function can be written as

$$p(y|\beta, h) = \frac{1}{(2\pi)^{\frac{N}{2}}} \left\{ h^{\frac{1}{2}} \exp\left[-\frac{h}{2}(\beta - \widehat{\beta})' X'X(\beta - \widehat{\beta}) \right] \right\} \left\{ h^{\frac{\nu}{2}} \exp\left[-\frac{h\nu}{2s^{-2}} \right] \right\} \tag{3.7}$$

3.4 THE PRIOR

The form of (3.7) suggests that the natural conjugate prior is Normal-Gamma, and this is indeed the case. In other words, if we elicit a prior for β conditional on h of the form

$$\beta|h \sim N(\underline{\beta}, h^{-1}\underline{V})$$

and a prior for h of the form

$$h \sim G(\underline{s}^{-2}, \underline{\nu})$$

then the posterior will also have these forms. In terms of the notation for the Normal-Gamma distribution, we have

$$\beta, h \sim NG(\underline{\beta}, \underline{V}, \underline{s}^{-2}, \underline{\nu}) \tag{3.8}$$

Note that (3.8) is identical to (2.7), except that $\underline{\beta}$ is now a k-vector containing the prior means for the k regression coefficients, $\beta_1, \ldots, \beta_k$, and $\underline{V}$ is now a $k \times k$ positive definite prior covariance matrix. The notation for the prior density is $p(\beta, h) = f_{NG}(\beta, h|\underline{\beta}, \underline{V}, \underline{s}^{-2}, \underline{\nu})$.

3.5 THE POSTERIOR

The posterior is derived by multiplying the likelihood in (3.7) by the prior in (3.8), and collecting terms (see Exercise 2). Doing so yields a posterior of the form

$$\beta, h|y \sim NG(\overline{\beta}, \overline{V}, \overline{s}^{-2}, \overline{\nu}) \tag{3.9}$$

where

$$\overline{V} = (\underline{V}^{-1} + X'X)^{-1} \tag{3.10}$$

$$\overline{\beta} = \overline{V}(\underline{V}^{-1}\underline{\beta} + X'X\widehat{\beta}) \tag{3.11}$$

$$\overline{\nu} = \underline{\nu} + N \tag{3.12}$$

and $\overline{s}^{-2}$ is defined implicitly through

$$\overline{\nu}\overline{s}^2 = \underline{\nu}\underline{s}^2 + \nu s^2 + (\widehat{\beta} - \underline{\beta})'[\underline{V} + (X'X)^{-1}]^{-1}(\widehat{\beta} - \underline{\beta}) \tag{3.13}$$

The previous expressions describe the joint posterior distribution. If we are interested in the marginal posterior for β, we can integrate out h as in Chapter 2 (see (2.13)). The result is a multivariate t distribution. In terms of the notation of Appendix B

$$\beta|y \sim t(\overline{\beta}, \overline{s}^2\overline{V}, \overline{\nu}) \tag{3.14}$$

and it follows from the definition of the t distribution that

$$E(\beta|y) = \overline{\beta} \tag{3.15}$$

and

$$var(\beta|y) = \frac{\overline{\nu}\overline{s}^2}{\overline{\nu} - 2}\overline{V} \tag{3.16}$$

The properties of the Normal-Gamma distribution imply immediately that:

$$h|y \sim G(\overline{s}^{-2}, \overline{\nu}) \tag{3.17}$$

and, hence, that

$$E(h|y) = \overline{s}^{-2} \tag{3.18}$$

and

$$var(h|y) = \frac{2\overline{s}^{-4}}{\overline{\nu}}. \tag{3.19}$$

These expressions are very similar to (2.8)–(2.18), except that now they are written in terms of matrices or vectors instead of scalars. For instance, $\widehat{\beta}$ is now a vector instead of a scalar, the matrix $(X'X)^{-1}$ plays the role that the scalar $\frac{1}{\sum x_i^2}$ did in Chapter 2, $\overline{V}$ is now a $k \times k$ matrix, etc. The interpretation of these formulae is also very similar. For instance, in Chapter 2 we said that the posterior mean of β, $\overline{\beta}$, was a weighted average of the prior mean, $\underline{\beta}$, and the OLS estimate, $\widehat{\beta}$ where the weights reflected the strength of information in the prior ($\underline{V}^{-1}$) and the data ($\sum x_i^2$). Here, the same intuition holds, except the posterior mean is a matrix-weighted average of prior and data information (see also Exercise 6).

The researcher must elicit the prior hyperparameters, $\underline{\beta}, \underline{V}, \underline{s}^{-2}, \underline{\nu}$. In many cases, economic theory, common sense, or a knowledge of previous empirical studies using different data sets will allow her to do so. The fact that the natural conjugate prior can be treated as though it comes from a fictitious data set generated from the same process as the actual data facilitates prior elicitation. Alternatively, the researcher may try a wide range of priors in a prior sensitivity analysis or work with a relatively noninformative prior. For instance, one could set $\underline{\nu}$ to a value much smaller than N and $\underline{V}$ to a 'large' value. When we are dealing with matrices, the interpretation of the term 'large' is not immediately obvious. The matrix generalization of the statement, $a > b$, where a and b are scalars, is usually taken to be $A - B$ is positive definite, where A and B are square matrices. One measure of the magnitude of a matrix is its determinant. Hence, when we say 'A should be large relative to B', we mean that $A - B$ should be a positive definite matrix with large determinant (see Appendix A, Definitions A.10 and A.14 for definitions of the determinant and positive definiteness).

Taking the argument in the previous paragraph to the limit suggests that we can create a purely noninformative prior by setting $\underline{\nu} = 0$ and setting $\underline{V}^{-1}$ to a small value. There is not a unique way of doing the latter (see Exercise 5). One common way is to set $\underline{V}^{-1} = cI_k$, where c is a scalar, and then let c go to zero. If we do this we find $\beta, h|y \sim NG(\overline{\beta}, \overline{V}, \overline{s}^{-2}, \overline{\nu})$, where

$$\overline{V} = (X'X)^{-1} \tag{3.20}$$

$$\overline{\beta} = \widehat{\beta} \tag{3.21}$$

$$\overline{\nu} = N \tag{3.22}$$

and

$$\overline{\nu s}^2 = \nu s^2 \tag{3.23}$$

As we found for the simpler model in the previous chapter, all of these formulae involve only data information, and are equal to ordinary least squares quantities.

As for the case with a single explanatory variable, this noninformative prior is improper and can be written as:

$$p(\beta, h) \propto \frac{1}{h} \tag{3.24}$$

3.6 MODEL COMPARISON

The linear regression framework with k explanatory variables allows for a wide variety of models to be compared. In this section, we consider two sorts of model comparison exercise. In the first, models are distinguished according to inequality restrictions on the parameter space. In the second, models are distinguished by equality restrictions.

3.6.1 Model Comparison Involving Inequality Restrictions

In some cases, interest might focus on regions of the parameter space. Consider, for instance, a marketing example where the dependent variable is sales of a product, and one of the explanatory variables reflects spending on a particular advertising campaign. In this case, the econometrician might be interested in finding out whether the advertising campaign increased sales (i.e. whether the coefficient on advertising was positive). In a production example, the econometrician could be interested in finding out whether returns to scale are increasing or decreasing. In terms of a regression model, increasing/decreasing returns to scale could manifest itself as particular combination of coefficients being greater/less than one. Both examples involve an inequality restriction involving one or more of the regression coefficients.

Suppose the inequality restrictions under consideration are of the form:

$$R\beta \geq r \tag{3.25}$$

where R is a known $J \times k$ matrix and r is a known J-vector. Equation (3.25) allows for any J linear inequality restrictions on the regression coefficients, β. To ensure that the restrictions are not redundant, we also must assume rank$(R) = J$. We can now define two models of the form:

$$M_1 : R\beta \geq r$$

and

$$M_2 : R\beta \ngeq r$$

where the notation in the equation defining M_2 means that one or more of the J inequality restrictions in M_1 are violated.

For models defined in this way, calculating posterior odds ratios is typically quite easy, and the use of noninformative priors is not a problem. That is,

$$PO_{12} = \frac{p(M_1|y)}{p(M_2|y)} = \frac{p(R\beta \geq r|y)}{p(R\beta \ngeq r|y)} \tag{3.26}$$

Since the posterior for β has a multivariate t distribution (see (3.14)), it follows that $p(R\beta|y)$ also has a t distribution (see Appendix B, Theorem B.14). Computer packages such as MATLAB allow for simple calculation of interval probabilities involving the t distribution and, hence, $p(R\beta \geq r|y)$ can easily be calculated. Alternatively, if $J = 1$, statistical tables for the univariate t distribution can be used.

3.6.2 Equality Restrictions

Model comparison involving equality restrictions is slightly more complicated, and additional issues arise with the use of noninformative priors. There are typically two types of model comparison exercise which fall into this category. First, the researcher might be interested in comparing M_1 which imposes $R\beta = r$ to

M_2, which does not have this restriction. M_1 is an example of a model that is *nested* in another model (i.e. M_1 is obtained from M_2 by placing the restrictions $R\beta = r$ on the latter's parameters). Secondly, the researcher might be interested in comparing $M_1 : y = X_1\beta_{(1)}+\varepsilon_1$ to $M_2 : y = X_2\beta_{(2)}+\varepsilon_2$, where X_1 and X_2 are matrices containing completely different explanatory variables. We use the notation $\beta_{(j)}$ to indicate the regression coefficients in the jth model (for $j = 1, 2$), since we have already used the notation β_j to indicate the scalar regression coefficients in (3.1). This is an example of *non-nested* model comparison.[1]

Both these categories involving equality restrictions can be dealt with by writing the two models to be compared as:

$$M_j : y_j = X_j\beta_{(j)} + \varepsilon_j \tag{3.27}$$

where $j = 1, 2$ indicates our two models, y_j will be defined below, X_j is an $N \times k_j$ matrix of explanatory variables, $\beta_{(j)}$ is a k_j-vector of regression coefficients and ε_j is an N-vector of errors distributed as $N(0_N, h_j^{-1}I_N)$.

The case of non-nested model comparison can be dealt with by setting $y_1 = y_2$. The case of nested model comparison involves beginning with the unrestricted linear regression model in (3.2). M_2 is simply this unrestricted model. That is, we set $y_2 = y$, $X_2 = X$ and $\beta_{(2)} = \beta$. M_1, which imposes $R\beta = r$, can be dealt with by imposing the restrictions on the explanatory variables. This may imply a redefinition of the dependent variable. A detailed discussion of how to do this at a high level of generality is given in Poirier (1995, pp. 540–541). However, a consideration of a few examples should be enough to show that restrictions of the form $R\beta = r$ can always be imposed on (3.2) by suitably redefining the explanatory and dependent variables. Restrictions of the form $\beta_m = 0$ imply that X_1 is simply X with the mth explanatory variable omitted. Restrictions of the form $\beta_m = r$ imply that X_1 is simply X with the mth explanatory variable omitted, and $y_1 = y - rx_m$ where x_m is the mth column of X. The restriction $\beta_2 - \beta_3 = 0$ can be handled by deleting the second and third explanatory variables, and inserting a new explanatory variable which is the sum of these deleted variables. Multiple and/or more complicated restrictions can be handled by generalizing the concepts illustrated by these simple examples in a straightforward way.

We denote the Normal-Gamma priors for the two models by:

$$\beta_{(j)}, h_j|M_j \sim NG(\underline{\beta}_j, \underline{V}_j, \underline{s}_j^{-2}, \underline{\nu}_j) \tag{3.28}$$

for $j = 1, 2$. The posteriors take the form

$$\beta_{(j)}, h_j|y_j \sim NG(\overline{\beta}_j, \overline{V}_j, \overline{s}_j^{-2}, \overline{\nu}_j) \tag{3.29}$$

[1]Non-nested model comparison problems can be put into the form of nested model comparison problems by defining M_3, which has explanatory variables $X = [X_1, X_2]$. If we did this, M_1 and M_2 would both be nested in M_3.

where

$$\overline{V}_j = (\underline{V}_j^{-1} + X_j'X_j)^{-1} \tag{3.30}$$

$$\overline{\beta}_j = \overline{V}_j(\underline{V}_j^{-1}\underline{\beta}_j + X_j'X_j\widehat{\beta}_j) \tag{3.31}$$

$$\overline{\nu}_j = \underline{\nu}_j + N \tag{3.32}$$

and $\overline{s}_j^{-2}$ is defined implicitly through

$$\overline{\nu}_j\overline{s}_j^2 = \underline{\nu}_j\underline{s}_j^2 + \nu_j s_j^2 + (\widehat{\beta}_j - \underline{\beta}_j)'[\underline{V}_j + (X_j'X_j)^{-1}]^{-1}(\widehat{\beta}_j - \underline{\beta}_j) \tag{3.33}$$

$\widehat{\beta}_j$, s_j^2 and ν_j are OLS quantities analogous to (3.4)–(3.6).

The derivation of the marginal likelihood for each model and, hence, the posterior odds ratio, proceeds along the same lines as in the previous chapter (see (2.31)–(2.34)). In particular, the marginal likelihood becomes

$$p(y_j|M_j) = c_j\left(\frac{|\overline{V}_j|}{|\underline{V}_j|}\right)^{\frac{1}{2}}(\overline{\nu}_j\overline{s}_j^2)^{-\frac{\overline{\nu}_j}{2}} \tag{3.34}$$

for $j = 1, 2$, where

$$c_j = \frac{\Gamma\left(\frac{\overline{\nu}_j}{2}\right)(\underline{\nu}_j\underline{s}_j^2)^{\frac{\underline{\nu}_j}{2}}}{\Gamma\left(\frac{\underline{\nu}_j}{2}\right)\pi^{\frac{N}{2}}} \tag{3.35}$$

The posterior odds ratio comparing M_1 to M_2 is

$$PO_{12} = \frac{c_1\left(\frac{|\overline{V}_1|}{|\underline{V}_1|}\right)^{\frac{1}{2}}(\overline{\nu}_1\overline{s}_1^2)^{-\frac{\overline{\nu}_1}{2}}p(M_1)}{c_2\left(\frac{|\overline{V}_2|}{|\underline{V}_2|}\right)^{\frac{1}{2}}(\overline{\nu}_2\overline{s}_2^2)^{-\frac{\overline{\nu}_2}{2}}p(M_2)} \tag{3.36}$$

The factors which affect the posterior odds ratio were discussed in Chapter 2. In particular, the posterior odds ratio depends upon the prior odds ratio, and contains rewards for model fit, coherency between prior and data information and parsimony.

The issue of the reward for parsimony relates closely to problems involved with use of noninformative priors. When discussing posterior inference, we considered a prior where $\underline{\nu} = 0$ and $\underline{V}^{-1} = cI_k$, where c was a scalar. We then defined a noninformative prior as one where c was set to zero. Loosely speaking, setting $\underline{\nu} = 0$ implies there is no prior information about the error precision, h, and letting c go to zero implies there is no prior information about the regression coefficients, β. In this section, we consider these two steps for becoming noninformative separately. An important result will be that it is reasonable to use noninformative priors for h_j for $j = 1, 2$, but it is not reasonable to use noninformative priors for $\beta_{(j)}$. The reason is that the error precision is a parameter which is common to both models, and has the same interpretation in each. However, $\beta_{(1)}$ and $\beta_{(2)}$

are not the same and, in cases where $k_1 \neq k_2$, the use of noninformative priors causes serious problems for Bayesian model comparison using a posterior odds ratio. These considerations motivate an important rule of thumb: *When comparing models using posterior odds ratios, it is acceptable to use noninformative priors over parameters which are common to all models. However, informative, proper priors should be used over all other parameters.* This rule of thumb is relevant not only for the regression model, but for virtually any model you might wish to use.

To justify the statements in the previous paragraph, consider first what happens if we set $\underline{\nu}_1 = \underline{\nu}_2 = 0$.[2] The formula for the posterior odds ratio in (3.36) simplifies substantially since $c_1 = c_2$. However, the posterior odds ratio still has a sensible interpretation involving model fit (i.e. s_j^2), the coherency between prior and data information (see the last term in (3.33)), etc. In short, using a noninformative prior for the error precisions in the two models is perfectly reasonable.

However, using noninformative priors for the $\beta_{(j)}$'s causes major problems which occur largely when $k_1 \neq k_2$. In the case of non-nested model comparison, we have two models which have different explanatory variables and it is clear that the dimension and interpretation of $\beta_{(1)}$ and $\beta_{(2)}$ can be different. For the case of nested model comparison, the restrictions imposed under M_1 will ensure that $\beta_{(1)}$ is of lower dimension than $\beta_{(2)}$ and, hence, $k_1 < k_2$. Thus, having $k_1 \neq k_2$ is quite common. The problem with interpreting posterior odds ratios in this case occurs because of the term $|\underline{V}_j|$. If we set $\underline{V}_j^{-1} = cI_{k_j}$, then $|\underline{V}_j| = \frac{1}{c^{k_j}}$. If we then let c go to zero, an examination of (3.36) should convince you that terms involving c will not cancel out. In fact, provided the prior odds ratio is positive and finite, if $k_1 < k_2$, PO_{12} becomes infinite, while if $k_1 > k_2$, PO_{12} goes to zero. In other words, the posterior odds ratio will always lend overwhelming support for the model with fewer parameters, regardless of the data. In the limit, the reward for parsimony becomes completely dominant and the more parsimonious model is always selected! Clearly, this is unreasonable and provides a strong argument for saying that informative priors should always be used for $\beta_{(1)}$ and $\beta_{(2)}$, at least for coefficients that are not common to both models.

You may think that you are safe when $k_1 = k_2$ as, in this case, the noninformative prior yields a posterior odds ratio of:

$$PO_{12} = \frac{(|X_1'X_1|)^{-\frac{1}{2}}(\nu_1 s_1^2)^{-\frac{N}{2}} p(M_1)}{(|X_2'X_2|)^{-\frac{1}{2}}(\nu_2 s_2^2)^{-\frac{N}{2}} p(M_2)}. \tag{3.37}$$

Note, however, that this expression depends upon units of measurement. For instance, if your explanatory variables in M_1 are measured in dollars and you decide to change this and measure them in thousands of dollars, leaving X_2 unchanged, your posterior odds ratio will change. This is a very undesirable feature which makes many Bayesians reluctant to use posterior odds based on

[2]To be mathematically precise, we should let them go to zero at the same rate.

noninformative priors, even in the case where $k_1 = k_2$. When the researcher elicits an informative prior, this problem does not arise. For instance, in the empirical illustration in the next section, the dependent variable is house price (measured in dollars) and one of the explanatory variables, x_2, is the lot size (measured in square feet). The coefficient, β_2, can be interpreted through a statement of the form: "An extra square foot of lot size will tend to add β_2 dollars to the price of a house, holding other house characteristics constant". The researcher would choose a prior for β_2 with this interpretation in mind. However, if the units of measurement of x_2 were changed to hundreds of square feet, then the interpretation would be based on a statement of the form: "An extra hundred square feet of lot size will tend to add β_2 dollars to the price of a house, holding other house characteristics constant". β_2 has a very different interpretation if the units of measurement of x_2 are changed and, hence, the researcher would use a very different prior. In other words, when the researcher elicits an informative prior, she is implicitly taking into account the units of measurement. However, with a noninformative prior the researcher does not take into account such considerations.

An important message of this section is that, when doing model comparison, it is important to elicit informative priors for parameters which differ or are restricted across models. With the other activities that an econometrician might do (i.e. estimation and prediction) noninformative priors are an acceptable path to take for the Bayesian who seeks to remain 'objective' and not introduce prior information. However, when calculating posterior odds ratios, a noninformative path may not be acceptable.

The ideas in this section have all been developed for the case of two models but can be extended to the case of many models in a straightforward way (see the discussion after (1.7) in Chapter 1). We also stress that posterior odds ratios can be used to form the posterior model probabilities which are necessary for Bayesian model averaging (see (2.42)).

3.6.3 Highest Posterior Density Intervals

Standard Bayesian model comparison techniques are based on the intuitively appealing idea that $p(M_j|y)$ summarizes all of our knowledge and uncertainty about M_j after seeing the data. However, as we have seen, calculating meaningful posterior model probabilities typically requires the elicitation of informative priors. For the Bayesian who wants to do model testing or comparison with a noninformative prior, there are some other techniques which can be used. However, these techniques are not as intuitively appealing as Bayesian model probabilities and have only *ad hoc* justifications. In later chapters, we discuss some of these techniques. In this subsection, we introduce the idea of a Highest Posterior Density Interval (HPDI), and show how it can be used in an *ad hoc* fashion to compare nested models.

Before discussing model comparison, we begin with some definitions of basic concepts. We define these concepts in the context of the parameter vector β in the Normal linear regression model, but they are quite general and can be used with the parameters of any model. Suppose that the elements of the vector of regression coefficients, β, can each lie anywhere in the interval $(-\infty, \infty)$, which is denoted by $\beta \in R^k$. Let $\omega = g(\beta)$ be some m-vector of functions of β which is defined over a region, Ω, where $m \leq k$. Let C be a region within Ω, denoted by $C \subseteq \Omega$.

Definition 3.1: Credible Sets
The set $C \subseteq \Omega$ is a $100(1-\alpha)\%$ credible set with respect to $p(\omega|y)$ if:

$$p(\omega \in C|y) = \int_C p(\omega|y)d\omega = 1 - \alpha$$

As an example, suppose $\omega = g(\beta) = \beta_j$, a single regression coefficient. Then a 95% credible interval for β_j is any interval, $[a, b]$ such that:

$$p(a \leq \beta_j \leq b|y) = \int_a^b p(\beta_j|y)d\beta_j = 0.95$$

There are typically numerous possible credible intervals. Suppose, for instance, that $\beta_j|y$ is $N(0, 1)$. Then, using statistical tables for the standard Normal, we find that $[-1.96, 1.96]$ is a 95% credible interval, as is $[-1.75, 2.33]$ and $[-1.64, \infty)$, etc. To choose from among the infinite number of credible intervals, it is common to choose the one with smallest area. In the standard Normal example, $[-1.96, 1.96]$ is the shortest credible interval. The name given for such a choice is a *Highest Posterior Density Interval*. This is formalized in the following definition.

Definition 3.2: Highest Posterior Density Intervals
A $100(1-\alpha)\%$ highest posterior density interval for ω is a $100(1-\alpha)\%$ credible interval for ω with the property that it has a smaller area than any other $100(1-\alpha)\%$ credible interval for ω.

It is common to present highest posterior density intervals in addition to point estimates when doing Bayesian estimation. For instance, the researcher might report a posterior mean plus a 95% HPDI of β_j. The researcher is 95% sure that β_j lies within the HPDI. HPDIs can also be used in an *ad hoc* manner to do model comparison. Consider, for instance, two Normal linear regression models as in (3.2), and that interest centers on deciding whether the jth explanatory variable should be included. Thus, the two models under consideration are

$$M_1 : \beta_j = 0$$

and

$$M_2 : \beta_j \neq 0$$

Posterior inference under M_2 can be performed as outlined in (3.28)–(3.33), and an HPDI can be calculated for β_j using the properties of the t distribution. If this HPDI does not include zero, then this is evidence against M_1. A finding that the HPDI does include zero is taken as evidence in favor of M_1. Such a strategy can be generalized in the obvious way to the case where we are interested in investigating whether $R\beta = r$.

The reader who knows frequentist econometrics will recognize the similarity of this approach with common hypothesis testing procedures. For instance, a frequentist test of the hypothesis that $\beta_j = 0$ can be done by calculating a confidence interval for β_j. If this confidence interval contains zero, then the hypothesis is accepted. If it does not, the hypothesis is rejected. We stress, however, that this similarity only holds far enough to provide some very crude intuition. Confidence intervals have a very different interpretation from HPDIs.

HPDIs are a very general tool in that they will exist any time the posterior exists. Thus, they can be used with the noninformative prior discussed previously. However, the justification for using them to compare models, although sensible, is an informal one which, in contrast to posterior odds, is not rooted firmly in probability theory.

3.7 PREDICTION

Prediction for the case of the Normal linear regression model with a single explanatory variable is outlined in Chapter 2 (2.36)–(2.40). The case of several explanatory variables is a simple extension of this material. Suppose we have a Normal linear regression model as in (3.2), with likelihood and prior given in (3.3) and (3.8). Posterior inference can be carried out using (3.9). We want to carry out predictive inference on T unobserved values of the dependent variable, which we denote by $y^* = (y_1^*, \ldots, y_T^*)'$, which are generated according to

$$y^* = X^*\beta + \varepsilon^* \tag{3.38}$$

where ε^* is independent of ε and is $N(0, h^{-1}I_T)$ and X^* is a $T \times k$ matrix analogous to X, containing the k explanatory variables for each of the T out-of-sample data points.

The steps in deriving the predictive density for y^* are simple generalizations of those outlined in (2.37)–(2.40). That is, for the Bayesian prediction is based on

$$p(y^*|y) = \int\int p(y^*|y, \beta, h)p(\beta, h|y)d\beta\, dh$$

The fact that ε^* is independent of ε implies that y and y^* are independent of one another and, hence, $p(y^*|y, \beta, h) = p(y^*|\beta, h)$. The latter term can be written as

$$p(y^*|\beta, h) = \frac{h^{\frac{S}{2}}}{(2\pi)^{\frac{S}{2}}} \exp\left[-\frac{h}{2}(y^* - X^*\beta)'(y^* - X^*\beta)\right] \tag{3.39}$$

Multiplying (3.38) by the posterior given in (3.9) and integrating yields a multivariate t predictive density of the form

$$y^*|y \sim t(X^*\overline{\beta}, \overline{s}^2\{I_T + X^*\overline{V}X^{*'}\}, \overline{\nu}) \tag{3.40}$$

This result can be used to carry out predictive inference in the Normal linear regression model with natural conjugate prior.

3.8 COMPUTATIONAL METHODS: MONTE CARLO INTEGRATION

Model comparison, prediction and posterior inference about β can all be done analytically using the results in previous sections. Furthermore, since the marginal posterior for β is a multivariate t distribution, linear combinations of β are also multivariate t (see Appendix B, Theorem B.14). Thus, if R is defined as in (3.25), posterior inference on $R\beta$ can be carried out using the multivariate t distribution. Since the marginal posterior of h is Gamma, the properties of this well-known distribution can be used to make inferences about the error precision.

However, there are some cases where interest centers not on β, nor on $R\beta$, but on some nonlinear function of β which we will call $f(\beta)$. We will assume $f(.)$ is a scalar function, but the techniques in this section can be extended to several functions by simply handling one function at a time.

In general, the posterior for $f(\beta)$ will not lie in the class of densities with well-known analytical properties. This, then, is a convenient place to start discussing posterior simulation. As described in Chapter 1, even if we do not know the properties (e.g. mean, standard deviation, etc.) of a density, it is possible to figure them out on the computer using simulation. The simplest algorithm for doing posterior simulation is called Monte Carlo integration. In the context of the Normal linear regression model, we can write the basic theorem underlying Monte Carlo integration (see Theorem 1.1) as:

Theorem 3.1: Monte Carlo Integration

Let $\beta^{(s)}$ for $s = 1, \ldots, S$ be a random sample from $p(\beta|y)$ and $g(.)$ be any function and define

$$\widehat{g}_S = \frac{1}{S}\sum_{r=1}^{S} g(\beta^{(s)}) \tag{3.41}$$

then $\widehat{g}_S$ converges to $E[g(\beta)|y]$ as S goes to infinity.

Do not be confused by the introduction of two functions $f(.)$ and $g(.)$. By setting $g(.) = f(.)$, we can obtain an estimate of $E[f(\beta)|y]$ for any $f(.)$. However, we may wish to calculate other posterior properties of $f(\beta)$ and this requires the introduction of the function $g(.)$. For instance, the calculation of $var[f(\beta)|y]$ involves setting $g(.) = f(.)^2$ and using (3.41) to calculate $E[f(\beta)^2|y]$. As described in Chapter 1, by suitably redefining $g(.)$ we can calculate a variety of posterior properties of our function of interest, $f(.)$.

Equation (3.41) says that, given random draws from the posterior for β, inference about any function of the parameters can be done. Here Monte Carlo integration requires computer code which takes random draws from the multivariate t distribution. This is available in many places. For instance, MATLAB code relating to the following empirical illustration is available on the website associated with this book. This shows how Monte Carlo integration is done in practice. The structure of the code is as follows:

Step 1: Take a random draw, $\beta^{(s)}$ from the posterior for β given in (3.14) using a random number generator for the multivariate t distribution.
Step 2: Calculate $g(\beta^{(s)})$ and keep this result.
Step 3: Repeat Steps 1 and 2 S times.
Step 4: Take the average of the S draws $g(\beta^{(1)}), \ldots, g(\beta^{(S)})$.

These steps will yield an estimate of $E[g(\beta)|y]$ for any function of interest.

It is worth stressing that Monte Carlo integration yields only an approximation for $E[g(\beta)|y]$ (since you cannot set $S = \infty$). However, by selecting S, the researcher can control the degree of approximation error. Furthermore, as described in Chapter 1 (see (1.13)), we can obtain a numerical measure of the approximation error using a central limit theorem. In particular, we obtain

$$\sqrt{S}\{\widehat{g}_S - E[g(\beta)|y]\} \to N(0, \sigma_g^2) \tag{3.42}$$

as S goes to infinity, where $\sigma_g^2 = var(g(\beta)|y)$. The latter quantity can itself be estimated using Monte Carlo integration, and we shall call such an estimate $\widehat{\sigma}_g^2$. Using this estimate, (3.42) and the properties of the Normal density we can write:

$$Pr\left\{E[g(\beta)|y] - 1.96\frac{\widehat{\sigma}_g}{\sqrt{S}} \leq \widehat{g}_S \leq E[g(\beta)|y] + 1.96\frac{\widehat{\sigma}_g}{\sqrt{S}}\right\} \approx 0.95 \tag{3.43}$$

We can then rearrange the probability statement in (3.43) to find an approximate 95% confidence interval for $E[g(\beta)|y]$ of the form $\left[\widehat{g}_S - 1.96\frac{\widehat{\sigma}_g}{\sqrt{S}}, \widehat{g}_S + 1.96\frac{\widehat{\sigma}_g}{\sqrt{S}}\right]$. The researcher can present this as a measure of how accurate her estimate of $E[g(\beta)|y]$ is or to use it as a guide for selecting S. Alternatively, the numerical standard error, $\frac{\widehat{\sigma}_g}{\sqrt{S}}$, can be reported as implicitly containing the same information in a more compact form.

3.9 EMPIRICAL ILLUSTRATION

To illustrate Bayesian inference in the multiple regression model, we use a data set containing the sales price of $N = 546$ houses sold in Windsor, Canada in 1987. Further details about this data set are provided in Anglin and Gencay

(1996). Interest centers on finding out which factors affect house prices and, hence, sales price is our dependent variable. We use four explanatory variables: the size of the lot the house is on, the number of bedrooms, number of bathrooms and number of storeys. Thus, we have:

- y_i = sales price of the ith house measured in Canadian dollars,
- x_{i2} = the lot size of the ith house measured in square feet,
- x_{i3} = the number of bedrooms in the ith house,
- x_{i4} = the number of bathrooms in the ith house,
- x_{i5} = the number of storeys in the ith house.

Presumably, a researcher doing work with this data set would have knowledge of the Windsor real estate market, and could use such knowledge to elicit a reasonable informative prior. Or, the researcher could ask a local real estate agent to help provide prior information. For instance, the researcher could ask the real estate agent a series of questions of the form: "How much would you expect a house with a lot of size 4000 square feet, with two bedrooms, one bathroom and one storey to cost?"; "How much would you expect a house with a lot of size 6000 square feet, with three bedrooms, two bathrooms and two storeys to cost?", etc. Since there are five unknown regression coefficients, the answers to five questions of this form would give the researcher five equations in five unknowns. She could then solve these equations to find the real estate agent's implicit guesses as to what the regression coefficients are. These guesses could be used as the prior mean for β.

For illustrative purposes, here we will use only a crudely elicited informative prior. House prices in Windsor in 1987 showed a wide variation, but most houses sold for prices in the \$50 000–\$150 000 region. A regression model which fits well might have errors that typically are of the order of magnitude of a few thousand dollars and maybe \$10 000 at most. This suggests that σ might be roughly 5000. That is, since the errors are Normally distributed with mean zero, if $\sigma = 5000$ then 95% of the errors will be less than $1.96 \times 5000 = \$9800$ in absolute value. Since $h = \frac{1}{\sigma^2}$, this suggests that a reasonable prior guess for h would be $\frac{1}{5000^2} = 4.0 \times 10^{-8}$. Thus, we set $\underline{s}^{-2} = 4.0 \times 10^{-8}$. However, this is a very crude guess and, hence, we want to attach little weight to it by setting $\underline{\nu}$ to a value which is much smaller than N. Since $N = 546$, setting $\underline{\nu} = 5$ is relatively noninformative. Loosely speaking, we are saying our prior information about h should have about 1% of the weight as the data information $\left(\text{i.e. } \frac{\underline{\nu}}{N} \approx 0.01\right)$.

For the regression coefficients, we set:

$$\underline{\beta} = \begin{bmatrix} 0.0 \\ 10 \\ 5000 \\ 10\,000 \\ 10\,000 \end{bmatrix}$$

Remember that regression coefficients can be interpreted as saying "if explanatory variable j is increased by one unit and all other explanatory variables are held constant, the price of the house tends to increase by β_j dollars". Hence, our prior mean implies statements of the form "if we compare two houses which are identical except the first house has one bedroom more than the second, then we expect the first house to be worth \$5000 more than the second" or "if the number of bathrooms is increases by one, holding all other house characteristics constant, we expect the price of the house go up by \$10 000", etc.

All these guesses about the regression coefficients are rather crude, so it makes sense to attach a relatively large prior variance to each of them. For instance, suppose our prior information about the intercept is very uncertain. In this case, we might want $var(\beta_1) = 10\,000^2$ (i.e. the prior standard deviation is 10 000 and, hence, we are attaching approximately 95% prior probability to the region $[-20\,000, 20\,000]$ which is a very wide interval).[3] If we think it highly probable that the effect of lot size would be between 0 and 20, we would choose $var(\beta_2) = 25$ (i.e. choose a prior standard deviation for β_2 of 5). For the other regression coefficients, we choose $var(\beta_3) = 2500^2$ and $var(\beta_4) = var(\beta_5) = 5000^2$. These hyperparameter values say, for instance, that our best prior guess of β_4 is 10 000 and we think it very likely that it lies in the interval [0, 20 000].

Given these choices, we can figure out the prior covariance matrix. The properties of the Normal-Gamma distribution imply that the prior covariance matrix for β has the form:

$$var(\beta) = \frac{\underline{\nu}\,\underline{s}^2}{\underline{\nu} - 2}\underline{V}$$

Since $\frac{\underline{\nu}\,\underline{s}^2}{\underline{\nu}-2} = 41\,666\,666\frac{2}{3}$, our choices for $var(\beta_j)$ for $j = 1, \ldots, 5$ imply:

$$\underline{V} = \begin{bmatrix} 2.40 & 0 & 0 & 0 & 0 \\ 0 & 6.0 \times 10^{-7} & 0 & 0 & 0 \\ 0 & 0 & 0.15 & 0 & 0 \\ 0 & 0 & 0 & 0.60 & 0 \\ 0 & 0 & 0 & 0 & 0.60 \end{bmatrix}$$

Note that we have set all the prior covariances to zero. This is commonly done, since it is often hard to make reasonable guesses about what they might be. It implies that your prior information about what plausible values for β_j might be are uncorrelated with those for β_i for $i \neq j$. In many cases, this is a reasonable assumption. This completes our specification of an informative natural conjugate prior for the parameters of our model.

The preceding paragraphs illustrate how prior elicitation might be done in practice. As you can see, prior elicitation can be a bit complicated and involve a lot of

[3]Here we are using a useful approximate rule-of-thumb that says that roughly 95% of the probability in a density is located within two standard deviations of its mean. This approximation works best for the Normal distribution or distributions which have a similar shape to the Normal (e.g. the t distribution).

guesswork. However, it is a very useful exercise to carry out, since it forces the researcher to think carefully about her model and how its parameters are interpreted. For the researcher who has no prior information (or does not wish to use it), is also possible to carry out a noninformative Bayesian analysis using (3.24).

Tables 3.1 and 3.2 present prior and posterior results for both the informative and noninformative priors. Posterior results based on the informative prior can be calculated using (3.9)–(3.19), and those based on the noninformative prior use (3.20)–(3.23). Table 3.1 confirms that our prior is relatively noninformative, since posterior results based on the informative prior are quite similar to those based on the noninformative prior. In the previous chapter, we saw that the posterior mean of the single regression coefficient using the informative prior lay between the prior mean and the OLS estimate. In Table 3.1, there is also a tendency for the posterior mean based on the informative prior to lie between the prior mean and the OLS estimate. Remember that the OLS estimate is identical to the posterior mean based on the noninformative prior (see (3.21)). However, not every posterior mean based on the informative prior lies between the prior mean and the OLS estimate (see results for β_1). This is because the posterior mean is a matrix weighted average of the prior mean and the OLS estimate (see (3.11)). The matrix weighting does not imply that every individual coefficient lies between its prior mean and OLS estimate.

Table 3.2 presents prior and posterior results for h. For this parameter, too, it can be seen that data information dominates prior information. That is, posterior results using the informative prior are quite similar to those using the noninformative prior.

A written summary of results in Tables 3.1 and 3.2 proceeds in the standard way, based on the interpretation of regression parameters. For instance, the researcher might write statements such as: "Regardless of whether we use the informative or noninformative priors, we find the posterior mean of β_4 to be roughly 17 000. Thus, our point estimate indicates that, if we compare two houses which are the same except the first house has one more bathroom than the second, we would expect the first house to be worth roughly \$17 000 more than the second." Or, more tersely, "the point estimate of the the marginal effect of bathrooms on house price is roughly \$17 000".

Table 3.3 contains results relating to the various methods of model comparison discussed in this chapter. All results can be used to shed light on the question of whether an individual regression coefficient is equal to zero. The column labelled $p(\beta_j > 0|y)$ uses (3.14) and the properties of the t distribution to calculate the probability that each individual coefficient is positive. The usefulness of such probabilities is described in Section 3.6.1. The column labelled 'Posterior Odds in Favor of $\beta_j = 0$' contains the posterior odds ratio comparing a model which restricts the appropriate element of β to be zero against the unrestricted alternative. That is, it uses the methods outlined in Section 3.6.2 to calculate the posterior odds ratio comparing two regression models:

Table 3.1 Prior and Posterior Means for β (standard deviations in parentheses)

	Prior	Posterior	
	Informative	Using Noninformative Prior	Using Informative Prior
β_1	0	−4009.55	−4035.05
	(10 000)	(3593.16)	(3530.16)
β_2	10	5.43	5.43
	(5)	(0.37)	(0.37)
β_3	5000	2824.61	2886.81
	(2500)	(1211.45)	(1184.93)
β_4	10 000	17 105.17	16 965.24
	(5000)	(1729.65)	(1708.02)
β_5	10 000	7634.90	7641.23
	(5000)	(1005.19)	(997.02)

Table 3.2 Prior and Posterior Properties of h

	Prior	Posterior	
	Informative	Using Noninformative Prior	Using Informative Prior
Mean	4.0×10^{-8}	3.03×10^{-9}	3.05×10^{-9}
St. Deviation	1.6×10^{-8}	3.33×10^{-6}	3.33×10^{-6}

$M_1 : \beta_j = 0$ to $M_2 : \beta_j \neq 0$. The restricted model uses an informative prior which is identical to the unrestricted prior, except that $\underline{\beta}$ and $\underline{V}$ become 4×1 and 4×4 matrices, respectively, with prior information relating to β_j omitted. A prior odds ratio of one is used. The last two columns of Table 3.3. present 99% and 95% Highest Posterior Density Intervals for each β_j using the noninformative prior. As described in Section 3.6.3, HPDIs can be used to carry out tests of equality restrictions. Remember that these have a sensible, but *ad hoc*, justification even when a noninformative prior is used. Don't forget that posterior odds ratios usually require the use of informative priors (at least over parameters which are common to the two models being compared). Hence, we do not present posterior odds ratios using the noninformative prior.

Table 3.3 Model Comparison Involving β

	Informative Prior			
	$p(\beta_j > 0 \mid y)$	95% HPDI	99% HPDI	Posterior Odds for $\beta_j = 0$
β_1	0.13	[−10 957, 2887]	[−13 143, 5073]	4.14
β_2	1.00	[4.71, 6.15]	[4.49, 6.38]	2.25×10^{-39}
β_3	0.99	[563.5, 5210.1]	[−170.4, 5944]	0.39
β_4	1.00	[13 616, 20 314]	[12 558, 21 372]	1.72×10^{-19}
β_5	1.00	[5686, 9596]	[5069, 10 214]	1.22×10^{-11}
	Noninformative Prior			
	$p(\beta_j > 0 \mid y)$	95% HPDI	99% HPDI	Posterior Odds for $\beta_j = 0$
β_1	0.13	[−11 055, 3036]	[−13 280, 5261]	—
β_2	1.00	[4.71, 6.15]	[4.48, 6.38]	—
β_3	0.99	[449.3, 5200]	[−301.1, 5950]	—
β_4	1.00	[13 714, 20 497]	[12 642, 21 568]	—
β_5	1.00	[5664, 9606]	[5041, 10 228]	—

The results in Table 3.3 are consistent with those in Table 3.1. In the latter table, we saw that the posterior means of β_2, β_4 and β_5 were all positive and very large relative to their posterior standard deviations, providing strong evidence that all these coefficients are non-zero and positive. Regardless of whether we use the informative or noninformative priors, Table 3.3 indicates $p(\beta_j > 0|y)$ is 1 (to several decimal places) for $j = 2, 4, 5$, and none of the HPDIs contains 0. For the informative prior, the posterior odds ratios comparing $M_1 : \beta_j = 0$ to $M_2 : \beta_j \neq 0$ for $j = 2, 4, 5$, are all very small, indicating that the unrestricted model receives massively more probability than the restricted model. Results for β_1 and β_3 are more mixed. For instance, most of the evidence indicates that $\beta_3 \neq 0$. However, the 99% HPDI for this parameter does include zero. Hence, if we were to use the model selection strategy outlined in Section 3.6.3, our results would depend upon precisely which HPDI we chose. A 95% HPDI would imply that $\beta_3 \neq 0$, whereas the 99% HPDI would imply $\beta_3 = 0$. This uncertainty is reflected in the posterior odds ratio, which indicates that the restricted model is 0.39 times as likely as the unrestricted model. If we use this posterior odds ratio to calculate a posterior model probability we find that $p(M_1 : \beta_3 = 0|y) = 0.28$. In words, there is a 28% chance that $\beta_3 = 0$ and 72% chance that it is not. When such uncertainty is present, it may make sense to consider Bayesian model averaging. The alternative is to choose either the unrestricted or the restricted model. In either case, there is a substantial probability that you are choosing the wrong model.

To illustrate how prediction can be done using the Normal linear regression model, we consider the case where the researcher is interested in predicting the sales price of a house with a lot size of 5000 square feet, two bedrooms, two bathrooms and one storey. Using (3.40), we can work out that the predictive distribution in the case of the informative prior is $t(70\,468, 3.33 \times 10^8, 551)$. For the noninformative prior, the predictive distribution is $t(70\,631, 3.35 \times 10^8, 546)$. The researcher might use either of these predictive densities to present information to a client wishing to sell a house with the characteristics listed above. For instance, she might say that her best guess of the sales price is slightly over $70 000, but that there is a large uncertainty associated with this guess (i.e. the predictive standard deviation is roughly $18 000).

Section 3.8 introduces Monte Carlo integration. As discussed in that section, Monte Carlo integration is not required for the Normal linear regression model with natural conjugate prior, unless interest centers on nonlinear functions of the regression coefficients. That is, we already know the posterior properties of β (see Table 3.1), so there is no need to do Monte Carlo integration here. However, to illustrate how Monte Carlo integration is carried out, we will use it to calculate the posterior mean and standard deviation of β_2. From Table 3.1, we know that these should be 5.43 and 0.37, respectively. This gives us a benchmark to see how well Monte Carlo integration works. For the sake of brevity, we calculate results only for the informative prior.

Monte Carlo integration can be implemented by taking random draws from the posterior distribution of β and then averaging appropriate functions of these draws (see (3.41)). From (3.14), we know that the $p(\beta|y)$ is a t density. Thus, we can write a program which repeatedly takes random draws from (3.14) and averages them.

Table 3.4 presents the posterior mean and standard deviation for β_2 calculated in various ways. The row labelled 'Analytical' is the exact result obtained using (3.14)–(3.16). The other rows present results calculated using Monte Carlo integration with different numbers of replications. These rows also present numerical standard errors (see the discussion at end of Section 3.8) which give insight into the accuracy of the Monte Carlo approximation of $E(\beta_2|y)$.

As expected, the accuracy of approximation of both the posterior mean and standard deviation gets better and better as the number of replications is increased.[4] In an empirical context, the exact choice of S will depend upon the accuracy desired by the researcher. For instance, if the researcher is doing a preliminary exploration of the data, then perhaps a rough estimate will do and setting $S = 10$ or 100 may be enough. However, to get highly accurate estimates (perhaps for the final results written up in a report), then the researcher may set $S = 10\,000$ or even 100 000. The numerical standard error does seem to give a good idea

[4]We remind the reader that the computer programs for calculating the results in the empirical illustrations are available on the website associated with this book. If you use these programs (or create your own programs), you should be able to exactly reproduce all tables up to and including Table 3.3. However, since Monte Carlo integration involves taking random draws, you will not be able to exactly reproduce Table 3.4. That is, your random draws will be different from mine and, hence, your results may differ slightly from mine. Formally, the random generator requires what is called a *seed* to get started. The seed is a number and it is usually taken from the computer's clock. Hence, programs run at different times will yield different random draws.

Table 3.4 Posterior Results for β_2 Calculated Various Ways

	Mean	Standard Deviation	Numerical Standard Error
Analytical	5.4316	0.3662	—
Number of Replications			
$S = 10$	5.3234	0.2889	0.0913
$S = 100$	5.4877	0.4011	0.0401
$S = 1000$	5.4209	0.3727	0.0118
$S = 10\,000$	5.4330	0.3677	0.0037
$S = 100\,000$	5.4323	0.3664	0.0012

of the accuracy of each approximation in that approximate posterior means are rarely much more than one numerical standard error from the true posterior mean given in the row labelled 'Analytical'.

It is also worth noting that, although increasing S will increase the accuracy of the Monte Carlo approximation of $E(\beta_2|y)$, the increase is not linear in S. For instance, Table 3.4 shows that results with $S = 100\,000$ are not ten times as accurate as those with $S = 10\,000$. Analytically, the numerical standard error, $\frac{\widehat{\sigma}_g}{\sqrt{S}}$, decreases at a rate of $\frac{1}{\sqrt{S}}$. Thus, results with $S = 100\,000$ should only be roughly $\sqrt{10} = 3.16$ times as accurate as those with $S = 10\,000$.

3.10 SUMMARY

In this chapter, we have gone through a complete Bayesian analysis (i.e. likelihood, prior, posterior, model comparison and prediction) for the Normal linear regression model with natural conjugate prior and k explanatory variables. This chapter is mostly the same as the previous one, except that matrix notation is used throughout to accommodate the complications caused by $k > 1$ explanatory variables. The concept of a highest posterior density interval was introduced. We also showed how Monte Carlo integration, a topic first discussed in Chapter 1, can be used to carry out posterior inference on nonlinear functions of the regression parameters.

3.11 EXERCISES

3.11.1 Theoretical Exercises

1. For the Normal linear regression model, show that the likelihood function in (3.3) can be written in terms of OLS quantities as in (3.7).

2. Assuming the prior for β and h is $NG(\underline{\beta}, \underline{V}, \underline{s}^{-2}, \underline{\nu})$, derive the posterior for β and h and, thus, show that the Normal-Gamma prior is a conjugate prior for this model.
3. Show that (3.13) can be written in the following alternative ways:

$$\overline{\nu s}^2 = \underline{\nu s}^2 + \nu s^2 + (\widehat{\beta} - \underline{\beta})'[(X'X)\overline{V}\underline{V}^{-1}](\widehat{\beta} - \underline{\beta})$$
$$= \underline{\nu s}^2 + (y - X\overline{\beta})'(y - X\overline{\beta}) + (\overline{\beta} - \underline{\beta})'\underline{V}^{-1}(\overline{\beta} - \underline{\beta})$$

4. Suppose you have a Normal linear regression model with *partially informative* natural conjugate prior, where prior information is available only on $J \leq k$ linear combinations of the regression coefficients, and the prior for h is the noninformative one of (3.24). Thus, $R\beta|h \sim N(r, h^{-1}\underline{V}_r)$, where R and r are defined after (3.25) and $\underline{V}_r$ is a $J \times J$ positive definite matrix. Show that the posterior is given by

$$\beta, h|y \sim NG(\widetilde{\beta}, \widetilde{V}, \widetilde{s}^{-2}, \widetilde{\nu})$$

where

$$\widetilde{V} = (R'\underline{V}_r^{-1}R + X'X)^{-1}$$
$$\widetilde{\beta} = \widetilde{V}(R'\underline{V}_r^{-1}\underline{\beta} + X'X\widehat{\beta})$$
$$\widetilde{\nu} = N$$

and

$$\widetilde{\nu}\widetilde{s}^2 = \underline{\nu s}^2 + (\widetilde{\beta} - \widehat{\beta})'X'X(\widetilde{\beta} - \widehat{\beta}) + (R\widehat{\beta} - r)'\underline{V}_r^{-1}(R\widehat{\beta} - r)$$

5. *Problems with Bayes Factors using Noninformative Priors.* Consider comparing two models using Bayes factors based on noninformative priors in the setup of Section 3.6.2. The Bayes factor can be obtained using (3.34) for the two models:
   (a) Consider a noninformative prior created by setting $\underline{\nu}_j = 0$, $\underline{V}_j^{-1} = cI_{k_j}$ and letting $c \to 0$ for $j = 1, 2$. Show that the Bayes factor comparing M_1 to M_2 reduces to:

$$\begin{cases} 0 & \text{if } k_1 > k_2 \\ \left[\dfrac{|X_1'X_1|}{|X_2'X_2|}\right]^{-\frac{1}{2}} \left(\dfrac{\nu_1 s_1^2}{\nu_2 s_2^2}\right)^{-\frac{N}{2}} & \text{if } k_2 = k_1 \\ \infty & \text{if } k_1 < k_2 \end{cases}$$

   (b) Consider a noninformative prior created by setting $\underline{\nu}_j = 0$, $\underline{V}_j^{-1} = \left(c^{\frac{1}{k_j}}\right) I_{k_j}$ and letting $c \to 0$ for $j = 1, 2$. Show that the Bayes factor reduces to:

$$\left[\frac{|X_1'X_1|}{|X_2'X_2|}\right]^{-\frac{1}{2}} \left(\frac{\nu_1 s_1^2}{\nu_2 s_2^2}\right)^{-\frac{N}{2}}$$

(c) Consider a noninformative prior created by setting $\underline{\nu}_j = 0$, $\underline{V}_j^{-1} = \left(c^{\frac{1}{k_j}}\right) X_j' X_j$ and letting $c \to 0$ for $j = 1, 2$. Show that the Bayes factor reduces to:

$$\left(\frac{\nu_1 s_1^2}{\nu_2 s_2^2}\right)^{-\frac{N}{2}}$$

6. *Ellipsoid Bound Theorems.* Consider the Normal linear regression model with natural conjugate prior, $\beta, h \sim NG(\underline{\beta}, \underline{V}, \underline{s}^{-2}, \underline{\nu})$:
   (a) Show that for any choice of $\underline{V}$, the posterior mean of β must lie in the ellipsoid:

$$(\overline{\beta} - \beta_{ave})' X' X (\overline{\beta} - \beta_{ave}) \leq \frac{(\widehat{\beta} - \underline{\beta})' X' X (\widehat{\beta} - \underline{\beta})}{4}$$

   where

$$\beta_{ave} = \frac{1}{2}(\widehat{\beta} + \underline{\beta})$$

   (b) For the case where $k = 1$, show that the result of (a) implies the posterior mean must lie between the prior mean and OLS estimate.
   (c) Suppose the prior covariance matrix is bounded between $\underline{V}_1 \leq \underline{V} \leq \underline{V}_2$, in the sense that $\underline{V}_2 - \underline{V}$ and $\underline{V} - \underline{V}_1$ are both positive definite matrices. Derive an ellipsoid bound analogous to that of (a).
   (d) Discuss how the results in (a) and (c) can be used to investigate the sensitivity of the posterior mean to prior assumptions.
   Note: You may wish to read Leamer (1982) or Poirier (1995, pp. 526–537) for help with answering this question or more details about ellipsoid bound theorems.
7. *Multicollinearity.* Consider the Normal linear regression model with likelihood, prior and posterior as described Sections 3.3, 3.4 and 3.5, respectively. Assume in addition that $Xc = 0$ for some non-zero vector of constants c. Note that this is referred to as a case of *perfect multicollinearity*. It implies the matrix X is not of full rank and $(X'X)^{-1}$ does not exist (see Appendix A on matrix algebra for relevant definitions):
   (a) Show that, despite this pathology, the posterior exists if $\underline{V}$ is positive definite. Define

$$\alpha = c' \underline{V}^{-1} \beta$$

   (b) Show that, given h, the prior and posterior distributions of α are both identical and equal to:

$$N(c' \underline{V}^{-1} \underline{\beta}, h^{-1} c' \underline{V}^{-1} c)$$

   Hence, although prior information can be used to surmount the problems caused by perfect multicollinearity, there are some combinations of the regression coefficients about which learning does not occur.

3.11.2 Computer-Based Exercises

Remember that some data sets and MATLAB programs are available on the website associated with this book. The house price data set is available on this website or in the *Journal of Applied Econometrics* Data Archive listed under Anglin and Gencay (1996)
(http://qed.econ.queensu.ca/jae/1996-v11.6/anglin-gencay/)

8. (a) Generate an artificial data set of size $N = 100$ for the Normal linear regression model with an intercept and one other explanatory variable. Set the intercept to 0, the slope coefficient to 1.0 and $h = 1.0$. Generate the explanatory variable by taking random draws from the $U(0, 1)$ distribution (see Chapter 2, Exercise 1 for more information on how to artificially generate data sets).
   (b) Calculate the posterior mean and standard deviation for this data set using a Normal-Gamma prior with $\underline{\beta} = (0, 1)'$, $\underline{V} = I_2$, $\underline{s}^{-2} = 1$, $\underline{\nu} = 1$.
   (c) Graph the posterior of β_2.
   (d) Calculate the Bayes factor comparing the model $M_1 : \beta_2 = 0$ with $M_2 : \beta_2 \neq 0$.
   (e) Plot the predictive distribution for a future observation with $x_2 = 0.5$.
   (f) Carry out a prior sensitivity analysis by setting $\underline{V} = cI_2$ and repeating parts (b), (d) and (e) for values of $c = 0.01, 1.0, 100.0, 1 \times 10^6$. How sensitive is the posterior to changes in prior information? How sensitive is the Bayes factor? How sensitive is the predictive distribution?
   (g) Calculate the posterior mean and standard deviation of β using a noninformative prior.
   (h) Calculate a 99% HPDI for β_2 using the noninformative prior and use it for investigating whether $\beta_2 = 0$. Compare your results with those of part (d).
9. Repeat Exercise 8 for different choices of N and h to investigate the role of sample size and error size in Bayesian estimation, model comparison and prior sensitivity for the Normal linear regression model.
10. Repeat Exercise 8 (b) using Monte Carlo integration for various values of S. How large does S have to be before you reproduce the results of Exercise 8 (b) to two decimal places?
11. Using the house price data set described in the Empirical Illustration, carry out a prior sensitivity analysis using various priors of your choice. Are the results presented in the text robust to reasonable changes in the prior?

4 The Normal Linear Regression Model with Other Priors

4.1 INTRODUCTION

In the previous chapter, we developed Bayesian methods for estimation, model comparison and prediction for the Normal linear regression model with natural conjugate prior. In this chapter, we go through the same steps for the Normal linear regression model with two different priors. We do this partly since the natural conjugate prior may not accurately reflect the prior information of a researcher in a particular application. Hence, it may be desirable to have available Bayesian methods using other priors. However, we also develop methods for these new priors to introduce some important concepts in Bayesian computation. It is hoped that, by introducing them in the context of the familiar Normal linear regression model, the basic concepts will become clear. These concepts will then be used repeatedly throughout the book for many different models.

Remember that, with the natural conjugate prior, β and h were not independent of one another (see the discussion around (2.7)). In the present chapter, we begin with the Normal linear regression model with an independent Normal-Gamma prior. As we shall see, the minor change from having β and h dependent to independent has major implications for Bayesian computation. In particular, the posterior, posterior odds ratio and predictive no longer have convenient analytical forms like (3.9), (3.36) and (3.40), respectively. Hence, we need to use posterior simulation methods. We introduce the concept of a *Gibbs sampler*, and show how it can be used to carry out posterior and predictive inference. We show how calculation of the posterior odds ratio for nested model comparison can be done using something called the *Savage–Dickey density ratio* and output from the Gibbs sampler.

The second prior introduced in this section is of great practical use in many contexts. It is one which imposes inequality restrictions on β. Economists are often interested in imposing such constraints. For instance, in the context of production function estimation, the restriction that increasing an input will increase output is typically an inequality constraint of the form $\beta_j > 0$. In this chapter, we show how such constraints can be imposed through the prior and a technique called *importance sampling* used to carry out Bayesian inference.

The likelihood function used with both these priors will be the same as that used in the previous chapter. Hence, unlike previous chapters, we will not have a separate section discussing the likelihood function. The reader is referred to (3.3)–(3.7) for a reminder of what it looks like.

4.2 THE NORMAL LINEAR REGRESSION MODEL WITH INDEPENDENT NORMAL-GAMMA PRIOR

4.2.1 The Prior

The Normal linear regression model is defined in Chapter 3 (see (3.2)–(3.7)), and depends upon the parameters β and h. In that chapter, we used a natural conjugate prior where $p(\beta|h)$ was a Normal density and $p(h)$ a Gamma density. In this section, we use a similar prior, but one which assumes prior independence between β and h. In particular, we assume $p(\beta, h) = p(\beta)p(h)$ with $p(\beta)$ being Normal and $p(h)$ being Gamma:

$$p(\beta) = \frac{1}{(2\pi)^{\frac{k}{2}}} |\underline{V}|^{-\frac{1}{2}} \exp\left[-\frac{1}{2}(\beta - \underline{\beta})'\underline{V}^{-1}(\beta - \underline{\beta})\right] \tag{4.1}$$

and

$$p(h) = c_G^{-1} h^{\frac{\underline{\nu}-2}{2}} \exp\left(-\frac{h\underline{\nu}}{2\underline{s}^{-2}}\right) \tag{4.2}$$

where c_G is the integrating constant for the Gamma p.d.f. given in Appendix B, Definition B.22. For simplicity, we are using the same notation as in the previous chapters. That is, $\underline{\beta} = E(\beta|y)$ is still the prior mean of β and the prior mean and degrees of freedom of h are still $\underline{s}^{-2}$ and $\underline{\nu}$, respectively. However, be careful to note that $\underline{V}$ is now simply the prior covariance matrix of β, whereas in the previous chapter we had $var(\beta|h) = h^{-1}\underline{V}$.

4.2.2 The Posterior

The posterior is proportional to the prior times the likelihood. Hence, if we multiply (3.3), (4.1) and (4.2) and ignore terms that do not depend upon β and h, we obtain:

$$p(\beta, h|y) \propto \left\{\exp\left[-\frac{1}{2}\left\{h(y - X\beta)'(y - X\beta) + (\beta - \underline{\beta})'\underline{V}^{-1}(\beta - \underline{\beta})\right\}\right]\right\}$$
$$h^{\frac{N+\underline{\nu}-2}{2}} \exp\left[-\frac{h\underline{\nu}}{2\underline{s}^{-2}}\right] \tag{4.3}$$

This joint posterior density for β and h does not take the form of any well-known and understood density and, hence, cannot be directly used in a simple way for posterior inference. For instance, a researcher might be interested in presenting the posterior mean and variance of β. Unfortunately, there is not a simple analytical formula for these posterior features which can be written down and, hence, posterior simulation is required.

If we treat (4.3) as a joint posterior density for β and h, it does not take a convenient form. The conditionals of the posterior are, however, simple. That is, $p(\beta|y, h)$ can be obtained by treating (4.3) as a function of β for a fixed value of h.[1] If we do matrix manipulations similar to those used in derivation of the posterior for the natural conjugate prior, we can write the key term in the first line of (4.3) as:

$$\begin{aligned} h(y - X\beta)'(y - X\beta) + (\beta - \underline{\beta})'\underline{V}^{-1}(\beta - \underline{\beta}) \\ = (\beta - \overline{\beta})'\overline{V}^{-1}(\beta - \overline{\beta}) + Q \end{aligned}$$

where

$$\overline{V} = (\underline{V}^{-1} + hX'X)^{-1} \tag{4.4}$$

$$\overline{\beta} = \overline{V}(\underline{V}^{-1}\underline{\beta} + hX'y) \tag{4.5}$$

and

$$Q = hy'y + \underline{\beta}'\underline{V}^{-1}\underline{\beta} - \overline{\beta}'\overline{V}^{-1}\overline{\beta}$$

Plugging this expression into (4.3) and ignoring the terms that do not involve β (including Q), we can write

$$p(\beta|y, h) \propto \exp\left[-\frac{1}{2}(\beta - \overline{\beta})'\overline{V}^{-1}(\beta - \overline{\beta})\right] \tag{4.6}$$

which is the kernel of a multivariate Normal density. In other words,

$$\beta|y, h \sim N(\overline{\beta}, \overline{V}) \tag{4.7}$$

$p(h|y, \beta)$ is obtained by treating (4.3) as a function of h. It can be seen that

$$p(h|y, \beta) \propto h^{\frac{N+\underline{\nu}-2}{2}} \exp\left[-\frac{h}{2}\left\{(y - X\beta)'(y - X\beta) + \underline{\nu}\underline{s}^2\right\}\right]$$

By comparing this with the definition of the Gamma density (see Appendix B, Definition B.22) it can be verified that

$$h|y, \beta \sim G(\overline{s}^{-2}, \overline{\nu}) \tag{4.8}$$

[1]Formally, the rules of probability imply $p(\beta|y, h) = \frac{p(\beta,h|y)}{p(h|y)}$. However, since $p(h|y)$ does not depend upon β, $p(\beta, h|y)$ gives the kernel of $p(\beta|y, h)$. Since a density is defined by its kernel, examination of the form of $p(\beta, h|y)$, treating h as fixed, will tell us what $p(\beta|y, h)$ is.

where

$$\overline{\nu} = N + \underline{\nu} \tag{4.9}$$

and

$$\overline{s}^2 = \frac{(y - X\beta)'(y - X\beta) + \underline{\nu}\underline{s}^2}{\overline{\nu}} \tag{4.10}$$

These formulae look quite similar to those for the Normal linear regression model with natural conjugate prior (compare with (3.9)–(3.13)). Indeed, at an informal level, the intuition for how the posterior combines data and prior information is quite similar. However, it must be stressed that (4.4)–(4.10) do not relate directly to the posterior of interest, $p(\beta, h|y)$, but rather to the conditional posteriors, $p(\beta|y, h)$ and $p(h|y, \beta)$. Since $p(\beta, h|y) \neq p(\beta|y, h)p(h|y, \beta)$, the conditional posteriors in (4.7) and (4.8) do not directly tell us everything about $p(\beta, h|y)$. Nevertheless, there is a posterior simulator, called the *Gibbs sampler*, which uses conditional posteriors like (4.7) and (4.8) to produce random draws, $\beta^{(s)}$ and $h^{(s)}$ for $s = 1, \ldots, S$, which can be averaged to produce estimates of posterior properties just as with Monte Carlo integration.

4.2.3 Bayesian Computation: The Gibbs Sampler

The Gibbs sampler is a powerful tool for posterior simulation which is used in many econometric models. We will motivate the basic ideas in a very general context before returning to the Normal linear regression model with independent Normal-Gamma prior. Accordingly, let us temporarily adopt the general notation of Chapter 1, where θ is a p-vector of parameters and $p(y|\theta)$, $p(\theta)$ and $p(\theta|y)$ are the likelihood, prior and posterior, respectively. In the linear regression model, $p = k + 1$ and $\theta = (\beta', h)'$. Furthermore, let θ be partitioned into various *blocks* as $\theta = (\theta'_{(1)}, \theta'_{(2)}, \ldots, \theta'_{(B)})'$, where $\theta_{(j)}$ is a scalar or vector, $j = 1, 2, \ldots, B$. In the linear regression model, it is convenient to set $B = 2$ with $\theta_{(1)} = \beta$ and $\theta_{(2)} = h$.

Remember that Monte Carlo integration involves taking random draws from $p(\theta|y)$ and then averaging them to produce estimates of $E[g(\theta)|y]$ for any function of interest $g(\theta)$ (see Chapter 3, Theorem 3.1). In many models, including the one discussed in the present chapter, it is not easy to directly draw from $p(\theta|y)$. However, it often is easy to randomly draw from $p(\theta_{(1)}|y, \theta_{(2)}, \ldots, \theta_{(B)})$, $p(\theta_{(2)}|y, \theta_{(1)}, \theta_{(3)}, \ldots, \theta_{(B)}), \ldots, p(\theta_{(B)}|y, \theta_{(1)}, \ldots, \theta_{(B-1)})$. The preceding distributions are referred to as *full conditional posterior distributions*, since they define a posterior for each block conditional on all the other blocks. In the Normal linear regression model with independent Normal-Gamma prior $p(\beta|y, h)$ is Normal and $p(h|y, \beta)$ is Gamma, both of which are simple to draw from. It turns out that drawing from the full conditionals will yield a sequence $\theta^{(1)}, \theta^{(2)}, \ldots, \theta^{(s)}$ which can be averaged to produce estimates of $E[g(\theta)|y]$ in the same manner as did Monte Carlo integration.

To motivate the Gibbs sampler, consider the case $B = 2$, and suppose that you have one random draw from $p(\theta_{(2)}|y)$. Call this draw $\theta_{(2)}^{(0)}$. Don't forget our notational convention, where we are using superscripts to indicate draws and subscripts to indicate blocks. Since $p(\theta|y) = p(\theta_{(1)}|y, \theta_{(2)})p(\theta_{(2)}|y)$, it follows that a random draw from $p\left(\theta_{(1)}|y, \theta_{(2)}^{(0)}\right)$ is a valid draw of $\theta_{(1)}$ from $p(\theta|y)$.[2] Call this draw $\theta_{(1)}^{(1)}$. Since $p(\theta|y) = p(\theta_{(2)}|y, \theta_{(1)})p(\theta_{(1)}|y)$, it follows that a random draw from $p\left(\theta_{(2)}|y, \theta_{(1)}^{(1)}\right)$ is a valid draw of $\theta_{(2)}$ from $p(\theta|y)$. Hence, $\theta^{(1)} = \left(\theta_{(1)}^{(1)\prime}, \theta_{(2)}^{(1)\prime}\right)'$ is a valid draw from $p(\theta|y)$. You can continue this reasoning indefinitely. That is, $\theta_{(1)}^{(2)}$, a random draw from $p\left(\theta_{(1)}|y, \theta_{(2)}^{(1)}\right)$, will be a valid draw of $\theta_{(1)}$ from $p(\theta|y)$; $\theta_{(2)}^{(2)}$, a random draw from $p\left(\theta_{(2)}|y, \theta_{(1)}^{(2)}\right)$ will be a valid draw of $\theta_{(2)}$ from $p(\theta|y)$, etc. Hence, if you can successfully find $\theta_{(2)}^{(0)}$, then sequentially drawing from the posterior of $\theta_{(1)}$ conditional on the previous draw for $\theta_{(2)}$, then $\theta_{(2)}$ given the previous draw for $\theta_{(1)}$, will yield a sequence of draws from the posterior. This strategy of sequentially drawing from the full conditional posterior distributions is called Gibbs sampling.

The problem with such a motivation is that it is typically not possible to find such an initial draw $\theta_{(2)}^{(0)}$. After all, if we knew how to easily take random draws from $p(\theta_{(2)}|y)$, we could use this and $p(\theta_{(1)}|\theta_{(2)}, y)$ to do Monte Carlo integration and have no need for Gibbs sampling. However, it can be shown that subject to weak conditions,[3] the initial draw $\theta_{(2)}^{(0)}$ does not matter in the sense that the Gibbs sampler will converge to a sequence of draws from $p(\theta|y)$. Hence, it is common to choose $\theta_{(2)}^{(0)}$ in some manner and then run the Gibbs sampler for S replications. However, the first S_0 of these are discarded as so-called *burn-in replications* and the remaining S_1 retained for the estimate of $E[g(\theta)|y]$, where $S_0 + S_1 = S$.

The preceding motivation for the Gibbs sampler was written for the case of two blocks, but can be extended in a simple manner to more blocks. Formally, the Gibbs sampler involves the following steps.

Step 0: Choose a starting value, $\theta^{(0)}$.
For $s = 1, \ldots, S$:
Step 1: Take a random draw, $\theta_{(1)}^{(s)}$ from $p\left(\theta_{(1)}|y, \theta_{(2)}^{(s-1)}, \theta_{(3)}^{(s-1)}, \ldots, \theta_{(B)}^{(s-1)}\right)$.

[2]This statement follows from the fact that $p(\theta_{(1)}, \theta_{(2)}|y) = p(\theta_{(1)}|y, \theta_{(2)})p(\theta_{(2)}|y)$ implies that first drawing from the marginal posterior density of $\theta_{(2)}$, then drawing from the posterior of $\theta_{(1)}$ conditional on the draw of $\theta_{(2)}$ is equivalent to directly drawing from the joint posterior of $\theta_{(1)}$ and $\theta_{(2)}$.

[3]In the interests of keeping this book focused on Bayesian econometric practice, we do not discuss the precise nature of these conditions. Geweke (1999) provides a description of them and further references for readers interested in more mathematical rigor. All the Gibbs samplers developed in this book will satisfy these weak conditions. The prime case where these conditions are not satisfied is if the posterior is defined over two different regions which are not connected with one another. Then the Gibbs sampler can provide draws from only one region. Of course, for common distributions like the Normal and Gamma, this is not a problem.

Step 2: Take a random draw, $\theta_{(2)}^{(s)}$ from $p\left(\theta_{(2)}|y, \theta_{(1)}^{(s)}, \theta_{(3)}^{(s-1)}, \ldots, \theta_{(B)}^{(s-1)}\right)$.

Step 3: Take a random draw, $\theta_{(3)}^{(s)}$ from $p\left(\theta_{(3)}|y, \theta_{(1)}^{(s)}, \theta_{(2)}^{(s)}, \theta_{(4)}^{(s-1)}, \ldots, \theta_{(B)}^{(s-1)}\right)$.

$\vdots$

Step B: Take a random draw, $\theta_{(B)}^{(s)}$ from $p\left(\theta_{(B)}|y, \theta_{(1)}^{(s)}, \theta_{(2)}^{(s)}, \ldots, \theta_{(B-1)}^{(s)}\right)$.

Following these steps will yield a set of S draws, $\theta^{(s)}$ for $s = 1, \ldots, S$. After dropping the first S_0 of these to eliminate the effect of $\theta^{(0)}$, the remaining S_1 draws can be averaged to create estimates of posterior features of interest. That is, just as with Monte Carlo integration, a weak law of large numbers can be invoked to say that, if $g(.)$ is a function of interest and

$$\widehat{g}_{S_1} = \frac{1}{S_1} \sum_{s=S_0+1}^{S} g(\theta^{(s)}) \tag{4.11}$$

then $\widehat{g}_{S_1}$ converges to $E[g(\theta)|y]$ as S_1 goes to infinity.

This strategy will work for any choice of blocking. However, for many econometric models, a natural choice of blocking suggests itself. In the Normal linear regression model with independent Normal-Gamma prior, $p(\beta|y, h)$ is Normal and $p(h|y, \beta)$ is Gamma. This suggests that the blocking mentioned previously, with $\theta_{(1)} = \beta$ and $\theta_{(2)} = h$ is a natural one. With this choice, a Gibbs sampler can be set up which involves sequentially drawing from the Normal and Gamma distributions using (4.7) and (4.8).

Any posterior simulation approach such as Gibbs sampling provides us with $\widehat{g}_{S_1}$ which is an estimate of $E[g(\theta)|y]$. By choosing S sufficiently large, the approximation error implicit in the estimate can be made as small as the researcher wants. In the case of Monte Carlo integration, we showed (see (3.42) and (3.43)) how a central limit theorem can be used to obtain the numerical standard error which is a sensible measure of the degree of approximation error. In the case of Gibbs sampling, we can do something similar. However, two issues arise with Gibbs sampling which did not arise previously. First, with Gibbs sampling we have to make sure the choice of $\theta^{(0)}$ is not having an effect on results. Secondly, unlike with Monte Carlo integration, the sequence of draws produced, $\theta^{(s)}$ for $s = 1, \ldots, S$, is not i.i.d. In particular, $\theta^{(s)}$ and $\theta^{(s-1)}$ will not be independent from one another. This can be seen most simply by considering the steps above. In general, the draw of $\theta_{(j)}^{(s)}$ depends upon $\theta_{(l)}^{(s-1)}$ for $j = 1, \ldots, B-1$ and $l > j$. The practical importance of these two differences is that, typically, more draws are required with Gibbs sampling than with Monte Carlo integration to achieve a given level of accuracy.

4.2.4 Bayesian Computation: Markov Chain Monte Carlo Diagnostics

Formally, the fact that the state of the Gibbs sampler at draw s (i.e. $\theta^{(s)}$) depends on its state at draw $s-1$ (i.e. $\theta^{(s-1)}$) means that the sequence is a Markov chain.

There are many other posterior simulators which have this property. Some of these we will discuss in later chapters. Such posterior simulators have the general name of *Markov Chain Monte Carlo* (MCMC) algorithms. Associated with these are numerous measures of the approximation error in the MCMC algorithm and various other diagnostics to see whether the estimated results are reliable. We will call all of these *MCMC diagnostics*, and discuss some of them here in the context of the Gibbs sampler. We stress that these diagnostics will also be useful in other MCMC algorithms to be discussed in future chapters. This is a fairly recent field of research in the Bayesian statistics literature, and many of the important results are available only in journal articles. Some of the more accessible sources for more detailed discussion of MCMC diagnostics include *Markov Chain Monte Carlo in Practice*, edited by Gilks *et al.* (see the papers Raftery and Lewis (1996) and Gelman (1996)) and Geweke (1999). Zellner and Min (1995) is also an important reference in this field. Computer programs for many of these diagnostics are available over the web. CODA is a set of S-Plus programs associated with the computer package BUGS which implements some of these (see Best *et al.*, 1995). MATLAB versions are available from James LeSage's Econometrics Toolbox (LeSage, 1999). BACC (McCausland and Stevens, 2001) also provides some MCMC diagnostics.

The first MCMC diagnostic is the numerical standard error which was discussed in previous chapters (see the discussion after (1.13) or (3.43)). Remember that the numerical standard error was derived through the use of a central limit theorem. In the context of MCMC methods, a numerical standard error can be derived, but the fact the draws are not independent means that a different central limit theorem must be used. The reader interested in precise details is referred to Geweke (1992). Briefly, under the weak conditions necessary for the Gibbs sampler to converge to a sequence of draws from $p(\theta|y)$, we obtain a central limit theorem of the familiar form:

$$\sqrt{S_1}\{\widehat{g}_{S_1} - E[g(\theta)|y]\} \rightarrow N(0, \sigma_g^2) \tag{4.12}$$

as S_1 goes to infinity. However, σ_g^2 has a more complicated form than in (3.43), and no fully-justifiable way of estimating it has been developed in the literature. Intuitively, σ_g^2 has to compensate for the fact that $\theta^{(s)}$ for $s = 1, \ldots, S$ is a correlated sequence. Geweke (1992) uses this intuition to draw on ideas from the time series literature to develop an estimate of σ_g^2 of the form

$$\widehat{\sigma}_g^2 = \frac{S(0)}{S_1} \tag{4.13}$$

The justification for this estimate is informal rather than rigorous, but it does seem to work well in practice. For the reader who knows time series methods, $S(0)$ is the spectral density of the sequence $\theta^{(s)}$ for $s = S_0 + 1, \ldots, S$ evaluated at 0. For the reader who does not know what this means, do not worry. The key point to stress here is that an estimate of σ_g^2 is available (and can be calculated using the computer programs discussed above). It is thus possible to calculate a numerical standard error, $\frac{\widehat{\sigma}_g}{\sqrt{S_1}}$. Its interpretation is the same as that described in the previous chapter.

Geweke (1992) suggests another diagnostic based on the intuition that, if a sufficiently large number of draws have been taken, the estimate of $g(\theta)$ based on the first half of the draws should be essentially the same as the estimate based on the last half. If these two estimates are very different, this indicates either than too few draws have been taken (and estimates are simply inaccurate), or that the effect of initial draw, $\theta^{(0)}$, has not worn off and is contaminating the estimate which uses the first draws. More generally, let us divide our S draws from the Gibbs sampler into an initial S_0 which are discarded as burn-in replications and the remaining S_1 draws which are included. These latter draws are divided into a set first set of S_A draws, a middle set of S_B draws and a last set of S_C draws. That is, we have $\theta^{(s)}$ for $s = 1, \ldots, S$ which are divided into subsets as $s = 1, \ldots, S_0, S_0 + 1, \ldots, S_0 + S_A, S_0 + S_A + 1, \ldots, S_0 + S_A + S_B, S_0 + S_A + S_B + 1, \ldots, S_0 + S_A + S_B + S_C$. In practice, it has been found that setting $S_A = 0.1S_1$, $S_B = 0.5S_1$ and $S_C = 0.4S_1$ works well in many applications. For the purposes of calculating the MCMC diagnostic, we drop out the middle set of S_B replications. By dropping out this middle set, we make it likely that our first and last set of draws are independent of one another. Let $\widehat{g}_{S_A}$ and $\widehat{g}_{S_C}$ be the estimates of $E[g(\theta)|y]$ using the first S_A replications after the burn-in and last S_C replications, respectively, using (4.11). Define $\frac{\widehat{\sigma}_A}{\sqrt{S_A}}$ and $\frac{\widehat{\sigma}_C}{\sqrt{S_C}}$ be the numerical standard errors of these two estimates. Then a central limit theorem analogous to (4.12) can be invoked to say

$$CD \to N(0, 1)$$

where CD is the convergence diagnostic given by

$$CD = \frac{\widehat{g}_{S_A} - \widehat{g}_{S_C}}{\frac{\widehat{\sigma}_A}{\sqrt{S_A}} + \frac{\widehat{\sigma}_C}{\sqrt{S_C}}} \tag{4.14}$$

In an empirical application involving the Gibbs sampler, this convergence diagnostic can be calculated and compared to critical values from a standard Normal statistical table. Large values of CD indicate $\widehat{g}_{S_A}$ and $\widehat{g}_{S_C}$ are quite different from one another and, hence, that you have not taken enough replications. If the convergence diagnostic indicates that a sufficiently large number of draws has been taken, then final results can be calculated based on the complete set of S_1 draws.

The previous MCMC diagnostics are likely to be quite informative in assessing whether your Gibbs sampler is working well and whether you have taken a sufficiently large number of replications to achieve your desired degree of accuracy. However, they are not foolproof and, in some unusual models it is possible that the MCMC diagnostics will indicate all is well when they should not. The leading example of such a case occurs where the posterior is bimodal. As a precise example, consider a case where the posterior involves is a mixture of two

Normal distributions which are located in different parts of the parameter space. It is possible that a Gibbs sampler, if started out near the mean of one of these Normals, will just stay there, yielding all replications from the region where the first of the Normals allocates appreciable probability. Numerical standard errors may look reasonable, the convergence diagnostic in (4.14) may indicate convergence has been achieved, but in reality all your results would be missing one of the two Normals that comprise the posterior. This cannot happen in the case of the Normal linear regression model with independent Normal-Gamma prior, nor for virtually all the models considered in this book. However, some of the mixtures of Normals models of Chapter 10 do allow for multi-modal posteriors.

A second case where the Gibbs sampler could yield misleading results and the MCMC diagnostics not warn you of the problem occurs when your initial replication, $\theta^{(0)}$, is extremely far away from the region of the parameter space where most of the posterior probability lies. If the degree of correlation in your Gibbs draws is very high, it might take an enormous number of draws for the Gibbs sampler to move to the region of higher posterior probability. In most cases, the convergence diagnostic CD will catch this problem, since $\widehat{g}_{S_A}$ and $\widehat{g}_{S_C}$ will tend to be different from one another as the Gibbs sampler gradually moves away from $\theta^{(0)}$, but in unusual cases it may not.

It is common to hear Bayesians refer to Gibbs samplers as 'wandering' or 'ranging' over the posterior distribution, taking most draws in regions of high probability and fewer in regions of low probability. In the previous two cases, the Gibbs sampler is not wandering over the entire posterior distribution and this will imply the MCMC diagnostics considered so far are unreliable. After all, the Gibbs sampler cannot provide us with meaningful diagnostics about regions of the parameter space it has never visited.

These two cases both occur because the effect of the initial replication has not worn off. Informally, a common practice is for the researcher to run the Gibbs sampler several times, each time using a different value for $\theta^{(0)}$. If these different runs of the Gibbs sampler all yield essentially the same answer, the researcher is reassured that sufficient replications have been taken (and enough burn-in replications discarded) for the effect of the initial replication to vanish. These ideas were first discussed and formalized in an MCMC convergence diagnostic described in Gelman and Rubin (1992). Gelman (1996), which is Chapter 8 of *Markov Chain Monte Carlo in Practice*, offers a detailed explanation of the derivations below. To explain the intuition underlying this and related diagnostics, let $\theta^{(0,i)}$ for $i = 1, \ldots, m$ denote m initial values which are taken from very different regions of the parameter space. In the jargon of this literature, these should be *overdispersed starting values*. Let $\theta^{(s,i)}$ for $s = 1, \ldots, S$ denote the S Gibbs sampler draws from the ith starting value and $\widehat{g}_{S_1}^{(i)}$ denote the corresponding estimate of $E[g(\theta)|y]$ using (4.11). Intuitively, if the effect of the starting value has been removed, each of these m sequences should be the same as one another. Hence, the variance calculated across the sequences should be not be too large relative to the variance within a sequence. A common estimate of the variance of a sequence is

$$s_i^2 = \frac{1}{S_1 - 1} \sum_{s=S_0+1}^{S} \left[g(\theta^{(s,i)}) - \widehat{g}_{S_1}^{(i)} \right]^2 \tag{4.15}$$

which is referred to as the within-sequence variance. We can now define the average of the within-sequence variances as

$$W = \frac{1}{m} \sum_{i=1}^{m} s_i^2 \tag{4.16}$$

Similarly, it can be shown (see Gelman, 1996) that the between-sequence variance can be estimated by

$$B = \frac{S_1}{m-1} \sum_{i=1}^{m} (\widehat{g}_{S_1}^{(i)} - \widehat{g})^2 \tag{4.17}$$

where

$$\widehat{g} = \frac{1}{m} \sum_{i=1}^{m} \widehat{g}_{S_1}^{(i)} \tag{4.18}$$

Note that W is an estimate of $var[g(\theta)|y]$. It can be shown that

$$\widehat{var[g(\theta)|y]} = \frac{S_1 - 1}{S_1} W + \frac{1}{S_1} B \tag{4.19}$$

is also an estimate of $var[g(\theta)|y]$. However, if the Gibbs sampler has not converged then W will underestimate $var[g(\theta)|y]$. Intuitively, if the Gibbs sampler has wandered over only part of the posterior then it will underestimate its variance. B, however, is based on sequences with overdispersed starting values. This overdispersion implies that, if the Gibbs sampler has converged, $\widehat{var[g(\theta)|y]}$ is an overestimate of $var[g(\theta)|y]$. Thus, a commonly-presented MCMC convergence diagnostic:

$$\widehat{R} = \frac{\widehat{var[g(\theta)|y]}}{W} \tag{4.20}$$

will tend to be greater than one, with values near one indicating that the Gibbs sampler has successfully converged. $\sqrt{\widehat{R}}$ is called the *estimated potential scale reduction*. It can be interpreted as a bound on how far off estimates of the standard deviation of $g(\theta)$ might be due to poor convergence. In Chapter 8 of *Markov Chain Monte Carlo in Practice* it is suggested that values of $\widehat{R}$ greater than 1.2 indicate poor convergence.

It would be nice to provide a deeper understanding of $\widehat{R}$, however this would involve lengthy derivations which are beyond the scope of this book. The interested reader is referred to the above references. For the practical Bayesian, this MCMC convergence diagnostic can easily be calculated using (4.15)–(4.20). Using the rule-of-thumb that $\widehat{R}$ should be less than 1.2 should not lead the practical Bayesian wrong.

4.2.5 Model Comparison: The Savage–Dickey Density Ratio

Just as posterior inference cannot be done analytically, no analytical form for the marginal likelihood exists for the Normal linear regression model with independent Normal-Gamma prior. That is, the marginal likelihood is given by

$$p(y) = \iint p(y|\beta, h)p(\beta, h)\, d\beta\, dh$$

where $p(\beta, h)$ is given in (4.1) and (4.2) and $p(y|\beta, h)$ is the likelihood given in (3.3). If you multiply prior and likelihood together and attempt to work out the integrals in the previous equation, you will find that this is impossible to do analytically. This suggests that posterior simulation methods should be investigated. In the next chapter, we will introduce a generic simulation method for calculating the marginal likelihood due to Gelfand and Dey (1994). This method is totally general in that it can be used for any model, including any variant on the linear regression model. However, the Gelfand–Dey method can be somewhat complicated. Hence, in this chapter, we describe a method which is much simpler, but is not as general. This simpler method is really just another way of writing the Bayes factor for comparing nested models. It is referred to as the *Savage–Dickey density ratio*. It can only be used when comparing nested models, and is only applicable with certain types of priors, but in cases where it is applicable it offers a very simple way of calculating the Bayes factor and, thus, the posterior odds ratio.

The Savage–Dickey density ratio is potentially applicable in a wide variety of applications and, hence, we derive the essential ideas using general notation before applying it to the regression model. Suppose the unrestricted version of a model, M_2, has a parameter vector $\theta = (\omega', \psi')'$. The likelihood and prior for this model are given by $p(y|\omega, \psi, M_2)$ and $p(\omega, \psi|M_2)$. The restricted version of the model, M_1, has $\omega = \omega_0$ where ω_0 is a vector of constants. The parameters in ψ are left unrestricted in each model. The likelihood and prior for this model are given by $p(y|\psi, M_1)$ and $p(\psi|M_1)$. Since ω is simply equal to ω_0 under M_1, we do not need to specify a prior for it. As usual when discussing model comparison, we include M_1 and M_2 as conditioning arguments in probability densities in order to be explicitly clear about which model is referred to.

Theorem 4.1: The Savage–Dickey Density Ratio

Suppose the priors in the two models satisfy:

$$p(\psi|\omega = \omega_0, M_2) = p(\psi|M_1) \tag{4.21}$$

then BF_{12}, the Bayes factor comparing M_1 to M_2, has the form

$$BF_{12} = \frac{p(\omega = \omega_0|y, M_2)}{p(\omega = \omega_0|M_2)} \tag{4.22}$$

where $p(\omega = \omega_0|y, M_2)$ and $p(\omega = \omega_0|M_2)$ are the unrestricted posterior and prior for ω evaluated at the point ω_0.

Equation (4.22) is referred to as the Savage–Dickey density ratio. The proof of this theorem and a more complicated expression for the Bayes factor in the case where the priors do not satisfy (4.21) is given in Verdinelli and Wasserman (1995); see also Exercise 1.

Note that the conditions on the prior under which the Savage–Dickey density ratio can be used to calculate the Bayes factor are very sensible in most situations. That is, in most cases it is reasonable to use the same prior for parameters which are the same in each model (i.e. $p(\psi|M_2) = p(\psi|M_1)$). Such a choice implies (4.21). In fact, (4.21) is a much weaker condition, saying the prior for ψ in the restricted and unrestricted models must be the same only at the point $\omega = \omega_0$.

The Savage–Dickey density ratio can be a big help in calculating the Bayes factor. For one thing, the Savage–Dickey density ratio involves only M_2 so you do not have to worry about developing methods for posterior inference using M_1. For another thing, (4.22) involves only prior and posterior densities and these are often easy to manipulate. Direct calculation of the marginal likelihood is not required. As we shall see in the next chapter, direct calculation of the marginal likelihood can be difficult.

Let us now return to the Normal linear regression model with independent Normal-Gamma prior. As an illustration of how the Savage–Dickey density ratio is useful, consider the case where the restricted model, M_1, imposes $\beta = \beta_0$. The case of other equality restrictions such as $R\beta = r$ is a simple extension. The unrestricted model, M_2, is the one discussed in the earlier part of this chapter, with likelihood given by (3.3) and prior given by (4.1) and (4.2). The Bayes factor comparing these two models is given by

$$BF_{12} = \frac{p(\beta = \beta_0|y, M_2)}{p(\beta = \beta_0|M_2)} \tag{4.23}$$

The denominator of this expression can be easily calculated, since the marginal prior for β is Normal. Using (4.1), the denominator is

$$p(\beta = \beta_0|M_2) = \frac{1}{(2\pi)^{\frac{k}{2}}}|\underline{V}|^{-\frac{1}{2}} \exp\left[-\frac{1}{2}(\beta_0 - \underline{\beta})'\underline{V}^{-1}(\beta_0 - \underline{\beta})\right] \tag{4.24}$$

which can be evaluated directly.

The numerator of (4.23) is slightly more difficult to evaluate since, although we know $p(\beta|y, h, M_2)$ is Normal, we do not know what $p(\beta|y, M_2)$ is. However, using the rules of probability and output from the Gibbs sampler, $p(\beta = \beta_0|y, M_2)$ can be estimated in a straightforward fashion. The Gibbs sampler will provide output, $\beta^{(s)}$ and $h^{(s)}$ for $s = S_0 + 1, \ldots, S$, and, it turns out, that simply averaging $p(\beta = \beta_0|y, h^{(s)}, M_2)$ across the draws $h^{(s)}$ will yield an estimate of

$p(\beta = \beta_0|y, M_2)$. To be precise,

$$\frac{1}{S_1}\sum_{s=S_0+1}^{S} p(\beta = \beta_0|y, h^{(s)}, M_2) \rightarrow p(\beta = \beta_0|y, M_2) \tag{4.25}$$

as S_1 goes to infinity. Remember that $S_1 = S - S_0$ is the number of draws retained after discarding S_0 initial draws. Since

$$p(\beta = \beta_0|, y, h^{(s)}, M_2) = \frac{1}{(2\pi)^{\frac{k}{2}}}|\overline{V}|^{-\frac{1}{2}} \exp\left[-\frac{1}{2}(\beta_0 - \overline{\beta})'\overline{V}^{-1}(\beta_0 - \overline{\beta})\right] \tag{4.26}$$

the average of the right-hand side of (4.25) can be calculated in a simple fashion.

To understand why (4.25) holds, note that the rules of probability imply

$$p(\beta = \beta_0|y, M_2) = \int p(\beta = \beta_0|y, h, M_2)p(h|y, M_2)\,dh$$

However, since $p(\beta = \beta_0|y, h, M_2)$ does not have β in it (since we have plugged in the value β_0), the only random variable inside the integral is h. Hence, we can write

$$p(\beta = \beta_0|y, M_2) = \int g(h)p(h|y)\,dh = E[g(h)|y]$$

where $g(h) = p(\beta = \beta_0|y, h, M_2)$. But as we have stressed before, posterior simulators are developed precisely to calculate things like $E[g(h)|y]$. Hence, (4.25) will give an estimate of $p(\beta = \beta_0|y, M_2)$ for exactly the same reason that (1.12) or (4.11) provided estimates of $E[g(\theta)|y]$ for any parameter vector θ and any function of interest $g(\theta)$.

It is worth noting that there is a myriad of models for which Gibbs sampling can be done in a straightforward manner. For such models, the Savage–Dickey density ratio is almost always easy to calculate by using a step like (4.25). The Savage–Dickey density ratio is, thus, a very powerful and widely-used tool for Bayes factor calculation.

4.2.6 Prediction

Prediction with the Normal linear regression model with natural conjugate prior is described in equations (3.38)–(3.40), and we will use the same notation as adopted there. That is, we want to carry out predictive inference on T unobserved values of the dependent variable, which we will denote by $y^* = (y_1^*, \ldots, y_T^*)'$, which are generated according to:

$$y^* = X^*\beta + \varepsilon^* \tag{4.27}$$

where ε^* is independent of ε and is $N(0, h^{-1}I_T)$ and X^* is an $T \times k$ matrix analogous to X, containing the k explanatory variables for each of the T out-of-sample data points.

The predictive density is calculated as

$$p(y^*|y) = \iint p(y^*|y, \beta, h)p(\beta, h|y)\,d\beta\,dh \tag{4.28}$$

The fact that ε^* is independent of ε implies that y and y^* are independent of one another and, hence, $p(y^*|y, \beta, h) = p(y^*|\beta, h)$, which can be written as

$$p(y^*|\beta, h) = \frac{h^{\frac{T}{2}}}{(2\pi)^{\frac{T}{2}}} \exp\left[-\frac{h}{2}(y^* - X^*\beta)'(y^* - X^*\beta)\right] \tag{4.29}$$

With the natural conjugate prior, the integral in (4.28) could be solved analytically and the predictive turned out to be a multivariate t density. Unfortunately, with the independent Normal-Gamma prior this integral cannot be solved analytically. Nevertheless, simulation methods allow for predictive inference to be carried out in a straightforward fashion.

Virtually any predictive feature of interest can be written in the form $E[g(y^*)|y]$ for some function $g(.)$. For instance, calculating the predictive mean of y_i^* implies $g(y^*) = y_i^*$, calculating the predictive variance requires knowledge of the predictive mean and $E[y_i^{*2}|y]$ and, hence, $g(y^*) = y_i^{*2}$ is of interest, etc.. Thus, interest centers on calculating

$$E[g(y^*)|y] = \int g(y^*)p(y^*|y)\,dy^* \tag{4.30}$$

Hopefully by now, equations like (4.30) will look familiar to you. That is, we have stressed that virtually anything the Bayesian may wish to calculate about the parameter vector θ will have the form

$$E[g(\theta)|y] = \int g(\theta)p(\theta|y)\,d\theta \tag{4.31}$$

for some $g(\theta)$. Except for the replacement of θ by y^*, (4.30) is identical to (4.31). Furthermore, our discussions of Monte Carlo integration and Gibbs sampling showed that, if $\theta^{(s)}$ for $s = 1, \ldots, S$ are draws from the posterior, then

$$\widehat{g}_S = \frac{1}{S}\sum_{s=1}^{S} g(\theta^{(s)})$$

will converge to $E[g(\theta)|y]$ as S increases.[4] This line of reasoning suggests that, if we can find $y^{*(s)}$ for $s = 1, \ldots, S$ which are draws from $p(y^*|y)$, then

$$\widehat{g}_Y = \frac{1}{S}\sum_{s=1}^{S} g(y^{*(s)}) \tag{4.32}$$

will converge to $E[g(y^*)|y]$. This is indeed the case.

The following strategy will provide the required draws of y^*. For every $\beta^{(s)}$ and $h^{(s)}$ provided by the Gibbs sampler, take a draw, $y^{*(s)}$ from $p(y^*|y, \beta^{(s)}, h^{(s)})$. Since the latter density is Normal (see (4.29)), such a strategy is quite simple. We now have $\beta^{(s)}, h^{(s)}$ and $y^{*(s)}$ for $s = 1, \ldots, S$. The rules of probability say that $p(\beta, h, y^*|y) = p(y^*|y, \beta, h)p(\beta, h|y)$ and, hence, the strategy of first drawing from the posterior, then drawing from $p(y^*|y, \beta, h)$ will yield draws from $p(\beta, h, y^*|y)$. Hence, our set of draws $\beta^{(s)}$, $h^{(s)}$ and $y^{*(s)}$ thus created can be used to evaluate any posterior feature of interest using (4.11) and any predictive feature of interest using (4.32).[5]

The strategy outlined in this section can be used for any model where a posterior simulator is used to provide draws from $p(\theta|y)$ and $p(y^*|y, \theta)$ has a form that is easy to work with. Almost all of the models discussed in the remainder of this book fall in this category. Hence, in future chapters you will often see a very brief discussion of prediction, including a sentence of the form 'Predictive inference in this model can be carried out using the strategy outlined in Chapter 4'.

4.2.7 Empirical Illustration

We use the house price data set introduced in Chapter 3 to illustrate the use of Gibbs sampling in the Normal linear regression model with independent Normal-Gamma prior. The reader is referred to Chapter 3 (Section 3.9) for a precise description of the dependent and explanatory variables for this data set. Section 3.9 discusses prior elicitation using a natural conjugate prior. This discussion implies that sensible values for the hyperparameters of the independent Normal-Gamma prior would be $\underline{\nu} = 5$ and $\underline{s}^{-2} = 4.0 \times 10^{-8}$, and

$$\underline{\beta} = \begin{bmatrix} 0.0 \\ 10 \\ 5000 \\ 10\,000 \\ 10\,000 \end{bmatrix}$$

[4] As discussed above, with Gibbs sampling you may wish to omit some initial burn-in draws and, hence, the summation would go from $S_0 + 1$ through S.

[5] This result uses the general rule that, if we have draws from the joint density $p(\theta, y^*|y)$, then the draws of θ considered alone are draws from the marginal $p(\theta|y)$ and the draws of y^* considered alone are draws from $p(y^*|y)$.

These values are identical to those used in the previous chapter, and have the same interpretation. However, we stress that $\underline{V}$ has a different interpretation in this chapter than the previous one. With the independent Normal-Gamma prior we have

$$var(\beta) = \underline{V}$$

while with the natural conjugate prior we had

$$var(\beta) = \frac{\underline{\nu}\underline{s}^2}{\underline{\nu} - 2}\underline{V}$$

Accordingly, to have a prior comparable to that used in the previous chapter, we set

$$\underline{V} = \begin{bmatrix} 10\,000^2 & 0 & 0 & 0 & 0 \\ 0 & 5^2 & 0 & 0 & 0 \\ 0 & 0 & 2500^2 & 0 & 0 \\ 0 & 0 & 0 & 5000^2 & 0 \\ 0 & 0 & 0 & 0 & 5000^2 \end{bmatrix}$$

Note that, with the independent Normal-Gamma prior, it is usually easy to elicit $\underline{V}$, since it is simply the prior variance of β. With the natural conjugate prior, the prior dependence between β and h means that the prior variance of β depends upon the prior you have elicited for h as well as $\underline{V}$.

Bayesian inference in this model can be done using Gibbs sampling. Most common Bayesian computer software (e.g. Jim LeSage's Econometrics Toolbox or BACC; see Section 1.3 of Chapter 1) allows for a thorough analysis of this model. The interested reader, at this stage, may wish to download and use this software. Alternatively, for the reader with some knowledge of computer programming, writing your own programs is a simple option. The website associated with this book contains MATLAB code for such a program (although for the MCMC convergence diagnostics a function from Jim LeSage's Econometrics Toolbox has been used). The structure of this program is very similar to the Monte Carlo integration program of the previous chapter, although it sequentially draws from $p(\beta|y, h)$ and $p(h|y, \beta)$, instead of simply drawing from $p(\beta|y)$.

Table 4.1 contains empirical results relating to β, including MCMC convergence diagnostics, for the Normal linear regression model with the independent Normal-Gamma prior specified above. We set the initial draw for the error precision to be equal to the inverse of the OLS estimate of σ^2 (i.e. $h^{(0)} = \frac{1}{s^2}$). We discard an initial $S_0 = 1000$ burn-in replications and include $S_1 = 10\,000$ replications. For the sake of brevity, we do not present results for h.

The posterior means and standard deviations are similar to those in Table 3.1, reflecting the fact that we have used similarly informative priors in the two chapters. The column labelled 'NSE' contains numerical standard errors for the

Table 4.1 Prior and Posterior Results for β (standard deviations in parentheses)

	Prior	Posterior	NSE	Geweke's CD	Post. Odds for $\beta_j = 0$
β_1	0 (10 000)	−4063.08 (3259.00)	28.50	−0.68	1.39
β_2	10 (5)	5.44 (0.37)	0.0029	0.11	6.69×10^{-42}
β_3	5000 (2500)	3214.09 (1057.67)	12.45	−0.57	0.18
β_4	10 000 (5000)	16 132.78 (1617.34)	15.56	0.55	2.06×10^{-19}
β_5	10 000 (5000)	7680.50 (979.09)	8.44	1.22	3.43×10^{-12}

approximation of $E(\beta_j|y)$ for $j = 1, \ldots, 5$, calculated using (4.13).[6] They can be interpreted as in the previous chapter, and indicate that our estimates are quite accurate. Of course, if a higher degree of accuracy is desired, the researcher can increase S_1. The column labeled 'Geweke's CD' is described in (4.14), and compares the estimate of $E(\beta_j|y)$ based on the first 1000 replications (after the burn-in replications) to that based on the last 4000 replications. If the effect of the initial condition has vanished and an adequate number of draws have been taken, then these two estimates should be quite similar. Noting that CD is asymptotically standard Normal, a common rule is to conclude that convergence of the MCMC algorithm has occurred if CD is less than 1.96 in absolute value for all parameters. Using this rule, Table 4.1 indicates that convergence of the MCMC algorithm has been achieved.

Table 4.1 also contains posterior odds ratios comparing two regression models: $M_1 : \beta_j = 0$ to $M_2 : \beta_j \neq 0$. As in Chapter 3, the restricted model uses an informative prior, which is identical to the unrestricted prior, except that $\underline{\beta}$ and $\underline{V}$ become 4×1 and 4×4 matrices, respectively, with prior information relating to β_j omitted. A prior odds ratio of one is used. The model comparison information in Table 4.1 is qualitatively similar to that in Table 3.1. That is, there is overwhelming evidence that β_2, β_4 and β_5 are non-zero, but some uncertainty as to whether β_1 and β_3 are zero. Note, however, that, if we compare empirical

[6]For the reader who knows spectral methods, $S(0)$ is calculated using a 4% autocovariance tapered estimate.

results in Table 4.1 with those in Chapter 3, the model comparison results are more different than the posterior means. This is a common finding. That is, prior information tends to have a bigger effect on posterior odds ratios than it does on posterior means and standard deviations. Hence, the slight difference in prior between Chapters 3 and 4 reveals itself more strongly in posterior odds ratios than in posterior means.

Working out the predictive density of the price of a house with given characteristics can be done using the methods outlined in Section 4.2.6. As in the previous chapter, we consider the case where the researcher is interested in predicting the sales price of a house with a lot size of 5000 square feet, two bedrooms, two bathrooms and one storey. Unlike with the natural conjugate prior, with the independent Normal-Gamma prior analytical results for the predictive distribution are unavailable. Nevertheless, the properties of the predictive can be calculated by making minor modifications to our posterior simulation program. That is, if we add one line of code which takes a random draw of $y^{*(s)}$ conditional on $\beta^{(s)}$ and $h^{(s)}$ using (4.29) and save the resulting draws, $y^{*(s)}$ for $s = S_0 + 1, \ldots, S$, we can then calculate any predictive property we wish using (4.32). Using these methods, we find that the predictive mean of a house with the specified characteristics is \$69 750 and the predictive standard deviation is 18 402. As expected, the figures are quite close to those obtained in the previous chapter.

For one (or at most two) dimensional features of interest, graphical methods can be a quite effective way of presenting empirical results. Figure 4.1 presents a plot of the predictive density. This figure is simply a histogram of all the draws, $y^{*(s)}$ for $s = S_0 + 1, \ldots, S$. The approximation arises since a histogram

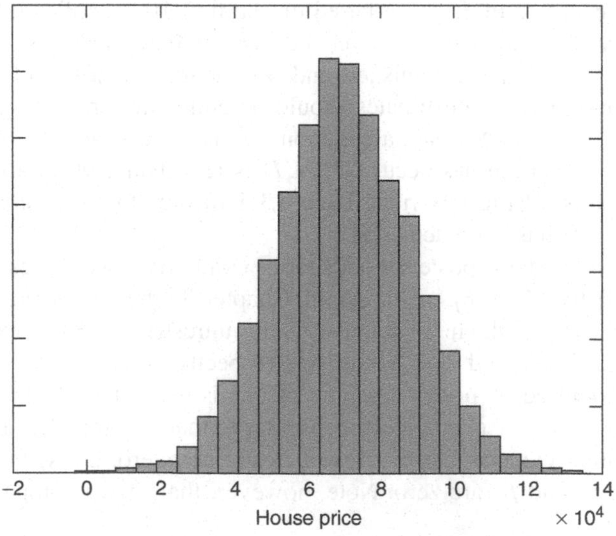

Figure 4.1 Predictive Density

is a discrete approximation to the continuous predictive density. This graph not only allows the reader to make a rough guess at the predictive mean, but also show the fatness of the tails of the predictive distribution. The graph shows that this data set does not allow for very precise predictive inference. Although the researcher's best prediction of the price of a house with a lot size of 5000 square feet, two bedrooms, two bathrooms and one storey is roughly $70 000, the predictive allocates non-negligible probability to the house price being less than $30 000 or more than $110 000.

4.3 THE NORMAL LINEAR REGRESSION MODEL SUBJECT TO INEQUALITY CONSTRAINTS

In this section, we discuss imposing inequality constraints on the coefficients in the linear regression model. This is something that the researcher may often wish to do. For instance, it may be desirable to impose concavity or monotonicity on a production function. In a model with autocorrelated errors (see Chapter 6, Section 6.5) the researcher may wish to impose stationarity. All such cases can be written in the form $\beta \in A$, where A is the relevant region. Bayesian analysis of the regression model subject to such restrictions is quite simple since we can simply impose them through the prior. To carry out posterior inference, we use something called 'importance sampling'. It is worth noting that, for some types of inequality constraints (e.g. linear inequality constraints such as $\beta_j > 0$), slightly simpler methods of posterior analysis are available. However, importance sampling is reasonably simple and works for any type of inequality constraint. Furthermore, importance sampling is a powerful tool which can be used with a wide variety of models, not only those with inequality constraints. Hence, we introduce the concept of importance sampling here, in the context of the familiar regression model. However, we stress it is a useful tool that works with many models. We remind the reader that the likelihood function for this model is the familiar one given in (3.3) or (3.7).

4.3.1 The Prior

It is convenient to introduce inequality restrictions through the prior. That is, saying $\beta \in A$ is equivalent to saying that a region of the parameter space which is not within A is *a priori* impossible and, hence, should receive a prior weight of 0. Such prior information can be combined with any other prior information. For instance, we can combine it with an independent Normal-Gamma or a natural conjugate prior. Here we combine it with the natural conjugate prior given in (3.8). Remember that a special case of this is the noninformative prior given in (3.24). Such a noninformative prior is useful in the common case where the researcher wishes to impose an inequality constraint on β, but has no other prior information.

Accordingly, our prior is given by

$$p(\beta, h) \propto f_{NG}(\beta, h|\underline{\beta}, \underline{V}, \underline{s}^{-2}, \underline{\nu})1(\beta \in A) \tag{4.33}$$

where $\underline{\beta}, \underline{V}, \underline{s}^{-2}, \underline{\nu}$ are prior hyperparameters to be selected by the researcher (see (3.8)) and $1(\beta \in A)$ is the indicator function, which equals 1 if $\beta \in A$ and 0 otherwise.

For future reference, note that, when the natural conjugate prior is used, the marginal prior for β has a t distribution for the same reasons that the marginal posterior does (see (3.14)). Hence, we can write the marginal prior for β as

$$p(\beta) \propto f_t(\beta|\underline{\beta}, \underline{s}^2\underline{V}, \underline{\nu})1(\beta \in A) \tag{4.34}$$

The noninformative variant of the natural conjugate prior involves setting $\underline{\nu} = 0$, $\underline{V}^{-1} = cI_k$ and letting c go to zero. This implies a prior of the form

$$p(\beta, h) \propto \frac{1}{h}1(\beta \in A) \tag{4.35}$$

4.3.2 The Posterior

The derivation of the posterior is the same as in Chapter 3, except we must impose the inequality constraints. Remember that, with the natural conjugate prior, we obtained a Normal-Gamma posterior for β and h (see (3.9)) and a multivariate t marginal posterior for β (see (3.14)). The noninformative prior was a special case of the natural conjugate prior. Here we obtain the same results, except these densities are truncated. Thus, $p(\beta, h|y)$ is Normal-Gamma truncated to the region $\beta \in A$ and $p(\beta|y)$ is multivariate t truncated to the region $\beta \in A$. Using the notation of Appendix B, Definition B.25, we obtain

$$p(\beta|y) \propto f_t(\beta|\overline{\beta}, \overline{s}^2\overline{V}, \overline{\nu})1(\beta \in A) \tag{4.36}$$

where $\overline{\beta}, \overline{s}^2, \overline{V}$ and $\overline{\nu}$ are defined in (3.10)–(3.13). The noninformative prior yields a posterior of the same form, except that $\overline{\beta}, \overline{s}^2, \overline{V}$ and $\overline{\nu}$ are defined in (3.20)–(3.23).

Combining inequality restrictions with the independent Normal-Gamma prior would imply that the formula for $p(\beta|y, h)$ in (4.6) would be multiplied by $1(\beta \in A)$.

4.3.3 Bayesian Computation: Importance Sampling

For some choices of A, analytical posterior results are available. For others, Gibbs sampling can be used, but for general choice of A, neither of these approaches work. Hence, we introduce a posterior simulation approach referred to as importance sampling. Because it is a generic method, we describe the basic ideas using our general notation, where θ is a vector of parameters and $p(y|\theta)$, $p(\theta)$ and $p(\theta|y)$ are the likelihood, prior and posterior, respectively.

Monte Carlo integration involved taking random draws from $p(\theta|y)$, but with many models this is not easy to do. Suppose instead, that random draws, $\theta^{(s)}$ for $s = 1, \ldots, S$, are taken from a density, $q(\theta)$, which is easy to draw from. This density is referred to as an *importance function*. Of course, simply taking draws from the importance function and averaging them in the familiar way is not appropriate. In particular,

$$\widehat{g}_S = \frac{1}{S}\sum_{r=1}^{S} g(\theta^{(s)})$$

will not converge to $E[g(\theta)|y]$ as $S \longrightarrow \infty$. For some intuition, consider the case where $q(\theta)$ and $p(\theta|y)$ have similar means, but $q(\theta)$ has a higher variance than $p(\theta|y)$. If random draws from $q(\theta)$ are taken, there will be too many of them out in the tails of $p(\theta|y)$ and too few near the mean of $p(\theta|y)$. What importance sampling does is correct for this by giving little weight to the former draws and giving more weight to the latter. In other words, instead of taking a simple average, importance sampling takes a weighted average.

Formally, the following theorem underlies importance sampling (see Geweke (1989) for more details and a proof; Kloek and van Dijk (1978) and Bauwens (1984) are important early references).

Theorem 4.2: Importance Sampling

Let $\theta^{(s)}$ for $s = 1, \ldots, S$ be a random sample from $q(\theta)$ and define

$$\widehat{g}_S = \frac{\sum_{s=1}^{S} w(\theta^{(s)})g(\theta^{(s)})}{\sum_{s=1}^{S} w(\theta^{(s)})} \tag{4.37}$$

where

$$w(\theta^{(s)}) = \frac{p(\theta = \theta^{(s)}|y)}{q(\theta = \theta^{(s)})} \tag{4.38}$$

then $\widehat{g}_S$ converges to $E[g(\theta)|y]$ as S goes to infinity (under weak conditions[7]).

In fact, because the weights appear in both the numerator and denominator of (4.37), we only need to be able to evaluate the kernels of $p(\theta|y)$ and $q(\theta)$. To be precise, if $p^*(\theta|y) \propto p(\theta|y)$ and $q^*(\theta) \propto q(\theta)$, (4.38) can be replaced by

$$w(\theta^{(s)}) = \frac{p^*(\theta = \theta^{(s)}|y)}{q^*(\theta = \theta^{(s)})} \tag{4.39}$$

and the theorem still holds.

---

[7]These conditions essentially amount to $q(\theta)$ having support which includes the support of $p(\theta|y)$ and $E[g(\theta)|y]$ existing.

At first glance, importance sampling seems a marvellous solution to any posterior simulation problem. After all, it seems to say that you can just randomly sample from any convenient density, $q(\theta)$, and simply weight using (4.37) to obtain an estimate of $E[g(\theta)|y]$. Unfortunately, in practice, things are not this easy. Unless $q(\theta)$ approximates $p(\theta|y)$ reasonably well, you can find cases where $w(\theta^{(s)})$ is virtually zero for almost every draw. This means the weighted average implicitly involves very few draws. Thus, S may have to be enormous to obtain reasonably accurate estimates of $E[g(\theta)|y]$. Thus, importance sampling may be infeasible unless $q(\theta)$ is very carefully chosen. Since selection of $q(\theta)$ can involve a lot of work and a different $q(\theta)$ is typically required for every class of models, most people use other strategies such as Gibbs sampling if at all possible. After all, once a particular blocking is chosen, Gibbs sampling simply involves drawing from the conditional posteriors (and monitoring convergence). Importance sampling involves hunting for and justifying a convenient class of important functions (e.g. the Normal class), and then fine tuning within this class (e.g. choosing the mean and variance of the Normal) to approximate $p(\theta|y)$. Particularly if θ is high-dimensional it can be extremely difficult to find a good importance function. Geweke (1989) discusses a reasonably general strategy for choosing importance functions and the interested reader is referred to this paper for more details. Richard and Zhang (2000) discuss a more generic strategy for getting good importance functions.

Fortunately, for the Normal linear regression model subject to inequality restrictions, an obvious importance function springs to mind and importance sampling can be carried out in a straightforward manner. Consider what happens if we set

$$q(\beta) = f_t(\beta|\overline{\beta}, \overline{s}^2\overline{V}, \overline{\nu}) \tag{4.40}$$

This importance function, since it is multivariate t, is easy to draw from. Furthermore, using (4.36) and (4.39), the weights can be calculated as

$$w(\beta^{(s)}) = 1(\beta^{(s)} \in A)$$

and (4.37) can be used to estimate $E[g(\beta)|y]$ for any function of interest, $g(.)$. Note that all these weights are either 1 (if $\beta^{(s)} \in A$) or 0 (if $\beta^{(s)} \notin A$). In other words, this strategy simply involves drawing from the unrestricted posterior and discarding draws which violate the inequality restrictions (i.e. the latter draws are given a weight of zero, which is equivalent to simply discarding them). Hence, it is very simple to implement (unless A is such a small region that virtually all the draws are discarded).

A numerical standard error can be calculated using a central limit theorem.

Theorem 4.3: A Numerical Standard Error

Using the setup and definitions of Theorem 4.2,

$$\sqrt{S}\{\widehat{g}_S - E[g(\theta)|y]\} \to N(0, \sigma_g^2) \tag{4.41}$$

as S goes to infinity, where σ_g^2 can be consistently approximated by:

$$\widehat{\sigma}_g^2 = \frac{\frac{1}{S}\sum_{s=1}^{S}\left[w(\theta^{(s)})\{g(\theta^{(s)}) - \widehat{g}_S\}\right]^2}{\left[\frac{1}{S}\sum_{s=1}^{S}w(\theta^{(s)})\right]^2}$$

Thus, a numerical standard error, $\frac{\widehat{\sigma}_g}{\sqrt{S}}$, can be calculated to guide the researcher in choosing S.

4.3.4 Model Comparison

The inequality restrictions usually make it impossible to calculate the marginal likelihood for this model directly. Depending upon the exact form of the models being compared, some of the model comparison techniques discussed above can be used. Alternatively, the generic method for marginal likelihood calculation which we will discuss in the next chapter can be used.

Here we discuss two particular sorts of model comparison. Consider first the case where M_1 is the model considered in this section where inequality restrictions are imposed on the Normal linear regression model with natural conjugate prior (i.e. $\beta \in A$). Let M_2 be the same model except that the inequality restrictions are violated (i.e. $\beta \notin A$). Since inequality restrictions are often implied by economic theory, comparing models of this form is often of interest. That is, a particular economic theory might imply $\beta \in A$ and, hence, $p(M_1|y)$ is the probability that the economic theory is correct.

A particular example of model comparison of this sort, for the case of linear inequality restrictions, is given in Chapter 3 (Section 6). As described in this previous material, model comparison involving such inequality restrictions is quite easy (and use of noninformative priors is not a problem). In practice, we can use the unrestricted Normal linear regression model with natural conjugate prior, and calculate $p(M_1|y) = p(\beta \in A|y)$ and $p(M_2|y) = 1 - p(M_1|y)$. If the inequality restrictions are linear, $p(\beta \in A|y)$ can be calculated analytically. Alternatively, the importance sampling strategy outlined in (4.40) allows for its simple calculation. That is, with the unrestricted model, $p(\beta \in A|y) = E[g(\theta)|y]$, where $g(\theta) = 1(\beta \in A)$. But as we have stressed, posterior simulation is designed precisely to evaluate such quantities. Hence, we can take random draws from the unrestricted posterior density (which is $f_t(\beta|\overline{\beta}, \overline{s}^2\overline{V}, \overline{\nu})$) and simply calculate the proportion which satisfy $\beta \in A$. This proportion is an estimate of $p(\beta \in A|y)$. But taking draws from $f_t(\beta|\overline{\beta}, \overline{s}^2\overline{V}, \overline{\nu})$ is precisely what we advocated in the importance sampling strategy outlined in (4.40). Hence, by doing importance sampling and keeping a record of how may draws are kept and how many are discarded (i.e. receive weight of zero), you can easily calculate $p(M_1|y)$ and $p(M_2|y)$.

The Savage–Dickey density ratio can be used to compare nested models where both have the same inequality restrictions imposed. That is, we now let M_2 be the model given in this section (i.e. the Normal linear regression model with natural conjugate prior with inequality restrictions imposed and posterior given by equation 4.36) and let M_1 be equal to M_2 except that $\beta = \beta_0$ is imposed. If the same prior is used for the error precision, h, in both models then the Savage–Dickey density ratio says that the Bayes factor can be calculated as

$$BF_{12} = \frac{p(\beta = \beta_0|y, M_2)}{p(\beta = \beta_0|M_2)}$$

Unfortunately, evaluating this Bayes factor is not as easy as it looks, since the results in (4.34) and (4.36) only provide the prior and posterior *kernels* (i.e. these equations have proportionality signs, not equality signs). Formally, the prior and posterior *densities* have the form

$$p(\beta) = \underline{c} f_t(\beta|\underline{\beta}, \underline{s}^2\underline{V}, \underline{\nu})1(\beta \in A)$$

and

$$p(\beta|y) = \overline{c} f_t(\beta|\overline{\beta}, \overline{s}^2\overline{V}, \overline{\nu})1(\beta \in A)$$

where $\underline{c}$ and $\overline{c}$ are prior and posterior integrating constants which ensure the densities integrate to one. The Savage–Dickey density ratio thus has the form

$$BF_{12} = \frac{\overline{c} f_t(\beta = \beta_0|\overline{\beta}, \overline{s}^2\overline{V}, \overline{\nu})}{\underline{c} f_t(\beta = \beta_0|\underline{\beta}, \underline{s}^2\underline{V}, \underline{\nu})} \tag{4.42}$$

Note that this involves evaluating two multivariate t densities at the point $\beta = \beta_0$ and calculating $\underline{c}$ and $\overline{c}$. For some hypotheses, it is easy to obtain $\underline{c}$ and $\overline{c}$. Consider, for instance, the case of univariate inequality restrictions such as $\beta_j > 0$. In this case, we can simply use statistical tables for the t distribution (or their computer equivalent) to obtain these integrating constants. For more general inequality restrictions, the method outlined in the previous paragraph can be used. That is, this method calculated $p(M_1|y)$ which was the probability that the restriction $\beta \in A$ held. But, $\overline{c} = \frac{1}{p(M_1|y)}$ since

$$\overline{c} = \frac{1}{\int f_t(\beta|\overline{\beta}, \overline{s}^2\overline{V}, \overline{\nu})1(\beta \in A)\, d\beta}$$

and $p(M_1|y) = \int f_t(\beta|\overline{\beta}, \overline{s}^2\overline{V}, \overline{\nu})1(\beta \in A)\, d\beta$. Calculation of $\underline{c}$ can be done in an analogous manner, except that importance sampling must be done on the prior instead of the posterior.

4.3.5 Prediction

The strategy outlined in (4.27) to (4.32) to carry out prediction can be implemented here with only slight modifications. With importance sampling, the draws from the importance function must be weighted as described in (4.37) and (4.38). In terms of our generic model notation, let $\theta^{(s)}$ be a random draw from an importance function, and $y^{*(s)}$ be a random draw from $p(y^*|y, \theta^{(s)})$ for $s = 1, \ldots, S$. Then

$$\widehat{g}_Y = \frac{\sum_{s=1}^{S} w(\theta^{(s)}) g(y^{*(s)})}{\sum_{s=1}^{S} w(\theta^{(s)})} \tag{4.43}$$

converges to $E[g(y^*)|y]$ as S goes to infinity, where $w(\theta^{(s)})$ is given in (4.38) or (4.39). This strategy for calculating predictive features of interest can be used anywhere importance sampling is done, including the Normal linear regression model with natural conjugate prior subject to inequality constraints.

4.3.6 Empirical Illustration

We continue with our empirical illustration using the house price data set. Remember that the dependent variable is the sales price of a house, and the explanatory variables are lot size, number of bedrooms, number of bathrooms and number of storeys. We would expect all of these explanatory variables to have a positive effect on the price of a house. Furthermore, let us suppose the researcher knows that $\beta_2 > 5$, $\beta_3 > 2500$, $\beta_4 > 5000$ and $\beta_5 > 5000$ and wishes to include this information in her prior. In terms of the terminology of (4.33), this defines the region A. The prior in (4.33) is the product of $1(\beta \in A)$ and a Normal-Gamma density and, hence, requires the elicitation of hyperparameters $\underline{\beta}$, $\underline{V}$, $\underline{s}^{-2}$ and $\underline{\nu}$. We choose the same values for these hyperparameters as in Chapter 3. That is, we set $\underline{s}^{-2} = 4.0 \times 10^{-8}$, $\underline{\nu} = 5$,

$$\underline{\beta} = \begin{bmatrix} 0.0 \\ 10 \\ 5000 \\ 10\,000 \\ 10\,000 \end{bmatrix}$$

and

$$\underline{V} = \begin{bmatrix} 2.40 & 0 & 0 & 0 & 0 \\ 0 & 6.0 \times 10^{-7} & 0 & 0 & 0 \\ 0 & 0 & 0.15 & 0 & 0 \\ 0 & 0 & 0 & 0.60 & 0 \\ 0 & 0 & 0 & 0 & 0.60 \end{bmatrix}$$

We use importance sampling to carry out inference in this model.[8] The computer code necessary to do this is a simple extension of the computer code used

[8]Note that, for simple restrictions of the sort considered in this empirical illustration, it would be possible to use Monte Carlo integration based on draws from the truncated Normal distribution.

to do Monte Carlo integration in the empirical illustration in Chapter 3. That is, we can use (4.40) as the importance function, but this importance function is precisely the same as the posterior in Chapter 3. The importance sampling weights are then calculated as (4.36). As described above (see the discussion after (4.38)), for this choice of importance function, the importance sampling weights are either equal to one (if the draw satisfies the constraints) or zero (if it does not). By taking weighted averages of the importance sampling draws, as in (4.35), we can calculate the posterior properties of β. Numerical standard errors can be calculated using the results in Theorem 4.3. Table 4.2 contains posterior means, standard deviations and NSEs of β along with a posterior odds ratio for comparing a model with $\beta_j = \underline{\beta}_j$ against the model with only the inequality restrictions imposed. This choice of models to compare is purely illustrative, and the posterior odds ratio is calculated using (4.42). Since $\beta_j = \underline{\beta}_j$ is a univariate restriction, $\underline{c}$ and $\overline{c}$ can be calculated using the properties of the univariate t distribution. Table 4.2 is based on 10 000 replications (i.e. $S = 10\,000$).

The results in Table 4.2 are quite close to those presented in Tables 3.1 or 4.1. Note that, for β_4 and β_5, the inequality restrictions we have imposed have little impact. That is, the unrestricted posterior means (standard deviations) for β_4 and β_5 in Table 3.1 are 16 965 (1708) and 7641 (997), respectively. Thus virtually all of the posterior probability is in the region where $\beta_4 > 5000$ and $\beta_5 > 5000$. Imposing the latter inequality restrictions through the prior thus has a minimal effect on the posterior. Intuitively, the data already tell us that $\beta_4 > 5000$ and $\beta_5 > 5000$, so incorporating these restrictions to the prior does not add any new information.

The inequality restrictions do, however, affect β_2 and β_3, increasing their posterior means somewhat. By cutting off the regions of the posterior with $\beta_2 < 5$ and $\beta_3 < 2500$, it is not surprising that the means increase. The posterior standard deviations in Table 4.2 are somewhat smaller than those in Table 3.1, indicating that the additional information provided in the prior decreases our posterior uncertainty about what the coefficients are.

The numerical standard errors indicate that we are achieving reasonably precise estimates and, as with any posterior simulator, if you wish more accurate estimates you can increase S. A careful comparison, however, with Table 3.4, indicates that NSEs (and, hence, approximation errors) are somewhat larger with

Table 4.2 Posterior Results for β

	Mean	Standard Deviation	NSE	Post. Odds for $\beta_j = \underline{\beta}_j$
β_1	−5645.47	2992.87	40.53	1.20
β_2	5.50	0.30	0.0041	1.36×10^{-29}
β_3	3577.58	782.58	10.60	0.49
β_4	16 608.02	1666.26	22.56	5.5×10^{-4}
β_5	7469.35	936.63	12.68	0.22

importance sampling than with Monte Carlo integration. For instance, with an identical number of replications, 10 000, the NSE relating to the estimation of $E(\beta_2|y)$ was 0.0037 with Monte Carlo integration and 0.0041 with importance sampling. Since Monte Carlo integration involves drawing directly from the posterior, and importance sampling involves drawing from an approximation to the posterior, the latter is less numerically efficient than the former.

The posterior odds ratios are in line with the evidence provided by posterior means and standard deviations. Except for the intercept, there is no strong evidence that $\beta_j = \underline{\beta}_j$. However, for β_3 and β_5 the posterior odds ratios attach a little bit of probability to the restrictions. Since, for these coefficients, the posterior means are not that far from $\underline{\beta}_j$ (relative to posterior standard deviations), the evidence of the posterior odds ratios is sensible.

The predictive density of the price of a house with given characteristics can be calculated as described in Section 4.3.5. That is, at each importance sampling draw for using the methods outlined in Section 4.2.6 can be used to take a random draw, $y^{*(s)}$ for $s = 1, \ldots, S$. These draws can then be averaged as described in (4.43) to obtain any predictive feature of interest. In the previous empirical illustration in Section 4.2.7, we took draws from $p(y^*|\beta^{(s)}, h^{(s)})$. This was simple to do since the latter density was Normal. It is straightforward to adopt the same strategy here, although we would have to extend our importance function to provide draws for $h^{(s)}$. The Normal-Gamma posterior in (3.9) would be a logical importance function for such a case. Alternatively, techniques analogous to those used to go from (3.39) to (3.40) imply that

$$p(y^*|y, \beta) = p(y^*|\beta) = f_t(y^*|X^*\beta, \overline{s}^2 I_T, \overline{\nu})$$

Hence, draws from $p(y^*|\beta^{(s)})$ can be taken from the t distribution. In case you are wondering where the inequality restrictions on β have gone to, note that our predictive draws taken from $p(y^*|\beta^{(s)})$ are conditional on the importance sampling draws from β. The latter draws already have the inequality restrictions imposed on them. If we use this method to work out the predictive density of the sales price of a house with a lot size of 5000 square feet, two bedrooms, two bathrooms and one storey, we find the predictive mean and standard deviation to be 69 408 and 18 246, respectively. These results are similar to those we have found in previous empirical illustrations using this data set.

4.4 SUMMARY

In this chapter, we have described Bayesian methods for posterior and predictive analysis and model comparison for the Normal linear regression model with two priors. The first of these is an independent Normal-Gamma prior and the second a natural conjugate prior subject to inequality restrictions. These priors were partly introduced because they are useful in many empirical settings. However, another reason for discussing them is that they allowed us to introduce important methods of computation in a familiar setting. The first of these computational methods is Gibbs sampling. In contrast to Monte Carlo integration, which involved drawing from the joint posterior distribution, Gibbs sampling involves sequentially drawing from the full posterior conditional distributions. Such draws can be treated as though they came from the joint posterior, although care has to be taken since Gibbs draws are not independent of one another, and can be dependent on the initial point chosen to start the Gibbs sampler. MCMC diagnostics are described which can be used to ensure that these two problems are overcome.

The second computational method introduced is importance sampling. This algorithm involves taking random draws from an importance function and then appropriately weighting the draws to correct for the fact that the importance function and posterior are not identical. This chapter also introduces the Savage–Dickey density ratio, which is a convenient way of writing the Bayes factor for nested model comparison.

At this stage, we have three posterior simulation algorithms: Monte Carlo integration, Gibbs sampling and importance sampling. The question of which one to use is a model-specific one. If it is easy to draw from the posterior, then Monte Carlo integration is the appropriate tool. If direct simulation of the posterior is difficult, but simulation from posterior conditionals is simple, then Gibbs sampling suggests itself. If neither Monte Carlo integration nor Gibbs sampling is easy, but a convenient approximation to the posterior suggests itself, then importance sampling is a sensible choice.

4.5 EXERCISES

4.5.1 Theoretical Exercises

1. *The Savage–Dickey density ratio.*
   (a) Prove Theorem 4.1. (Hint: If you are having trouble with this problem, the proof is provided in Verdinelli and Wasserman, 1995.)
   (b) How would your answer change if the condition $p(\psi|\omega = \omega_0, M_2) = p(\psi|M_1)$ did not hold?
2. For the Normal linear regression model with natural conjugate prior, the Bayes factor for comparing $M_1: \beta_i = 0$ to $M_2: \beta_i \neq 0$ (where the β_i is a single regression coefficient and the same prior is used for h in both models) can be obtained from Chapter 3 (3.34). Alternatively, this Bayes factor can be derived using the Savage–Dickey density ratio. Show that these two approaches lead to the same result.

4.5.2 Computer-Based Exercises

Remember that some data sets and MATLAB programs are available on the website associated with this book.

3. The purpose of this question is to learn about the properties of the Gibbs sampler in a very simple case. Assume that you have a model which yields a bivariate Normal posterior,

$$\begin{pmatrix} \theta_1 \\ \theta_2 \end{pmatrix} \sim N\left(\begin{bmatrix} 0 \\ 0 \end{bmatrix}, \begin{bmatrix} 1 & \rho \\ \rho & 1 \end{bmatrix}\right)$$

where $|\rho| < 1$ is the (known) posterior correlation between θ_1 and θ_2.
   (a) Write a program which uses Monte Carlo integration to calculate the posterior means and standard deviations of θ_1 and θ_2.
   (b) Write a program which uses Gibbs sampling to calculate the posterior means and standard deviations of θ_1 and θ_2. (Hint: Use the properties of the multivariate Normal in Appendix B, Theorem B.9 to work out the relevant conditional posterior distributions.)
   (c) Set $\rho = 0$ and compare the programs from parts (a) and (b). How many replications from each posterior simulator are necessary to estimate posterior means and standard deviations of θ_1 and θ_2 to two decimal places?
   (d) Repeat part (c) of this question for $\rho = 0.5, 0.9, 0.95, 0.99$ and 0.999. Discuss how the degree of correlation between θ_1 and θ_2 affects the performance of the Gibbs sampler.
   (e) Modify your Monte Carlo and Gibbs sampling programs to include numerical standard errors and (for the Gibbs sampling program) Geweke's convergence diagnostic. Repeat the parts (c) and (d) of this question. Do the numerical standard errors provide a correct view of the accuracy of approximation of the posterior simulators? Does the convergence diagnostic accurately indicate when convergence of the Gibbs sampler has been achieved?
4. The purpose of this question is to learn about the properties of importance sampling in a very simple case. Assume you have a model which a single parameter, θ, and its posterior is $N(0, 1)$.
   (a) Write a program which calculates the posterior mean and standard deviation of θ using Monte Carlo integration.
   (b) Write a program which calculates the posterior mean and standard deviation of θ using importance sampling, calculates a numerical standard error using Theorem 4.3 and calculates the mean and standard deviation of the importance sampling weights. Use the $f_t(\theta|0, 1, \nu_\theta)$ density as an importance function.
   (c) Carry out Monte Carlo integration and importance sampling with $\nu_\theta = 1, 3, 5, 10, 20, 50$ and 100 for a given number of replications (e.g. $S = 1000$). Compare the accuracy of the estimates across different algorithms and choices for ν_θ. What happens to the mean and standard deviation of the importance sampling weights as ν_θ increases?
   (d) Redo part (c) using $f_t(\theta|2, 1, \nu_\theta)$ as an importance function.

5 The Nonlinear Regression Model

5.1 INTRODUCTION

In previous chapters, we worked with the linear regression model:

$$y_i = \beta_1 + \beta_2 x_{i2} + \cdots + \beta_k x_{ik} + \varepsilon_i$$

where data was available on $i = 1, \ldots, N$ individuals. This model is useful not only when the relationship between the dependent and explanatory variables is a linear one, but also in cases where it can be transformed to linearity. For instance, the Cobb–Douglas production function relating an output, y, to inputs $x_2, \ldots, x_k$ is of the form

$$y = \alpha_1 x_2^{\beta_2}, \ldots, x_k^{\beta_k}$$

If we take logs of both sides of this equation and add an error term, we obtain a regression model:

$$\ln(y_i) = \beta_1 + \beta_2 \ln(x_{i2}) + \cdots + \beta_k \ln(x_{ik}) + \varepsilon_i$$

where $\beta_1 = \ln(\alpha_1)$. This specification is now linear in logs of the dependent and explanatory variables and, with this small difference, all the techniques of the previous chapters apply. The translog production function is another example of a nonlinear relationship which can be transformed to linearity.

There are, however, some functional forms which cannot be transformed to linearity. An example of an *intrinsically nonlinear* functional form is the constant elasticity of substitution (CES) production function, which is of the form

$$y_i = \left(\sum_{j=1}^{k} \gamma_j x_{ij}^{\gamma_{k+1}} \right)^{\frac{1}{\gamma_{k+1}}}$$

In this chapter, we consider Bayesian inference in regression models where the explanatory variables enter in an intrinsically nonlinear way. The empirical illustration will focus on the CES production function and, for this case, our nonlinear regression model will have the form:

$$y_i = \left(\sum_{j=1}^{k} \gamma_j x_{ij}^{\gamma_{k+1}} \right)^{\frac{1}{\gamma_{k+1}}} + \varepsilon_i \tag{5.1}$$

We use the same notation as before (e.g., see the discussion at the beginning of Chapter 3), and let ε and y be N-vectors stacking the errors and observations of the dependent variable, respectively, and let X be an $N \times k$ matrix stacking the observations of the k explanatory variables. We make the standard assumptions that:

1. ε is $N(0_N, h^{-1}I_N)$.
2. All elements of X are either fixed (i.e. not random variables) or, if they are random variables, they are independent of all elements of ε with a probability density function $p(X|\lambda)$, where λ is a vector of parameters that does not include any of the other parameters in the model.

The basic ideas discussed in this chapter will hold for the general nonlinear regression model:

$$y_i = f(X_i, \gamma) + \varepsilon_i$$

where X_i is the ith row of X, $f(\cdot)$ is a function which depends upon X_i and a vector of parameters, γ. With some abuse of notation, we write this model in matrix form as:

$$y = f(X, \gamma) + \varepsilon \tag{5.2}$$

where $f(X, \gamma)$ is now an N-vector of functions with ith element given by $f(X_i, \gamma)$. The exact implementation of the posterior simulation algorithm will depend upon the form of $f(\cdot)$ and, hence, we discuss basic concepts using (5.2) before discussing (5.1).

The nonlinear regression model is an important one in its own right. However, we also discuss it here, since it will give us a chance to introduce a number of techniques which are applicable in virtually any model. The linear regression model was a very special one in the sense that it was possible, in some cases, to obtain analytical posterior results (see Chapters 2 and 3). Even with priors which preclude the availability of analytical results, some special techniques are available for the Normal linear regression model (e.g. Gibbs sampling and the Savage–Dickey density ratio discussed in Chapter 4). However, many models do not allow for such specialized methods to be used and it is important to develop generic methods which can be used in any model. The nonlinear regression model allows us to introduce such generic methods in a context which is only a slight extension on our familiar linear regression model. With regards to posterior simulation we introduce a very important class of posterior simulators called the *Metropolis–Hastings* algorithms. These algorithms will be used in later chapters.

We will also introduce a generic method for calculating the marginal likelihood developed in Gelfand and Dey (1994), and a metric for evaluating the fit of a model called *the posterior predictive p-value*.

5.2 THE LIKELIHOOD FUNCTION

Using the definition of the multivariate Normal density, we can write the likelihood function of the nonlinear regression model as

$$p(y|\gamma, h) = \frac{h^{\frac{N}{2}}}{(2\pi)^{\frac{N}{2}}}\left\{\exp\left[-\frac{h}{2}\{y - f(X,\gamma)\}'\{y - f(X,\gamma)\}\right]\right\} \quad (5.3)$$

With the linear regression model, we were able to write this expression in terms of OLS quantities which suggested a form for a natural conjugate prior (see (3.7)). Here, no such simplification exists unless $f(\cdot)$ takes very specific forms.

5.3 THE PRIOR

Prior choice will depend upon what $f(\cdot)$ is and how γ is interpreted. For instance, with the CES production function in (5.1), γ_{k+1} is related to the elasticity of substitution between inputs. The researcher would likely have prior information about what plausible values for this parameter might be. Hence, prior elicitation is likely to be very dependent on the particular empirical context. In this section, some of the discussion proceeds at a completely general level, with the prior simply denoted by $p(\gamma, h)$, and some of the discussion uses a prior which was noninformative for the linear regression model:

$$p(\gamma, h) \propto \frac{1}{h} \quad (5.4)$$

This prior is Uniform for γ and $\ln(h)$. In many cases, this might be a sensible noninformative prior for the parameters of the nonlinear regression model.

5.4 THE POSTERIOR

The posterior density is proportional to the likelihood times prior and can be written as

$$p(\gamma, h|y) \propto p(\gamma, h)\frac{h^{\frac{N}{2}}}{(2\pi)^{\frac{N}{2}}}\left\{\exp\left[-\frac{h}{2}\{y - f(X,\gamma)\}'\{y - f(X,\gamma)\}\right]\right\} \quad (5.5)$$

In general, there is no way to simplify this expression, which will depend upon the precise forms for $p(\gamma, h)$ and $f(\cdot)$ and does not take the form of any well-known density. When the noninformative prior in (5.4) is used, the error precision, h, can be integrated out analytically in a step analogous to that required to derive (3.14). The resulting marginal posterior for γ is

$$p(\gamma|y) \propto [\{y - f(X,\gamma)\}'\{y - f(X,\gamma)\}]^{-\frac{N}{2}} \tag{5.6}$$

In the case where $f(\cdot)$ was linear, this expression could be rearranged to be put in the form of the kernel of a t distribution, but here it does not take any convenient form.

5.5 BAYESIAN COMPUTATION: THE METROPOLIS–HASTINGS ALGORITHM

The lack of analytical results relating to the posterior suggests that a posterior simulator is required. For some forms of $f(\cdot)$ it may be possible to derive a Gibbs sampler. In some cases, a convenient approximation to $p(\gamma|y)$ may suggest itself and, if so, importance sampling may be used. Here we introduce a third possibility, the Metropolis–Hastings algorithm. This can actually be thought of as a whole class of algorithms which can be used to create posterior simulators for a wide variety of models. As we have done previously when introducing a new algorithm, we adopt the general notation where θ is a vector of parameters and $p(y|\theta)$, $p(\theta)$ and $p(\theta|y)$ are the likelihood, prior and posterior, respectively.

The Metropolis–Hastings algorithm has some similarities with importance sampling. Both are useful in cases where the posterior itself is hard to take random draws from, but another convenient possibility exists. In importance sampling, we called the latter an importance function, with the Metropolis–Hastings algorithm we call it the *candidate generating density*. Let θ^* indicate a draw taken from this density, which we denote as $q(\theta^{(s-1)};\theta)$. This notation should be interpreted as saying that a candidate draw, θ^*, is taken of the random variable θ whose density depends upon $\theta^{(s-1)}$. In other words, as with the Gibbs sampler (but unlike importance sampling), the current draw depends on the previous draw. Thus, the Metropolis–Hastings algorithm, like the Gibbs sampler, is a Markov Chain Monte Carlo (MCMC) algorithm and the drawn values (i.e. $\theta^{(s)}$ for $s = 1, \ldots, S$) are often referred to as a chain.

With importance sampling, we corrected for the fact that the importance function differed from the posterior by weighting the draws differently from one another. With Metropolis–Hastings, we weight all draws equally, but not all the candidate draws are accepted. In other words, if $g(\cdot)$ is our function of interest, we can obtain an estimate of $E[g(\theta)|y]$, which we label $\widehat{g}_S$, by simply averaging the draws in the familiar way:

$$\widehat{g}_S = \frac{1}{S}\sum_{r=1}^{S} g(\theta^{(s)}) \tag{5.7}$$

The Metropolis–Hastings algorithm always takes the following form:

Step 0: Choose a starting value, $\theta^{(0)}$.
Step 1: Take a candidate draw, θ^* from the candidate generating density, $q(\theta^{(s-1)};\theta)$.
Step 2: Calculate an acceptance probability, $\alpha(\theta^{(s-1)},\theta^*)$.
Step 3: Set $\theta^{(s)}=\theta^*$ with probability $\alpha(\theta^{(s-1)},\theta^*)$ and set $\theta^{(s)}=\theta^{(s-1)}$ with probability $1-\alpha(\theta^{(s-1)},\theta^*)$.
Step 4: Repeat Steps 1, 2 and 3 S times.
Step 5: Take the average of the S draws $g(\theta^{(1)}),\ldots,g(\theta^{(S)})$.

These steps will yield an estimate of $E[g(\theta)|y]$ for any function of interest.

As with Gibbs sampling, the Metropolis–Hastings algorithm usually requires the choice of a starting value, $\theta^{(0)}$. To make sure that the effect of this starting value has vanished, it is usually wise to discard S_0 initial draws. The MCMC diagnostics presented for the Gibbs sampler can be used with the Metropolis–Hastings algorithm to make sure an adequate number of draws are taken and enough initial draws discarded (see Chapter 4, Section 4.2.4, for details).

We give a precise formula for the acceptance probability, $\alpha(\theta^{(s-1)},\theta^*)$, shortly. But first, it is useful to discuss the properties that a good acceptance probability should have.

In the previous chapter we introduced the intuition that one can interpret an MCMC algorithm as wandering over the posterior, taking most draws in areas of high posterior probability and proportionately fewer in areas of low posterior probability. The candidate generating density is not identical to the posterior and, hence, if left to wander freely, will not take the right number of draws in each area of the parameter space. What the Metropolis–Hastings algorithm does is correct for this by not accepting every candidate draw. It derives an acceptance probability which is highest in areas where posterior probability is highest and lowest in areas where posterior probability is lowest. Intuitively, if $\theta^{(s-1)}$ is in an area of low posterior probability the algorithm will tend to move quickly away from $\theta^{(s-1)}$ (i.e. the current location of the chain is in a region of low probability so candidate draws, which move you away from the current location, are likely to be accepted). However, if $\theta^{(s-1)}$ is in an area of high posterior probability the algorithm will tend to stay there (see Step 3, and note that the algorithm can stay at a particular point by setting $\theta^{(s)}=\theta^{(s-1)}$). By staying at this point of high posterior probability the algorithm is implicitly giving it more weight in an intuitively similar manner as importance sampling weights draws. Similar considerations hold with respect to the candidate draw, θ^*. For a given $\theta^{(s-1)}$, we want a candidate draw, θ^*, to be accepted with high probability if it is in a region of higher posterior probability than $\theta^{(s-1)}$. Candidate draws of θ^* in lower probability areas we want to be rejected with high probability.

The previous paragraph has provided intuition for an acceptance probability which depends on θ^* and $\theta^{(s-1)}$ in a manner which tends to move the chain away from areas of low posterior probability towards higher. Of course, we stress that the word 'tends' in the previous sentence is quite important. You do not want a chain which always stays in regions of high posterior probability, you want it to visit areas of low probability as well (but proportionately less of the time). The way the acceptance probability is constructed means that the chain will usually, but not always, move from areas of low posterior probability to high.

Chib and Greenberg (1995) is an excellent introduction to the Metropolis–Hastings algorithm and includes a derivation of the acceptance probability which ensures that the Metropolis–Hastings algorithm converges to the posterior. The reader is referred to this paper or Gilks, Richardson and Speigelhalter (1996a)[1] for more details. It turns out that the acceptance probability has the form

$$\alpha(\theta^{(s-1)}, \theta^*) = \min\left[\frac{p(\theta = \theta^*|y)q(\theta^*; \theta = \theta^{(s-1)})}{p(\theta = \theta^{(s-1)}|y)q(\theta^{(s-1)}; \theta = \theta^*)}, 1\right] \tag{5.8}$$

Remember that $p(\theta = \theta^*|y)$ is our notation for the posterior density evaluated at the point $\theta = \theta^*$, while $q(\theta^*; \theta)$ is a density for random variable θ and, hence, $q(\theta^*; \theta = \theta^{(s-1)})$ is this density evaluated at the point $\theta = \theta^{(s-1)}$. It can be verified that this acceptance probability has the intuitively desirable properties discussed above. The 'min' operator in (5.8) is included to ensure that the acceptance probability is not greater than 1.

Like importance sampling, the Metropolis–Hastings algorithm, at first glance, seems a marvellous solution to any posterior simulation problem. After all, it seems to say that you can just randomly sample from any convenient density, $q(\theta^{(s-1)}; \theta)$, and accept or reject candidate draws using (5.8) to obtain a sequence of draws, $\theta^{(s)}$ for $s = 1, \ldots, S$, which can be used to produce an estimate of $E(g(\theta)|y)$. Unfortunately, in practice, things are not this easy. If the candidate generating density is not well-chosen, virtually all of the candidate draws will be rejected and the chain will remain stuck at a particular point for long periods of time. Hence, care must be taken when choosing the candidate generating density and the MCMC diagnostics described in Chapter 4 should always be used to verify convergence of the algorithm. There is a myriad of possible strategies which can be used for choosing candidate generating densities. In the following, we discuss two of the more common ones.

5.5.1 The Independence Chain Metropolis–Hastings Algorithm

As its name suggests, the *Independence Chain Metropolis–Hastings algorithm* uses a candidate generating density which is independent across draws. That is, $q(\theta^{(s-1)}; \theta) = q^*(\theta)$ and the candidate generating density does not depend upon $\theta^{(s-1)}$. Such an implementation is useful in cases where a convenient

[1] Tierney (1996) is also relevant for the reader interested in mathematical rigor.

approximation exists to the posterior. This convenient approximation can be used as a candidate generating density. If this is done, the acceptance probability simplifies to

$$\alpha(\theta^{(s-1)}, \theta^*) = \min\left[\frac{p(\theta = \theta^*|y)q^*(\theta = \theta^{(s-1)})}{p(\theta = \theta^{(s-1)}|y)q^*(\theta = \theta^*)}, 1\right] \tag{5.9}$$

The Independence Chain Metropolis–Hastings algorithm is closely related to importance sampling. This can be seen by noting that, if we define weights analogous to the importance sampling weights (see Chapter 4 (4.38)):

$$w(\theta^A) = \frac{p(\theta = \theta^A|y)}{q^*(\theta = \theta^A)}$$

the acceptance probability in (5.9) can be written as

$$\alpha(\theta^{(s-1)}, \theta^*) = \min\left[\frac{w(\theta^*)}{w(\theta^{(s-1)})}, 1\right]$$

In words, the acceptance probability is simply the ratio of importance sampling weights evaluated at the old and candidate draws.

In terms of the nonlinear regression model, the usefulness of this algorithm depends on whether $f(\cdot)$ is of a form such that a convenient approximating density can be found. There is no completely general procedure for choosing an approximating density. However one is chosen, the MCMC diagnostics described in Chapter 4 should be used to verify that the resulting algorithm has converged.

One common procedure involves the use of frequentist maximum likelihood results as a basis for finding a good $q^*(\theta)$. The purely Bayesian reader may wish to skip the remainder of this paragraph and go to the bottom line recommendation for empirical practice given in the next. The frequentist econometrician will know that (subject to mild regularity conditions) the maximum likelihood estimator, $\widehat{\theta}_{ML}$, is asymptotically Normal with asymptotic covariance matrix given by

$$var(\widehat{\theta}_{ML}) = I(\theta)^{-1}$$

where $I(\theta)$ is the information matrix defined as the negative of the expected value of the second derivative of the log of the likelihood function (where the expectation is taken with respect to y):

$$I(\theta) = -E\left[\frac{\partial^2 \ln(p(y|\theta))}{\partial\theta\partial\theta'}\right]$$

In words, if the sample size is large, the inverse of the information matrix is going to give you a good idea of the shape of $p(y|\theta)$. Even if the information matrix cannot be directly calculated, $\frac{\partial^2 \ln(p(y|\theta))}{\partial\theta\partial\theta'}$ can be calculated (either by hand or through numerical differentiation procedures which are available in computer packages such as MATLAB) and used to obtain an approximation to $var(\widehat{\theta}_{ML})$ which we shall denote by $\widehat{var}(\widehat{\theta}_{ML})$. To introduce some additional jargon $\frac{\partial^2 \ln(p(y|\theta))}{\partial\theta\partial\theta'}$ is referred to as the *Hessian*, and you will often hear the phrase 'negative inverse Hessian' to indicate the estimate $\widehat{var}(\widehat{\theta}_{ML})$.

For the Bayesian, the discussion in the previous paragraph implies that, if sample size is reasonably large and the prior relatively noninformative, then the posterior might be approximately Normal with mean $\widehat{\theta}_{ML}$ and covariance matrix approximately $\widehat{var}(\widehat{\theta}_{ML})$. For some models, computer software exists which allows for direct calculation of such maximum likelihood quantities. Alternatively, computer packages such as MATLAB have associated routines which can optimize a user-specified function. These can be used to calculate maximum likelihood quantities. If you need to program up the routine which maximizes the likelihood function and finds $\widehat{var}(\widehat{\theta}_{ML})$, you may wish to maximize the posterior (i.e. find $\widehat{\theta}_{\max}$, the posterior mode) and take second-derivatives of the posterior to find an approximation to $var(\widehat{\theta}_{\max})$, instead. If you are using an informative prior, such a strategy will approximate your posterior a bit better. That is, asymptotic results suggest that the posterior might be approximately $f_N(\theta|\widehat{\theta}_{\max}, \widehat{var}(\widehat{\theta}_{\max}))$. In the following material, we will base our approximation on maximum likelihood results, but they can be replaced by $\widehat{\theta}_{\max}$ and $\widehat{var}(\widehat{\theta}_{\max})$ if these are available.

Setting $q^*(\theta) = f_N(\theta|\widehat{\theta}_{ML}, \widehat{var}(\widehat{\theta}_{ML}))$ can work well in some cases. However, it is more common to use the t distribution as a candidate generating density and set $q^*(\theta) = f_t(\theta|\widehat{\theta}_{ML}, \widehat{var}(\widehat{\theta}_{ML}), \nu)$. The reason for this is that, in practice, researchers have found that it is quite important for the candidate generating density to have tails which are at least as fat as those of the posterior. Geweke (1989) motivates why this is so for the case of importance sampling, but the same reasoning holds for the Independence Chain Metropolis–Hastings algorithm. The interested reader is referred to this paper for more detail.

The Normal density has very thin tails. The t-density has much fatter tails, especially for small values of ν. Useful properties of the t distribution are that, as $\nu \to \infty$, it approaches the Normal, and as ν gets small its tails get very fat. In fact, the t distribution with $\nu = 1$ is the Cauchy, a distribution with such fat tails that its mean is infinite (even though its median and mode are finite). In some cases it is possible, through examination of the posterior, to find a value for ν which ensures that the tails of the candidate generating density dominate those of the posterior. In general, however, a careful researcher will choose a small value for ν and use MCMC diagnostics to ensure the algorithm has converged.

It is important to stress that there are cases where use of a t-density to generate candidate draws is inappropriate. For instance, if the posterior is multi-modal then the unimodal t-density will usually not work well. Also, if the posterior is defined over a limited range (e.g. the Gamma density is defined only over the positive real numbers), then the t-density (which is defined over all the real numbers) might not work well (unless the posterior is sharply defined in the interior of its range).

In the case of the nonlinear regression model, maximizing the likelihood function (posterior) would involve writing a program which evaluates (5.3) (or (5.5)). Obtaining $var(\widehat{\theta_{ML}})$ would either involve second differentiating (5.3) or using a numerical differentiation subroutine available with most relevant computer packages. These steps would be dependent upon the precise form of $f(\cdot)$, so we will not say anything more about the nonlinear regression model at this stage.

From the previous discussion, you can see that finding an approximating density for use with either an Independence Chain Metropolis–Hastings algorithm or with importance sampling can be a bit of an art. Nevertheless, for most models asymptotic results exist which say that, as sample size goes to infinity, the posterior becomes Normal. For models in this class, if your sample size is reasonably large, $f_t(\theta|\widehat{\theta_{ML}}, var(\widehat{\theta_{ML}}), \nu)$ should approximate the posterior reasonably well. The empirical illustration below uses this strategy to estimate the CES model given in (5.1).

5.5.2 The Random Walk Chain Metropolis–Hastings Algorithm

The *Random Walk Chain Metropolis–Hastings algorithm* is useful when you cannot find a good approximating density for the posterior. Intuitively, with the Independence Chain Metropolis–Hastings algorithm (like importance sampling), you take draws from a density which is similar to the posterior and the acceptance probability (or weighting, in the case of importance sampling) is used to correct for the difference between the posterior and approximating densities. With the Random Walk Chain Metropolis–Hastings algorithm, no attempt is made to approximate the posterior, rather the candidate generating density is chosen to wander widely, taking draws proportionately in various regions of the posterior.

Formally, the Random Walk Chain Metropolis–Hastings algorithm generates candidate draws according to

$$\theta^* = \theta^{(s-1)} + z \tag{5.10}$$

where z is called the *increment random variable*. For the reader familiar with time series methods, the assumptions in (5.10) imply that candidates are generated according to a random walk. That is, candidates are drawn in random directions from the current point. The acceptance probability ensures that the chain moves in the appropriate direction. Note that θ^* and $\theta^{(s-1)}$ enter symmetrically in (5.10) and, hence, it will always be the case that $q(\theta^*; \theta = \theta^{(s-1)}) = q(\theta^{(s-1)}; \theta = \theta^*)$. This means that the acceptance probability simplifies to

$$\alpha(\theta^{(s-1)}, \theta^*) = \min\left[\frac{p(\theta = \theta^*|y)}{p(\theta = \theta^{(s-1)}|y)}, 1\right] \tag{5.11}$$

and it can be clearly seen that the random walk chain tends to move towards regions of higher posterior probability.

The choice of density for z determines the precise form of the candidate generating density. A common and convenient choice is the multivariate Normal. In this case, (5.10) determines the mean of the Normal (i.e. $\theta^{(s-1)}$ is the mean) and the researcher must choose the covariance matrix, which we shall denote by Σ. In terms of our notation for the Normal density:

$$q(\theta^{(s-1)}; \theta) = f_N(\theta|\theta^{(s-1)}, \Sigma) \tag{5.12}$$

With this approach, all that the researcher needs to select is Σ. This should be selected so that the acceptance probability tends to be neither too high nor too low. If the acceptance probability is usually very small, then candidate draws are almost always rejected and the chain tends to move only rarely. This is not a good situation as it will imply S must be huge if the chain is to move over the entire posterior. Small acceptance probabilities indicate that Σ is too large and most candidate draws are being generated far out in the tails of the posterior in regions which the posterior indicates are quite improbable. At the other extreme, it is not good to have the acceptance probability always being near one as this indicates that Σ is too small. That is, if Σ is very small, then θ^* and $\theta^{(s-1)}$ will tend to be very close to one another the acceptance probability will be near one (see (5.11)). In this case, too, S may have to be unfeasibly large to ensure that the chain explores the entire posterior.

There is no general rule which gives the optimal acceptance rate. In the special case where the posterior and candidate generating density are both Normal, the optimal acceptance rate has been calculated to be 0.45 in one-dimensional problems with slightly lower values in higher dimensional problems. As the number of dimensions approaches infinity, the optimal acceptance probability approaches 0.23. Another rule of thumb often-mentioned is that the acceptance probability should be roughly 0.5. In general, if you choose Σ to ensure that your acceptance probability is roughly in this region, you are unlikely to go too far wrong. However, you should always use the MCMC diagnostics discussed in Chapter 4 to verify that your algorithm has converged.

Saying that Σ should be chosen to ensure that the average acceptance probability is in the region 0.2 to 0.5 is enough information to select it in the case where θ (and, hence, Σ) is a scalar. In this case, you can simply experiment, running your Random Walk Metropolis–Hastings algorithm with various values for Σ until you find one which yields reasonable acceptance probabilities. In cases where θ is p-dimensional such an approach can be much harder since Σ will have $\frac{p(p+1)}{2}$ elements. In such cases, it usually works well to set $\Sigma = c\Omega$ where c is a scalar and Ω is an estimate of the posterior covariance matrix of θ. Then you can experiment with different values of c until you find one which yields a reasonable acceptance probability. This approach requires you to find Ω, an estimate of $var(\theta|y)$. This can be done in two ways. From the point of view of the researcher, the simplest way is to begin with $\Sigma = cI_p$ and try and find a value for c which does not imply completely useless values for the acceptance probabilities (i.e. if they are such that candidate draws are accepted at a rate of 0.000001 or 0.99999 they are likely to be completely useless, unless you have enough computing power to easily take billions of draws). This value of c can then be used to get a very crude estimate of Ω. You can then set $\Sigma = c\Omega$ and try and find a new value for c which yields a slightly more reasonable acceptance probability. Results for this case can be used to get a better Ω which can then be used to find a better value for Σ, etc. This process can be repeated until a good value of Σ is found. Such an approach is simple for the researcher since, once the basic computer code for the Random Walk Metropolis–Hastings algorithm has been created, no additional programming is required. However, it can be quite demanding in terms of the computer's time.

An alternative approach is to set Ω equal to $\widehat{var(\widehat{\theta}_{ML})}$, the estimate of the variance of the maximum likelihood estimate described above in the discussion of the Independence Chain Metropolis–Hastings algorithm. Such an approach, of course, requires the calculation of $\widehat{var(\widehat{\theta}_{ML})}$ which will require additional programming by the researcher.

In the empirical illustration we compare the Random Walk to Independence Chain Metropolis–Hastings algorithm in an application involving the CES production function.

5.5.3 Metropolis-within-Gibbs

The Metropolis–Hastings algorithm provides a posterior simulator for $p(\theta|y)$. In Chapter 4, we introduced another posterior simulator called the Gibbs sampler which, in the case of two blocks where $\theta = (\theta'_{(1)}, \theta'_{(2)})'$, involved sequentially drawing from $p(\theta_{(1)}|y, \theta_{(2)})$ and $p(\theta_{(2)}|y, \theta_{(1)})$. For the Normal linear regression model with independent Normal-Gamma prior, a Gibbs sampler was simple to implement since $p(\beta|y, h)$ was a Normal density and $p(h|y, \beta)$ was a Gamma density. In the nonlinear regression model, a noninformative or independent Gamma prior for h will imply that $p(h|y, \gamma)$ is a Gamma density. However, $p(\gamma|y, h)$ will be proportional to (5.5) and, hence, will not take the form of a density which is convenient to draw from. At first sight, this seems to imply that a Gibbs sampler involving $p(h|y, \gamma)$ and $p(\gamma|y, h)$ cannot be set up for the nonlinear regression model. However, it can be shown that, if we use a Metropolis–Hastings algorithm for $p(\gamma|y, h)$, the resulting simulated draws, $\gamma^{(s)}$ and $h^{(s)}$ for $s = 1, \ldots, S$, are valid posterior simulator draws. Formally, using a Metropolis–Hastings algorithm for either (or both) of the posterior conditionals used in the Gibbs sampler, $p(\theta_{(1)}|y, \theta_{(2)})$ and $p(\theta_{(2)}|y, \theta_{(1)})$, is perfectly acceptable. This statement is also true if the Gibbs sampler involves more than two blocks. Such *Metropolis-within-Gibbs* algorithms are fairly common since many models have posteriors where most of the conditionals are easy to draw from, but one or two conditionals do not have a convenient form. For the latter conditionals, Metropolis–Hastings algorithms can be used. The empirical illustration below will show how this can be done in the case of the nonlinear regression model.

5.6 A MEASURE OF MODEL FIT: THE POSTERIOR PREDICTIVE P-VALUE

The typical Bayesian method of model comparison is the posterior odds ratio, which is the relative probability of two completely specified models. However, there are some cases where the researcher is interested in investigating the performance of a model in some absolute sense, not relative to a specific alternative model. Also, there are many cases in which the researcher might want to use an improper, noninformative, prior and, as discussed in Chapter 3, posterior odds can be meaningless if such a prior is used on parameters which are not common to all models. In both such situations, the posterior predictive p-value approach is a sensible alternative to the posterior odds ratio. The reader interested in more details about this approach is referred to Gelman and Meng (1996). For some innovative extensions on the basic approach outlined below, the reader is referred to Bayarri and Berger (2000).

To motivate the posterior predictive p-value approach, it is important to distinguish between y, the data actually observed, and $y^\dagger$, observable data which could be generated from the model under study (i.e. $y^\dagger$ is an $N \times 1$ random vector with p.d.f. $p(y^\dagger|\theta)$ where the latter is the likelihood function without y plugged in). Let $g(\cdot)$ be some function of interest. $p(g(y^\dagger)|y)$ summarizes everything our model says about $g(y^\dagger)$ after seeing the data. In other words, it tells us the types of data sets that our model can generate. For the observed data we can directly calculate $g(y)$. If $g(y)$ is in the extreme tails of $p(g(y^\dagger)|y)$, then the model cannot do a good job of explaining $g(y)$ (i.e. $g(y)$ is not the sort of data characteristic that can plausibly be generated by the model). Formally, we can obtain tail area probabilities in a manner similar to frequentist p-value calculations. In particular, the posterior predictive p-value is the probability of a model yielding a data set with more extreme properties than that actually observed (i.e. analogous to a frequentist p-value. You may wish to present either a one-tailed or two-tailed p-value).

$p(g(y^\dagger)|y)$ can be calculated using simulation methods in a manner which is very similar to the one we used for predictive inference. That is, analogous to (4.28) and the discussion which follows it, we can write

$$p(g(y^\dagger)|y) = \int p(g(y^\dagger)|\theta, y)p(\theta|y)\,d\theta = \int p(g(y^\dagger)|\theta)p(\theta|y)d\theta \quad (5.13)$$

where the last equality follows from the fact that, conditional on θ, the actual data provides no additional information about $y^\dagger$. The posterior simulator provides draws from $p(\theta|y)$ and we can simulate from $p(g(y^\dagger)|\theta)$ by merely simulating artificial data from the model for a given parameter value in a manner identical to that used for prediction (see (4.30)–(4.32) of Chapter 4).

Posterior predictive p-values can be used in two different ways. First, they can be used as a measure of fit, of how likely the model was to have generated the data in an absolute sense. Secondly, they can be used to compare different models.

That is, if one model yields posterior predictive p-values which are much lower than another, this is evidence against the former model. However, most Bayesians prefer posterior odds ratios for the latter unless the use of noninformative priors makes the posterior odds ratios meaningless or difficult to interpret.

The posterior predictive p-value approach requires the selection of a function of interest, $g(\cdot)$. The exact choice of $g(\cdot)$ will vary depending upon the empirical application. To take a practical example, let us return to the nonlinear regression model. For this model, we have

$$y_i^\dagger = f(X_i, \gamma) + \varepsilon_i$$

for $i = 1, \ldots, N$. Alternatively, given the assumptions we have made about the errors,

$$p(y^\dagger|\gamma, h) = f_N(y^\dagger|f(X, \gamma), h^{-1}I_N) \tag{5.14}$$

where $f(X, \gamma)$ is the N-vector defined in (5.2). Note that, for given values of the parameters of the model, simulating values for $y^\dagger$ is quite simple, involving only taking draws from the multivariate Normal. This simplicity is common to many models, making the posterior predictive p-value easy to calculate in a wide variety of cases.

For the nonlinear regression model with noninformative prior given in (5.4), (5.14) can be simplified even further, since h can be integrated out. In particular, using a derivation virtually identical to that required to go from (5.5) to (5.6), it can be shown that

$$p(y^\dagger|\gamma) = f_t(y^\dagger|f(X, \gamma), \overline{s}^2 I_N, N) \tag{5.15}$$

where

$$\overline{s}^2 = \frac{[y - f(X, \gamma)]'[y - f(X, \gamma)]}{N} \tag{5.16}$$

Hence, conditional on γ, draws of $y^\dagger$ can be taken using the multivariate t distribution. These draws can be interpreted as reflecting the sorts of data sets that this model can generate. The posterior predictive p-value approach uses the idea that, if the model is a reasonable one, the actual observed data set should be of the type which is commonly generated by the model. Finding out at what percentile the point $g(y)$ lies in the density $p(g(y^\dagger)|y)$ is the formal metric used.

To make things more concrete, let us digress briefly to motivate a few choices for $g()$. It is common to evaluate the fit of a model through *residual analysis*. The frequentist econometrician might calculate OLS estimates of the errors, ε_i, and call these residuals. The properties of these residuals can then be investigated to shed light on whether the assumptions underlying the model are reasonable. In the Bayesian context, the errors are given, for $i = 1, \ldots, N$, by

$$\varepsilon_i = y_i - f(X_i, \gamma)$$

We have assumed these errors to have various properties. In particular, we have assumed them to be i.i.d. $N(0, h^{-1})$. These assumptions might be unreasonable in a particular data set and, hence, the researcher may wish to test them. The brief statement 'the errors are i.i.d. Normal' involves many assumptions (e.g. the assumption that the errors are independent of one another, that they have a common variance, etc.), and the researcher might choose to investigate any of them. Here we will focus on aspects of the Normality assumption. Two of the properties of the Normal are that it is symmetric, and its tails have a particular shape. In terms of statistical jargon, the Normal distribution does not exhibit skewness and its tails have a particular kurtosis (see Appendix B, Definition B.8). Skewness and kurtosis are measured in terms of the third and fourth moments of the distribution and, for the standard Normal (i.e. the $N(0, 1)$) the third moment is zero and fourth moment is three. The Normality assumption thus implies that the following commonly-used measures of skewness and excess kurtosis should both be zero:

$$Skew = \frac{\sqrt{N}\sum_{i=1}^{N}\varepsilon_i^3}{\left[\sum_{i=1}^{N}\varepsilon_i^2\right]^{\frac{3}{2}}} \tag{5.17}$$

and

$$Kurt = \frac{N\sum_{i=1}^{N}\varepsilon_i^4}{\left[\sum_{i=1}^{N}\varepsilon_i^2\right]^{2}} - 3 \tag{5.18}$$

These measures of skewness and excess kurtosis cannot be calculated directly since ε_i is unobserved. The frequentist econometrician would replace ε_i by the appropriate OLS residual in the preceding formulae and use the result to carry out a test for skewness or excess kurtosis. A finding that either skewness or excess kurtosis in the residuals indicates that the Normality assumption is an inappropriate one.

A Bayesian analogue to this frequentist procedure would be to calculate the expected value of either (5.17) and (5.18) and see whether they are reasonable. Formally,

$$E[Skew|y] = E\left\{\left.\frac{\sqrt{N}\sum_{i=1}^{N}[y_i - f(X_i, \gamma)]^3}{\left[\sum_{i=1}^{N}[y_i - f(X_i, \gamma)]^2\right]^{\frac{3}{2}}}\right| y\right\}$$

is something that we can calculate in a straightforward fashion once we have a posterior simulator. That is, *Skew* is simply a function of the model parameters (and the data) and, hence, its posterior mean can be calculated in the same way as the posterior mean of any function of interest can be calculated (e.g. see (5.7)). $E[Kurt|y]$ can be calculated in the same fashion. If the Normality assumption is a reasonable one, $E[Kurt|y]$ and $E[Skew|y]$ should both be roughly zero.

Let us now return to the topic of posterior predictive p-values which can be used to formalize the ideas of the previous paragraph. As stressed in the previous paragraph, $E[Skew|y]$ and $E[Kurt|y]$ are functions of the observed data and can be calculated using the posterior simulator. For any observable data, $y^\dagger$, $E[Skew|y^\dagger]$ and $E[Kurt|y^\dagger]$ can be calculated in the same fashion. If we calculate these latter functions for a wide variety of observable data sets, we can obtain distributions of values for skewness and excess kurtosis, respectively, that this model is able to generate. If either $E[Skew|y]$ or $E[Kurt|y]$ lie far out in the tails of the distribution of $E[Skew|y^\dagger]$ and $E[Kurt|y^\dagger]$ this is evidence against the assumption of Normality. It is worth stressing that $E[Skew|y]$ or $E[Kurt|y]$ are both simply numbers whereas $E[Skew|y^\dagger]$ and $E[Kurt|y^\dagger]$ are both random variables with probability distributions calculated using (5.13). In terms of our previous notation, we are setting $g(y) = E[Skew|y]$ or $E[Kurt|y]$ and $g(y^\dagger) = E[Skew|y^\dagger]$ or $E[Kurt|y^\dagger]$.

In practice, a program which calculates posterior predictive p-values for skewness for the nonlinear regression model using the noninformative prior would involve the following steps. The case of excess kurtosis (or any other function of interest) can be done in the same manner. These steps assume that you have derived a posterior simulator (i.e. a Metropolis–Hastings algorithm which is producing draws from the posterior). Details for how such a posterior simulator can be programmed up are given in the previous section.

Step 1: Take a draw, $\gamma^{(s)}$, using the posterior simulator.
Step 2: Generate a representative data set, $y^{\dagger(s)}$, from $p(y^\dagger|\gamma^{(s)})$ using (5.15).
Step 3: Set $\varepsilon_i^{(s)} = y_i - f(X_i, \gamma^{(s)})$ for $i = 1, \ldots, N$ and evaluate (5.17) at this point to obtain $Skew^{(s)}$.
Step 4: Set $\varepsilon_i^{\dagger(s)} = y_i^{\dagger(s)} - f(X_i, \gamma^{(s)})$ for $i = 1, \ldots, N$ and evaluate (5.17) at this point to obtain $Skew^{\dagger(s)}$.
Step 5: Repeat Steps 1, 2, 3 and 4 S times.
Step 6: Take the average of the S draws $Skew^{(1)}, \ldots, Skew^{(S)}$ to give an estimate of $E[Skew|y]$.
Step 7: Calculate the proportion of the S draws $Skew^{\dagger(1)}, \ldots, Skew^{\dagger(S)}$ which are smaller than your estimate of $E[Skew|y]$ from Step 6. If this number is less than 0.5 then it is your estimate of the posterior predictive p-value. Otherwise the posterior predictive p-value is one minus this number.

There is no hard and fast rule for exactly what value of the posterior predictive p-value should be taken as evidence against a model. A useful rule of thumb is to take a posterior predictive p-value of less than 0.05 (or 0.01) as evidence against a model. Remember that, if the posterior predictive p-value for skewness is equal to 0.05, then we can say "This model generates measures of skewness greater than the one actually observed only five percent of the time. Hence, it is fairly unlikely that this model generated the observed data."

5.7 MODEL COMPARISON: THE GELFAND–DEY METHOD

If it is sensible to elicit informative priors and there are two or more competing models you wish to investigate, then posterior odds ratios remain the preferred method of model comparison. In the case of the nonlinear regression model, you may wish to compare different choices for $f(\cdot)$. This would typically involve non-nested model comparison. Alternatively, you may wish to compare a nonlinear to a linear regression model. After all, the linear regression model is, as we have seen in previous chapters, much easier to work with. Hence, you may only wish to work with a nonlinear model if the extra complexity is really worth it. In the case of the CES production function given in (5.1), the model reduces to linearity if $\gamma_{k+1} = 1$ and the linear model would be nested in the nonlinear one. That is, we would have $M_1 : \gamma_{k+1} = 1$ and $M_2 : \gamma$ unrestricted. When comparing nested models of this sort, the Savage–Dickey density ratio is often a convenient tool (see Chapter 4, Theorem 4.1). For the nonlinear regression model, if the prior for γ is of a simple form, then the Savage–Dickey density ratio can often be easily calculated (i.e. the posterior for γ, which must be evaluated at the point $\gamma_{k+1} = 1$ in the Savage–Dickey density ratio, will typically be easy to work with so the ease of calculation depends crucially on the prior).[2] For instance, if the prior is of the form:

$$p(\gamma, h) \propto \frac{p(\gamma)}{h}$$

then the posterior for γ will be of the form given in (5.6) times $p(\gamma)$. However, if the prior is more complicated then it may be impossible to easily evaluate the Savage–Dickey density ratio.

In sum, for non-nested model comparison, or for nested model comparison where the Savage–Dickey density ratio cannot be easily evaluated, there is a need for another, more general, method for calculating the posterior odds ratio. A method due to Gelfand and Dey (1994) can be used in such cases. Indeed, it can be used in virtually any model, including those in previous and future chapters in this book. Computer software to implement this algorithm is available in BACC

[2]It is also worth noting in passing that there exist algorithms which, given posterior simulator output, $\gamma^{(s)}$ for $s = 1, \ldots, S$, can approximate $p(\gamma|y)$. Such an approximation can be used to calculate the numerator of the Savage–Dickey density ratio. However, such algorithms, which fall in the general area of *nonparametric density estimation*, typically require specialized knowledge beyond the scope of the present book and, hence, we do not discuss their use further. However, the reader familiar with nonparametric methods may wish to consider their use.

(see McCausland and Stevens, 2001) and, hence, researchers using this software can use the Gelfand–Dey method in an automatic fashion.

The Gelfand–Dey method is based on the fact that the inverse of the marginal likelihood for a model, M_i, which depends upon parameter vector, θ, can be written as $E[g(\theta)|y, M_i]$ for a particular choice of $g(\cdot)$. In addition, as we have stressed throughout this book, posterior simulators such as the Metropolis–Hastings algorithm are designed precisely to estimate such quantities. The following theorem provides the necessary choice of $g(\cdot)$.

Theorem 5.1: The Gelfand–Dey Method of Marginal Likelihood Calculation
Let $p(\theta|M_i)$, $p(y|\theta, M_i)$ and $p(\theta|y, M_i)$ denote the prior, likelihood and posterior, respectively, for model M_i defined on the region Θ. If $f(\theta)$ is any p.d.f. with support contained in Θ, then

$$E\left[\frac{f(\theta)}{p(\theta|M_i)p(y|\theta, M_i)}\middle| y, M_i\right] = \frac{1}{p(y|M_i)} \tag{5.19}$$

Proof

$$\begin{aligned}
E\left[\frac{f(\theta)}{p(\theta|M_i)p(y|\theta, M_i)}\middle| y, M_i\right] &= \int \frac{f(\theta)}{p(\theta|M_i)p(y|\theta, M_i)} p(\theta|y, M_i)\, d\theta \\
&= \int \frac{f(\theta)}{p(\theta|M_i)p(y|\theta, M_i)} \frac{p(\theta|M_i)p(y|\theta, M_i)}{p(y|M_i)} d\theta \\
&= \frac{1}{p(y|M_i)} \int f(\theta) d\theta \\
&= \frac{1}{p(y|M_i)}
\end{aligned}$$

This theorem seems remarkably powerful in that it says for *any* p.d.f. $f(\theta)$, we can simply set:

$$g(\theta) = \frac{f(\theta)}{p(\theta|M_i)p(y|\theta, M_i)} \tag{5.20}$$

and use posterior simulator output to estimate $E[g(\theta)|y, M_i]$ in the same way as for any posterior feature of interest. However, for the Gelfand–Dey method to work well, $f(\theta)$ must be chosen very carefully. For instance, as discussed in Geweke (1999), the asymptotic theory underlying the Gelfand–Dey method implies that $\frac{f(\theta)}{p(\theta|M_i)p(y|\theta, M_i)}$ must be bounded from above (i.e. it must be finite for every possible value of θ). Just as importance sampling required careful selection of an importance function and the Metropolis–Hastings algorithm required careful selection of a candidate generating density, the Gelfand–Dey method requires careful selection of $f(\theta)$.

Geweke (1999) recommends the following strategy for choosing $f(\theta)$, which has been found to work well in practice. This strategy involves letting $f(\cdot)$ be a Normal density with the tails chopped off. The justification for this truncation is that it is often difficult to verify that $\frac{f(\theta)}{p(\theta|M_i)p(y|\theta,M_i)}$ is finite out in the tails of the Normal density. By cutting off the tails, $f(\theta)$ is set to zero in these potentially problematic regions. Formally, let $\widehat{\theta}$ and $\widehat{\Sigma}$ be estimates of $E(\theta|y, M_i)$ and $var(\theta|y, M_i)$ obtained from the posterior simulator. Furthermore, for some probability, $p \in (0, 1)$, let $\widehat{\Theta}$ denote the support of $f(\theta)$ which is defined by

$$\widehat{\Theta} = \{\theta : (\widehat{\theta} - \theta)'\widehat{\Sigma}^{-1}(\widehat{\theta} - \theta) \leq \chi^2_{1-p}(k)\} \tag{5.21}$$

where $\chi^2_{1-p}(k)$ is the $(1 - p)$th percentile of the Chi-squared distribution with k degrees of freedom and k is the number of elements in θ (see Appendix B, Definition B.22). Then Geweke (1999) recommends letting $f(\theta)$ be the multivariate Normal density truncated to the region $\widehat{\Theta}$,

$$f(\theta) = \frac{1}{p(2\pi)^{\frac{k}{2}}}|\widehat{\Sigma}|^{-\frac{1}{2}} \exp\left[-\frac{1}{2}(\widehat{\theta} - \theta)'\widehat{\Sigma}^{-1}(\widehat{\theta} - \theta)\right] 1(\theta \in \widehat{\Theta}) \tag{5.22}$$

where $1(\,)$ is the indicator function. We would expect small values of p (e.g. $p = 0.01$) to work best, since then more draws will be included when estimating the marginal likelihood. However, as Geweke (1999) points out, the additional cost of trying several different values for p is very low. As with any posterior feature of interest estimated using a posterior simulator, a numerical standard error can be calculated in the standard way (see, for instance, Chapter 4, (4.12) and (4.13)) and used to evaluate the accuracy of the Gelfand–Dey marginal likelihood estimate. BACC contains a subroutine which implements this version of the Gelfand–Dey method and provides numerical standard errors for one or more values of p chosen by the researcher.

We stress that the general Gelfand–Dey method for calculating the marginal likelihood outlined in Theorem 5.1, works for virtually any model. In practice, the only real requirements are that a posterior simulator is available, and that $p(\theta|M_i)$ and $p(y|\theta, M_i)$ are known. The latter are non-trivial requirements, since in some cases, you may only know the kernels of the prior and/or likelihood, not the full p.d.f.'s. In such cases, the Gelfand–Dey method cannot be used. Geweke (1999)'s implementation of the Gelfand–Dey method works only if the support of the posterior contains the region defined in (5.21) (i.e. $\widehat{\Theta} \in \Theta$). If this is not the case, Geweke (1999) provides some suggestions for how this implementation can be changed in minor ways. The reader is referred to Geweke (1999) for additional details. The Geweke implementation of the Gelfand–Dey method is, thus, a little less generic, but still can be used for a wide array of models (and virtually any model discussed in this book).

5.8 PREDICTION

Predictive inference in this model can be carried out using the strategy outlined in Chapter 4. That is, draws from $p(y^*|y)$ are provided by the Metropolis–Hastings algorithm and, conditional on these, draws from $p(y^*|y, \gamma^{(s)})$ can be taken and averaged as in (4.32) to provide an estimate of any predictive function of interest. For instance, for the noninformative prior given in (5.2), and using techniques analogous to those used to derive (3.40), it can be verified that

$$p(y^*|y, \gamma) = f_t(y^*|f(X^*, \gamma), \overline{s}^2 I_T, N)$$

where $\overline{s}^2$ is defined in (5.16). Hence, draws from y^* conditional on γ can easily be obtained.

5.9 EMPIRICAL ILLUSTRATION

We use a microeconomic application to illustrate Bayesian inference in a nonlinear regression model. We use data from $N = 123$ companies on output, y, and the inputs labor, x_1, and capital, x_2. So as not to have to worry about units of measurement, the dependent and explanatory variables are all standardized to have standard deviation one. That is, we divide each variable by its standard deviation. Such standardization is sometimes done since the interpretation of coefficients is now in terms of standard deviations of variables. For instance, in a linear regression context a coefficient, β_j, can be interpreted as saying: "If explanatory variable j increases by one standard deviation, the dependent variable will tend to increase by β_j standard deviations."

We assume a production function of the CES form given in (5.1):

$$y_i = \gamma_1 + (\gamma_2 x_{i1}^{\gamma_4} + \gamma_3 x_{i2}^{\gamma_4})^{\frac{1}{\gamma_4}} + \varepsilon_i$$

The choice of an additive intercept is motivated by the consideration that a simple restriction, $\gamma_4 = 1$, yields the linear regression model.

We begin by comparing posterior inference in this model using Independence Chain and Random Walk Chain Metropolis–Hastings algorithms using the noninformative prior given in (5.4). We focus on the marginal posterior for $\gamma = (\gamma_1, \ldots, \gamma_4)'$, given in (5.8). To construct the candidate generating densities for these two algorithms, we first use an optimization subroutine to find $\widehat{\gamma}_{\max}$, the mode of $p(\gamma|y)$. The optimization subroutine (taken from Jim LeSage's Econometrics Toolbox, although most optimization subroutines will provide similar information) provides an estimate of the Hessian which can be used to construct $\widehat{var(\widehat{\gamma}_{\max})}$, as described in Section 5.5.1.

Table 5.1 presents results based on a fairly crude implementation of both algorithms. A computer program for implementing these algorithms follows the steps spelled out in Section 5.5. The two algorithms differ only in their candidate generating densities and, hence, in their acceptance probabilities (see (5.9) and (5.11)). The Independence Chain algorithm takes candidate draws from the $f_t(\gamma|\widehat{\gamma}_{\max}, \widehat{var(\widehat{\gamma}_{\max})}, 10)$ density and the Random Walk algorithm takes candidate draws from the $f_N(\gamma|\gamma^{(s-1)}, \widehat{var(\widehat{\gamma}_{\max})})$ density. These implementations

Table 5.1 Posterior Properties Using Two M-H Algorithms

	Independence Chain		Random Walk	
	Mean	St. Dev	Mean	St. Dev.
γ_1	1.04	0.05	1.02	0.06
γ_2	0.73	0.08	0.72	0.12
γ_3	0.97	0.12	1.00	0.16
γ_4	1.33	0.23	1.36	0.29

can, undoubtedly, be improved upon by experimenting with different covariance matrices in either candidate generating density. For instance, we could replace $\widehat{var(\widehat{\gamma}_{\max})}$ by an approximation of $var(\gamma|y)$ obtained using an initial run of either Metropolis–Hastings algorithm. Or we could use $c\widehat{var(\widehat{\gamma}_{\max})}$ in either candidate generating density and then experiment with various values of the scalar c to improve algorithm performance. For the Independence Chain Metropolis–Hastings algorithm, we could experiment with different values of the degrees of freedom parameter in the candidate generating density. In both cases, we set $S = 25\,000$.

For the Independence Chain and Random Walk algorithms, 7.4% and 20.6%, respectively, of the candidate draws were accepted. Table 5.1 indicates that even these crude implementations of two different algorithms yield roughly the same results. Relative to the posterior standard deviations, the estimated posterior means are very close to one another. The estimated posterior standard deviations differ a bit more across algorithms, but are still fairly close. Of course, if the researcher wishes more accurate estimates, she could either fine tune the candidate generating densities or could increase the number of replications.[3]

The results in Table 5.1 are obtained using an improper, noninformative prior and, hence, we cannot use posterior odds ratios to compare this model to another. Instead, we will use posterior predictive p-values to provide some indication of how well our model is fitting the data. Section 5.6 describes how to calculate them and includes a list of steps that the computer program follows. Briefly, at each replication, an artificial data set is drawn from (5.15) and the measures of

[3]Using a 500 MHz Pentium computer, the program which produced all the results in Table 5.1 took under five minutes to run. We could easily take several hundred thousand draws (i.e. start the program running and come back after lunch to find it completed) or even several million draws (i.e. start the program running before leaving the office in the evening and coming back the next morning to find it completed). As computers get faster and faster, it is becoming easier and easier to take large numbers of replications. This has implications for empirical research, in that the researcher who wants to achieve a certain degree of accuracy often has a choice between improving her posterior simulation algorithm or simply make do with a less efficient algorithm, but take more replications. When making such a choice, it is worth noting that improving an algorithm usually involves the researcher's time, while taking more replications involves the computer's time. Most researchers value their own time more than their computers' and, hence, it is common to see a researcher presenting results which are based on an algorithm which is 'good enough', rather than one which is 'the best'.

skewness and excess kurtosis calculated using (5.17) and (5.18). The posterior predictive density for these measures can then be obtained. Formally, we are calculating $E[Skew|y^\dagger]$ and $E[Kurt|y^\dagger]$ which are random variables, since $y^\dagger$ is random. We can also calculate the measures and skewness and kurtosis for our observed data, $E[Skew|y]$ and $E[Kurt|y]$, which are not random variables since y is not random. The location of $E[Skew|y]$ and $E[Kurt|y]$ in the posterior predictive densities for $E[Skew|y^\dagger]$ and $E[Kurt|y^\dagger]$ provides insight into how well the model is fitting the data.

Figures 5.1 and 5.2 graph approximations to these posterior predictive densities. These figures are simply histograms of the draws we called $Skew^{\dagger(1)}, \ldots, Skew^{\dagger(S)}$ and $Kurt^{\dagger(1)}, \ldots, Kurt^{\dagger(S)}$ in Section 5.6. We have also labeled $E[Skew|y]$ and $E[Kurt|y]$ on these graphs as 'Observed skewness' and 'Observed kurtosis', respectively. In words, the posterior predictive density for skewness (or kurtosis) tells you what values for skewness (or kurtosis) the model under study tends to generate. For instance, Figure 5.1 indicates that the nonlinear regression model used in this empirical illustration virtually always generates data sets with skewness measures that are less than one in absolute value. If the actual data had generated a skewness value which is greater than one in absolute value, this would be evidence that the model is not suitable for use with this data set. In fact, $E[Skew|y] = 0.17$ which lies near the center of the density plotted in Figure 5.1. The corresponding p-value is 0.37, indicating that 37% of artificial data sets generated with this model will exhibit a greater degree of skewness than the actual data. Thus, our actual data set exhibits a degree of skewness which is

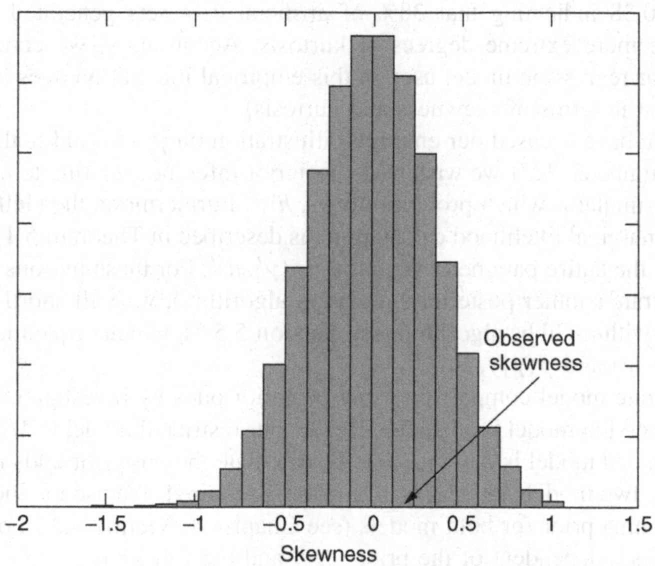

Figure 5.1 Posterior Predictive Density for Skewness

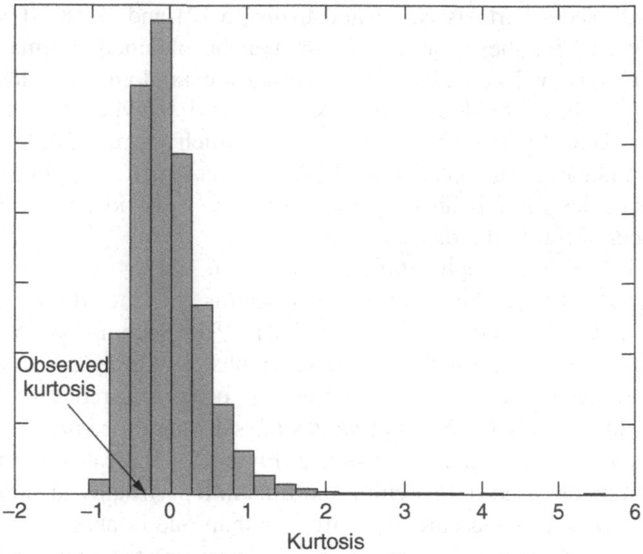

Figure 5.2 Posterior Predictive Density for Kurtosis

consistent with the nonlinear regression model. Similar considerations hold for the posterior predictive density for excess kurtosis. The observed data yields a kurtosis measure of $E[Kurt|y] = -0.37$ which is consistent with the types of kurtosis measures that this model can generate. In fact, the posterior predictive p-value is 0.38 indicating that 38% of artificial data sets generated from this model have more extreme degrees of kurtosis. Accordingly, we conclude that the nonlinear regression model used in this empirical illustration does fit the data well (at least in terms of skewness and kurtosis).

So far, we have focused our empirical illustration on γ and said nothing about the other parameter, h. If we wish to do posterior inference relating to h, we need a posterior simulator which provides draws, $h^{(s)}$. Furthermore, the Gelfand–Dey method of marginal likelihood calculation (as described in Theorem 5.1), requires draws from the entire parameter vector, $\theta = (\gamma', h)'$. For these reasons (and simply to illustrate another posterior simulation algorithm), we will shortly derive a Metropolis-within-Gibbs algorithm (see Section 5.5.3), which sequentially draws from $p(h|y, \gamma)$ and $p(\gamma|y, h)$.

We illustrate model comparison using posterior odds by investigating whether a linear regression model is adequate. Hence, our restricted model is $M_1 : \gamma_4 = 1$. Our unrestricted model is $M_2 : \gamma_4 \neq 1$. To calculate the posterior odds ratio comparing these two models an informative prior is required. We use the independent Normal-Gamma prior for both models (see Chapter 4, Section 4.2). For M_2, our prior for γ is independent of the prior for h and the former is

$$\gamma \sim N(\underline{\gamma}, \underline{V}_{\gamma})$$

while the latter is

$$h \sim G(\underline{s}^{-2}, \underline{\nu})$$

We choose $\underline{\gamma} = (1, 1, 1, 1)'$, $\underline{V} = 0.25I_4$, $\underline{\nu} = 12$ and $\underline{s}^{-2} = 10.0$. Given the likely magnitudes of the marginal products of labor and capital and the way we have standardized the data, these choices are sensible, but relatively noninformative.

For M_1 we use the same prior, except that γ now has only three elements and, thus, $\underline{\gamma}$ and $\underline{V}$ have their last row and row and column, respectively, omitted.

For the linear regression model, M_1, Chapter 4, Section 4.2, provides all the necessary details for posterior inference, including development of a Gibbs sampler for posterior simulation. For the nonlinear regression model with this prior we will derive a Metropolis-within-Gibbs algorithm. To construct such an algorithm, we must first derive $p(h|y, \gamma)$ and $p(\gamma|y, h)$. Using the same techniques as for (4.8) through (4.10), we find that

$$h|y, \gamma \sim G(\overline{s}^{-2}, \overline{\nu}) \tag{5.23}$$

where

$$\overline{\nu} = N + \underline{\nu}$$

and

$$\overline{s}^2 = \frac{[y - f(X, \gamma)]'[y - f(X, \gamma)] + \underline{\nu}\underline{s}^2}{\overline{\nu}}$$

Using (5.5) and noting that $p(\gamma|y, h) \propto p(\gamma, h|y)$, it can be seen that

$$p(\gamma|y, h) \propto \exp\left[-\frac{h}{2}\{y - f(X, \gamma)\}'\{y - f(X, \gamma)\}\right]$$
$$\exp\left[-\frac{1}{2}(\gamma - \underline{\gamma})'\underline{V}^{-1}(\gamma - \underline{\gamma})\right] \tag{5.24}$$

This conditional posterior density does not have a form that can be directly drawn from in a simple manner. However, we can use a Metropolis–Hasting algorithm for $p(\gamma|y, h)$ which, combined with draws from the Gamma distribution given in (5.23), provides us with a Metropolis-within-Gibbs algorithm. In this empirical application, we use a Random Walk Chain Metropolis–Hastings algorithm to draw from (5.24). This algorithm is identical to that used to create the numbers in the last two columns of Table 5.1, except that the acceptance probability is calculated using (5.24).

Given output from posterior simulators such as we have developed for M_1 and M_2, it is relatively straightforward to calculate the marginal likelihoods using the method of Gelfand and Dey. Section 5.7 provides the details. Briefly, once draws from the posterior simulator have been obtained, we must evaluate the likelihood and prior at these draws (see (5.20)). Since, for both models, the priors are independent Normal-Gamma the prior evaluation component is straightforward to do. For M_1 the likelihood function is given in (3.3) and for M_2 it is given in (5.3). It is not hard to write computer code which evaluates these at the posterior draws. The function labeled $f(\theta)$ in (5.20) is also simple to program up, involving only (5.21) and (5.22). Once the prior, likelihood and $f(\cdot)$ are evaluated at every posterior draw, we can calculate $g(\cdot)$ as given in (5.20) and average the result to provide us with an estimate of the marginal likelihood. As with every empirical illustration, computer code for obtaining the marginal likelihoods for the linear and nonlinear regression models is available on the website associated with this book.

The two posterior simulators (i.e. the Gibbs sampler for the linear model and the Metropolis-within-Gibbs for the nonlinear model) were run using $S_0 = 2500$ burn-in replications and $S_1 = 25\,000$ included replications. The marginal likelihoods were calculated using $p = 0.01, 0.05$ and 0.10 (see (5.21)). The resulting estimates of the Bayes factors for these three truncation values were 1.067, 1.075 and 1.065, respectively. The fact that these three different estimates are so close to one another provides informal evidence that our posterior simulators are giving reliable results. Of course, the MCMC convergence diagnostics discussed in Chapter 4 (see Section 4.2.4) can be used to provide more formal evidence of MCMC convergence.

The Bayes factors are very near one, indicating that the linear and nonlinear regression models are receiving roughly equal support. Given that the posterior mean of γ_4 is a little more than one standard deviation from 1.0 (see Table 5.1), this finding is quite reasonable.

5.10 SUMMARY

In this chapter, Bayesian methods for inference in the nonlinear regression model are described. The nonlinear aspect means that analytical results are unavailable, even when simple noninformative priors are used. It is also not possible, in general, to develop a Gibbs sampler for posterior simulation. These features are used to motivate the introduction of a new class of very general algorithms for posterior simulation: the Metropolis–Hastings class of algorithms. Two commonly-used variants, the Independence Chain and Random Walk Chain Metropolis–Hastings algorithms are described in detail. The former works well in cases where a convenient approximation to the posterior exists, while the latter can be used if such an approximation cannot be found. We also use the nonlinear regression model as a framework for introducing a new method for evaluating a model or models, the posterior predictive p-value. This is a general tool for measuring the fit of a model. A general tool for marginal likelihood calculation, the Gelfand–Dey method, is also described.

At this stage, we have introduced virtually all of the basic tools and concepts used in the remainder of this book. In terms of posterior simulators, we have covered Monte Carlo integration, importance sampling, Gibbs sampling and the Metropolis–Hastings algorithm. For model comparison, we have covered the Gelfand–Dey method for marginal likelihood calculation as well as easier, but less generic, methods such as the Savage–Dickey density ratio. The posterior predictive p-value approach has also been described. This approach is best interpreted as a way of measuring model fit, but it can also be used to compare models. The remaining chapters of this book are mostly concerned with adapting these general tools and concepts to particular models. It is worth stressing that we now have a wide variety of tools and, for some models, more than one can be used. For instance, in the nonlinear regression model, it is possible to use either importance sampling or a Metropolis–Hastings algorithm. For model comparison, it is possible to use a posterior odds ratio or use posterior predictive p-values. The precise tool chosen can be a matter of preference or convenience and, in this book, we make particular choices. The choices we make are not intended to be the only (or even the best) ones for a given model.

5.11 EXERCISES

5.11.1 Theoretical Exercises

1. Which of the following regression functions is intrinsically linear? For those which are intrinsically linear, write out a linear regression model in terms of (possibly transformed) dependent and explanatory variables and discuss what the properties of the regression errors are. Assume y_i and x_i are scalars and ε_i is $N(0, h)$ for $i = 1, \ldots, N$.

   (a)
   $$y_i = \beta_1 x_i^{\beta_2} \varepsilon_i.$$

   (b)
   $$y_i = \left(\beta_1 + \beta_2 \frac{1}{x_i}\right) \varepsilon_i.$$

   (c)
   $$\frac{1}{y_i} = \beta_1 + \beta_2 \frac{1}{x_i} + \varepsilon_i.$$

   (d)
   $$y_i = \exp\left(\beta_1 + \beta_2 \frac{1}{x_i} + \varepsilon_i\right).$$

   (e)
   $$y_i = \exp\left(\beta_1 + \beta_2 \frac{1}{x_i}\right) + \varepsilon_i.$$

(f)

$$y_i = \frac{1}{1 + \exp\,(\beta_1 + \beta_2 x_i + \varepsilon_i)}.$$

2. Consider the Normal regression model which is partly linear and partly nonlinear:

$$y_i = \beta_1 + \beta_2 x_{i2} + \cdots + \beta_k x_{ik} + f(x_{i,k+1}, \ldots, x_{i,k+p}, \gamma) + \varepsilon_i$$

   (a) Using a noninformative prior, derive the marginal posterior $p(\gamma|y)$.
   (b) Discuss how you might derive a posterior simulator if you wished to carry out Bayesian inference solely on γ.
   (c) Discuss how you might derive a posterior simulator if you wished to carry out Bayesian inference on γ and $\beta_1, \ldots, \beta_k$.

5.11.2 Computer-Based Exercises

Remember that some data sets and MATLAB programs are available on the website associated with this book.

3. Use the house price data set and the Normal linear regression model with natural conjugate prior described in the Empirical Illustration to Chapter 3.[4] Use the posterior predictive p-value approach (see Section 5.6) to see whether this model can reasonably capture the skewness and kurtosis in the data.
4. Use the house price data set and the Normal linear regression model which is partly nonlinear and partly linear described in Exercise 2. You may use whatever prior you wish for β and γ provided it is informative. Normal priors will be the most convenient.
   (a) Using your results from Exercise 2, write and run a program for posterior simulation which allows for the explanatory variables lot size and number of bedrooms to have a nonlinear effect on house price. Assume the other explanatory variables enter linearly and $f()$ has a CES form (see (5.1)).
   (b) Do posterior results indicate that lot size and number of bedrooms have a nonlinear effect on house price?
   (c) Use the Savage–Dickey density ratio to calculate the Bayes factor for testing whether there is nonlinearity in the regression relationship (see the discussion at the beginning of Section 5.7).
   (d) Re-do part (d) using the Gelfand–Dey method.
5. Re-do this chapter's empirical illustration using importance sampling using the $f_t(\theta|\widehat{\theta}_{ML}, \widehat{var(\widehat{\theta}_{ML})}, \nu)$ as an importance function.

[4]The house price data set is available on the website associated with this book or in the *Journal of Applied Econometrics* Data Archive listed under Anglin and Gencay (1996) (http://qed.econ.queensu.ca/jae/1996-v11.6/anglin-gencay/).

6. Investigate the performance of Independence Chain and Random Walk Chain Metropolis–Hastings algorithms using artificial data. The artificial data should be generated from (5.1) using various values for γ and h. You should experiment with different candidate generating densities and different numbers of replications. This chapter's empirical illustration contains a simple example of how you might compare algorithms.

6 The Linear Regression Model with General Error Covariance Matrix

6.1 INTRODUCTION

In this chapter we return to the linear regression model

$$y = X\beta + \varepsilon \tag{6.1}$$

In previous chapters, we assumed ε to be $N(0_N, h^{-1}I_N)$. This statement is really a combination of several assumptions, some of which we might want to relax. The assumption that the errors have mean zero is an innocuous one. If a model has errors with a non-zero mean, this non-zero mean can be incorporated into the intercept. To be precise, a new model, which is identical to the old except for the intercept, can be created which does have mean zero errors. However, the assumption that the covariance matrix of the errors is $h^{-1}I_N$ might not be innocuous in many applications. Similarly, the assumption that the errors have a Normal distribution is one which might be worth relaxing in many cases. In this chapter, we consider several empirically-relevant ways of relaxing these assumptions and describe Bayesian inference in the resulting models.

All the models in this chapter are based on (6.1) and the following assumptions:

1. ε has a multivariate Normal distribution with mean 0_N and covariance matrix $h^{-1}\Omega$, where Ω is an $N \times N$ positive definite matrix.
2. All elements of X are either fixed (i.e. not random variables) or, if they are random variables, they are independent of all elements of ε with a probability density function, $p(X|\lambda)$, where λ is a vector of parameters that does not include β and h.

Note that these assumptions are identical to those made in Chapters 2, 3 and 4, except for the assumption about the error covariance matrix. However, as we shall show in this chapter, assumptions about this error covariance matrix are closely related to distributional assumptions. Hence, we can use this framework to free up the assumption that the errors are Normally distributed.

The various models we discuss differ in the precise form that Ω takes. After discussing some general theory which is relevant for any choice of Ω, we examine several specific choices which arise in many applications. We begin by considering *heteroskedasticity*, which is the name given for cases where the error variances differ across observations. We consider two types of heteroskedasticity: one where its form is known and one where it is unknown. The latter case allows us to free up the Normality assumption, and we discuss, in particular, how a certain model with heteroskedasticity of unknown form is equivalent to a linear regression models with Student-t errors. This model allows us to introduce the concept of a *hierarchical prior*, which will be used extensively in the remainder of this book. Subsequently, we consider a case where the errors are correlated with one another. In particular, we discuss the Normal linear regression model with *autoregressive* or *AR* errors. In addition to being of interest in and of themselves, AR models are important time series models and provide us with a convenient starting point for an introduction to time series methods. The final model considered in this chapter is the *seemingly unrelated regressions* or *SUR* model. This is a model which has several equations corresponding to multiple dependent variables and is a component of models considered in future chapters.

6.2 THE MODEL WITH GENERAL Ω

6.2.1 Introduction

Before discussing the likelihood function, prior, posterior and computational methods, we present a general result which has implications for both interpretation and computation for this model. Since Ω is a positive definite matrix, it follows from Appendix A, Theorem A.10 that an $N \times N$ matrix P exists with the property that $P\Omega P' = I_N$. If we multiply both sides of (6.1) by P, we obtain a transformed model

$$y^* = X^*\beta + \varepsilon^* \tag{6.2}$$

where $y^* = Py$, $X^* = PX$ and $\varepsilon^* = P\varepsilon$. It can be verified that ε^* is $N(0_N, h^{-1}I_N)$. Hence, the transformed model given in (6.2) is identical to the Normal linear regression model discussed in Chapters 2, 3 and 4. This has two important implications. First, if Ω is known, Bayesian analysis of the Normal linear regression model with nonscalar error covariance matrix is straightforward. The researcher can transform her data and carry out Bayesian inference using the methods of earlier chapters. Secondly, if Ω is unknown, (6.2) suggests methods for Bayesian computation. That is, conditional on Ω, (6.2) implies that the posteriors of β and h will be of the same form as in previous chapters and, hence, these earlier results can be used for derivations relating to β and h. If the prior for β and h is $NG(\underline{\beta}, \underline{V}, \underline{s}^{-2}, \underline{\nu})$, then all the results of Chapters 2 and 3 are applicable *conditional upon* Ω and we can draw upon these results to derive a posterior simulator. For instance, (3.14) can be used to verify that $p(\beta|y, \Omega)$ is a multivariate t distribution and this, combined with a posterior simulator for $p(\Omega|y)$ can be used to carry out posterior inference on β and Ω. This is done in Griffiths (2001) for the noninformative limiting case of the natural conjugate prior. In this chapter we use a prior of the independent Normal-Gamma form of Chapter 4, Section 4.2, and a Gibbs sampler which sequentially draws from $p(\beta|y, h, \Omega)$, $p(h|y, \beta, \Omega)$ and $p(\Omega|y, \beta, h)$ can be set up. The first two of these posterior conditionals will be Normal and Gamma, as in Section 4.2.2 of Chapter 4, while $p(\Omega|y, \beta, h)$ depends upon the precise form of Ω. Hence, the only new derivations which are required relate to this latter distribution. Similar considerations hold for priors which impose inequality constraints (see Chapter 4, Section 4.3).

6.2.2 The Likelihood Function

Using the properties of the multivariate Normal distribution, the likelihood function can be seen to be:

$$p(y|\beta, h, \Omega) = \frac{h^{\frac{N}{2}}}{(2\pi)^{\frac{N}{2}}} |\Omega|^{-\frac{1}{2}} \left\{ \exp\left[-\frac{h}{2}(y - X\beta)'\Omega^{-1}(y - X\beta) \right] \right\} \tag{6.3}$$

or, in terms of the transformed data,

$$p(y^*|\beta, h, \Omega) = \frac{h^{\frac{N}{2}}}{(2\pi)^{\frac{N}{2}}} \left\{ \exp\left[-\frac{h}{2}(y^* - X^*\beta)'(y^* - X^*\beta) \right] \right\} \tag{6.4}$$

In Chapter 3, we showed how the likelihood function could be written in terms of OLS quantities (see (3.4)–(3.7)). Here an identical derivation using the transformed model yields a likelihood function written in terms of Generalized Least Squares[1] (GLS) quantities:

$$\nu = N - k \tag{6.5}$$

$$\widehat{\beta}(\Omega) = (X^{*\prime}X^*)^{-1}X^{*\prime}y^* = (X'\Omega^{-1}X)^{-1}X'\Omega^{-1}y \tag{6.6}$$

and

$$s^2(\Omega) = \frac{(y^* - X^*\widehat{\beta}(\Omega))'(y^* - X^*\widehat{\beta}(\Omega))}{\nu} \tag{6.7}$$

$$= \frac{(y - X\widehat{\beta}(\Omega))'\Omega^{-1}(y - X\widehat{\beta}(\Omega))}{\nu}$$

[1]For the reader unfamiliar with the concept of a Generalized Least Squares estimator, any frequentist econometrics textbook such as Green (2000) will provide a detailed discussion. Knowledge of this material is not necessary to understand the material in this chapter.

as:

$$p(y|\beta, h, \Omega) = \frac{1}{(2\pi)^{\frac{N}{2}}} \times \left\{h^{\frac{1}{2}} \exp\left[-\frac{h}{2}(\beta - \widehat{\beta}(\Omega))' X' \Omega^{-1} X (\beta - \widehat{\beta}(\Omega))\right]\right\} \times \left\{h^{\frac{\nu}{2}} \exp\left[-\frac{h\nu}{2s(\Omega)^{-2}}\right]\right\} \tag{6.8}$$

6.2.3 The Prior

Here we use an independent Normal-Gamma prior for β and h (see Chapter 4, Section 4.2.1), and use the general notation, $p(\Omega)$, to indicate the prior for Ω. In other words, the prior used in this section is

$$p(\beta, h, \Omega) = p(\beta)p(h)p(\Omega)$$

where

$$p(\beta) = f_N(\beta|\underline{\beta}, \underline{V}) \tag{6.9}$$

and

$$p(h) = f_G(h|\underline{\nu}, \underline{s}^{-2}) \tag{6.10}$$

6.2.4 The Posterior

The posterior is proportional to the prior times the likelihood and is of the form

$$p(\beta, h, \Omega|y) \propto p(\Omega) \times \left\{\exp\left[-\frac{1}{2}\{h(y^* - X^*\beta)'(y^* - X^*\beta) + (\beta - \underline{\beta})'\underline{V}^{-1}(\beta - \underline{\beta})\}\right]\right\} \times h^{\frac{N+\underline{\nu}-2}{2}} \exp\left[-\frac{h\underline{\nu}}{2\underline{s}^{-2}}\right] \tag{6.11}$$

This posterior is written based on the likelihood function expressed as in (6.4). Alternative expressions based on (6.3) or (6.8) can be written out. However, we do not do this, since this joint posterior density for β, h and Ω does not take the form of any well-known and understood density and, hence, cannot be directly used in a simple way for posterior inference. At least some of the conditionals of the posterior are, however, simple. Proceeding in the same manner as in Chapter 4 (see (4.4)–(4.10) and surrounding discussion), it can be verified that the posterior of β, conditional on the other parameters of the model is multivariate Normal:

$$\beta|y, h, \Omega \sim N(\overline{\beta}, \overline{V}) \tag{6.12}$$

where

$$\overline{V} = (\underline{V}^{-1} + hX'\Omega^{-1}X)^{-1} \tag{6.13}$$

and

$$\overline{\beta} = \overline{V}(\underline{V}^{-1}\underline{\beta} + hX'\Omega^{-1}X\widehat{\beta}(\Omega)) \tag{6.14}$$

The posterior for h conditional on the other parameters in the model is Gamma:

$$h|y, \beta, \Omega \sim G(\overline{s}^{-2}, \overline{\nu}) \tag{6.15}$$

where

$$\overline{\nu} = N + \underline{\nu} \tag{6.16}$$

and

$$\overline{s}^2 = \frac{(y - X\beta)'\Omega^{-1}(y - X\beta) + \underline{\nu}\underline{s}^2}{\overline{\nu}} \tag{6.17}$$

The posterior for Ω conditional on β and h has a kernel of the form

$$p(\Omega|y, \beta, h) \propto p(\Omega)|\Omega|^{-\frac{1}{2}}\left\{\exp\left[-\frac{h}{2}(y - X\beta)'\Omega^{-1}(y - X\beta)\right]\right\} \tag{6.18}$$

In general, this conditional posterior does not take any easily recognized form. In future sections of this chapter we consider particular forms for Ω and derive appropriate posterior simulators. At this stage, we only note that, if we could take posterior draws from $p(\Omega|y, \beta, h)$, then a Gibbs sampler for this model could be set up in a straightforward manner, since $p(\beta|y, h, \Omega)$ is Normal and $p(h|y, \beta, \Omega)$ is Gamma.

6.3 HETEROSKEDASTICITY OF KNOWN FORM

6.3.1 Introduction

Heteroskedasticity is said to occur if the error variances differ across observations. The models in previous chapters all had error variances which were identical across observations and were, thus, *homoskedastic*. A couple of examples will serve to motivate how heteroskedasticity might arise in practice. Consider first a microeconomic example where the dependent variable is company sales. If errors are proportionate to firm size, then errors for small firms will tend to smaller than those for large firms. Secondly, heteroskedasticity might arise in a study involving data from many countries. Since developed countries have better agencies for collecting statistics than developing countries, it might be the case that errors are smaller in the former countries.

In terms of our regression model, heteroskedasticity occurs if

$$\Omega = \begin{bmatrix} \omega_1 & 0 & \cdot & \cdot & 0 \\ 0 & \omega_2 & 0 & \cdot & \cdot \\ \cdot & 0 & \cdot & \cdot & \cdot \\ \cdot & \cdot & \cdot & \cdot & 0 \\ 0 & \cdot & \cdot & 0 & \omega_N \end{bmatrix} \tag{6.19}$$

In other words, the Normal linear regression model with heteroskedastic errors is identical to that studied in Chapters 2–4, except that we now assume $var(\varepsilon_i) = h^{-1}\omega_i$ for $i = 1, \ldots, N$.

The examples above indicate that we often know (or at least suspect), what form this heteroskedasticity will take. For instance, ω_i might depend upon whether firm i is small or large or whether country i is developing or developed. Here we will assume that

$$\omega_i = h(z_i, \alpha) \tag{6.20}$$

where $h()$ is a positive function which depends on parameters α and a p-vector of data, z_i. z_i may include some or all of the explanatory variables, x_i. A common choice for $h()$, which ensures that the error variances are positive is:

$$h(z_i, \alpha) = (1 + \alpha_1 z_{i1} + \alpha_2 z_{i2} + \cdots + \alpha_p z_{ip})^2 \tag{6.21}$$

but the discussion of this section works for other choices.

The prior, likelihood and posterior for this model are simply those in Section 6.2 with the expression for Ω given in (6.19) plugged in. Accordingly, we do not write them out here. Note, however, that in the present section Ω depends upon α and, hence, the formulae below are written as depending on α.

To carry out Bayesian inference in the present heteroskedastic model, a posterior simulator is required. The previous discussion suggests that a Metropolis-within-Gibbs algorithm (see Section 5.5.3) might be appropriate. In particular, as noted in (6.12) and (6.15), $p(\beta|y, h, \alpha)$ is Normal and $p(h|y, \beta, \alpha)$ is Gamma, and we require only a method for taking draws from $p(\alpha|y, \beta, h)$ to have a complete posterior simulator. Unfortunately, if we plug (6.19) and (6.20) into (6.18) to obtain an expression for $p(\alpha|y, \beta, h)$ the result does not take the form of any convenient density. Nevertheless, a Metropolis–Hastings algorithm can be developed. In the empirical illustration which follows, a Random Walk Chain Metropolis–Hastings algorithm (see Section 5.5.2 of Chapter 5) is used although other algorithms are possible. Bayes factors for any hypothesis of interest (e.g. $\alpha_1 = \cdots = \alpha_p = 0$ which is the hypothesis that heteroskedasticity does not exist) can be calculated using the Gelfand–Dey approach. Alternatively, posterior predictive p-values or HPDIs can be calculated to shed light on the fit and appropriateness of the model. Predictive inference in this model can be carried out using the strategy outlined in Chapter 4, Section 4.2.6.

6.3.2 Empirical Illustration: Heteroskedasticity of a Known Form

We use the house price data set introduced in Chapter 3, to illustrate the use of Gibbs sampling in the Normal linear regression model with heteroskedasticity of known form. The reader is referred to Section 3.9 of Chapter 3 for a precise description of the dependent and explanatory variables for this data set. We assume the heteroskedasticity takes the form given in (6.21) and that $z_i = (x_{i2}, \dots, x_{ik})'$. The priors for β and h are given in (6.9) and (6.10) and we use the same values for hyperparameters as in Chapter 4, Section 4.2.7. We use a noninformative prior for α of the form

$$p(\alpha) \propto 1$$

Note that this prior is improper and, hence, we cannot calculate meaningful Bayes factors for hypotheses involving the elements of α. Accordingly, we present 95% HPDIs along with posterior means and standard deviations in Table 6.1.

The posterior simulator is a Metropolis-within-Gibbs algorithm, with draws of β and h taken from (6.12) and (6.15), respectively. Draws from $p(\alpha|y, \beta, h)$ are taken using a Random Walk Chain Metropolis–Hastings algorithm with a Normal increment random variable (see Chapter 5, (5.10)). $p(\alpha|y, \beta, h)$ is given in (6.18) with (6.21) providing the precise form for Ω. Equation (6.18), evaluated at old and candidate draws, is used to calculate the acceptance probability (see Chapter 5, (5.11)). The variance of the proposal density, labelled Σ in (5.12), is chosen by first setting $\Sigma = cI$ and experimenting with different values of the scalar c until a value is found which yields reasonable acceptance probabilities. The posterior simulator is then run using this value to yield an estimate of the posterior variance of α, $\widehat{var}(\alpha|y)$. We then set $\Sigma = c\widehat{var}(\alpha|y)$ and experiment with different values of c until we find one which yields an average acceptance probability of roughly 0.50. Then a final long run of 30 000 replications, with 5000 burn-in replications

Table 6.1 Posterior Results for β, h and α

	Mean	Standard Deviation	95% HPDI
β_1	−5453.92	2976.04	[−10 310, 557]
β_2	6.12	0.40	[5.42, 6.82]
β_3	3159.52	1025.63	[1477, 4850]
β_4	14 459.34	1672.43	[11 742, 17 224]
β_5	7851.11	939.34	[6826, 9381]
h	1.30×10^{-7}	4.05×10^{-8}	$[7 \times 10^{-8}, 2 \times 10^{-7}]$
α_1	5.49×10^{-4}	1.36×10^{-4}	$[3 \times 10^{-4}, 8 \times 10^{-4}]$
α_2	0.68	0.32	[0.21, 1.26]
α_3	0.70	0.42	[0.08, 1.40]
α_4	−0.35	0.33	[−0.89, 0.18]

discarded, is taken. MCMC diagnostics indicate convergence of the Metropolis-within-Gibbs algorithm and numerical standard errors indicate an approximation error which is small relative to posterior standard deviations of all parameters.

Table 6.1 indicates that heteroskedasticity does seem to exist for this data set. That is, the 95% HPDIs do not include zero for α_1, α_2 and α_3 indicating that lot size, number of bedrooms and number of bathrooms have significant explanatory power in the equation for heteroskedasticity. The fact that all of these coefficients are positive indicates that the error variance for large houses tends to be bigger than for small houses. In previous chapters, we ignored heteroskedasticity when working with this data set. To see what effect this omission had, you may wish to compare the results in Table 6.1 with those in Table 4.1. The latter table contains results for the homoskedastic version of the model, but uses the same data and the same prior for β and h. It can be seen that including heteroskedasticity has some effect on the posterior of β. For instance, the posterior mean of β_4 was 16 133 in the homoskedastic model and is 14 459 in the heteroskedastic one. However, for many purposes, such differences might be fairly small and the researcher might conclude that the incorporation of heteroskedasticity has not had an enormous effect on results relating to β.

6.4 HETEROSKEDASTICITY OF AN UNKNOWN FORM: STUDENT-t ERRORS

6.4.1 General Discussion

In the previous section, we assumed that the heteroskedasticity was of a form given by (6.20). The question arises as to how to proceed if you suspect heteroskedasticity is present, but of unknown form. In other words, you are willing to assume (6.19), but unwilling to assume a functional form as in (6.20). With N observations and $N + k + 1$ parameters to estimate (i.e. β, h and $\omega = (\omega_1, \ldots, \omega_N)'$), treatment of heteroskedasticity of unknown form may sound like a difficult task. However, as we shall see, it is not too difficult to extend the techniques of the previous sections of this chapter to be applicable to this model. Furthermore, the method developed to handle this case is quite important for two reasons. Firstly, the method involves the use of a *hierarchical prior*. This is a concept we will use again and again throughout the remainder of this book. Hierarchical priors have played a big role in many recent developments in Bayesian statistical theory and are gradually becoming more popular in econometrics as well. They are commonly used as a way of making flexible, parameter-rich models more amenable to statistical analysis.[2] Secondly, this model also allows us to

[2]Frequentist econometricians also work with models that are hierarchical in structure and very similar to ones discussed in this book. However, the frequentist statistical theory surrounding these models is often quite difficult. Accordingly, Bayesian methods are particularly popular in this area of the statistical literature.

introduce concepts relating to flexible econometric modelling (see Chapter 10) and, in particular, allows us to free up the assumption of Normal errors that we have used so far.

We begin by eliciting $p(\omega)$, the prior for the N-dimensional vector ω. As in previous chapters, it proves convenient to work with error precisions rather than variances and, hence, we define $\lambda \equiv (\lambda_1, \lambda_2, \ldots, \lambda_N)' \equiv (\omega_1^{-1}, \omega_2^{-1}, \ldots, \omega_N^{-1})'$. Consider the following prior for λ:

$$p(\lambda) = \prod_{i=1}^{N} f_G(\lambda_i | 1, \nu_\lambda) \tag{6.22}$$

Note that the prior for λ depends upon a hyperparameter, ν_λ, which is chosen by the researcher and assumes each λ_i comes from the same distribution. In other words, (6.22) implies that the λ_is are i.i.d. draws from the Gamma distribution. This assumption (or something similar) is necessary to deal with the problems caused by the high-dimensionality of λ. Intuitively, if we were to simply treat $\lambda_1, \ldots, \lambda_N$ as N completely independent and unrestricted parameters, we would not have enough observations to estimate each one of them. Equation (6.22) puts some structure which allows for estimation. It allows for all the error variances to be different from one another, but says they are all drawn from the same distribution. Thus, we can have a very flexible model, but enough structure is still imposed to allow for statistical inference.

You may be wondering why we chose the particular form given in (6.22). For instance, why should the λ_is be i.i.d. draws from the Gamma distribution with mean 1.0? Rather remarkably, it turns out that this model, with likelihood given by (6.3) and prior given by (6.9), (6.10) and (6.22) is *exactly the same* as the linear regression model with i.i.d. Student-t errors with ν_λ degrees of freedom. In other words, if we had begun by assuming

$$p(\varepsilon_i) = f_t(\varepsilon_i | 0, h^{-1}, \nu_\lambda) \tag{6.23}$$

for $i = 1, \ldots, N$, derived the likelihood and used (6.9) and (6.10) as priors for β and h, respectively, we would have ended up with exactly the same posterior. We will not formally prove this statement and the interested reader is referred to Geweke (1993) for proofs and further explanation. Note, however, the power and convenience of this result. The Student-t distribution is similar to the Normal, but has fatter tails and is more flexible. In fact, the Normal distribution is a special case of the Student-t which occurs as $\nu_\lambda \to \infty$. Thus, we have a model that allows for a more flexible error distribution, but we have achieved this result without leaving our familiar Normal linear regression model framework. Furthermore, we can draw on the computational methods derived above to develop a posterior simulator for the linear regression model with independent Student-t errors. For this reason, an explicit statement of the likelihood function for this model is not given here.

In Chapter 10 we discuss several ways of making models more flexible. However, it is worthwhile briefly noting that the model discussed here involves a *mixture of Normals* distribution of a particular sort. Intuitively, if a Normal distribution is too restrictive, you can create a more flexible distribution by taking a weighted average of more than one Normal distribution. As more and more Normals are mixed, the distribution becomes more and more flexible and, as discussed in Chapter 10, can approximate any distribution to a high degree of accuracy. Thus, mixtures of Normals models are a powerful tool for use when economic theory does not suggest a particular form for the likelihood function and you wish to be very flexible. Our treatment of heteroskedasticity of an unknown form is equivalent to a *scale mixture of Normals*. This means that the assumption that ε_i are independent $N(0, h^{-1}\lambda_i^{-1})$ with prior for λ_i given in (6.22) is equivalent to the assumption that the error distribution is a weighted average (or mixture) of different Normal distributions which have different variances (i.e. different scales) but the same means (i.e. all errors have mean zero). When this mixing is done using $f_G(\lambda_i|1, \nu_\lambda)$ densities, the mixture of Normals ends up being equivalent to the t distribution. However, using densities other than the $f_G(\lambda_i|1, \nu_\lambda)$ will yield other distributions more flexible than the Normal. See Chapter 10 for further details.

The previous discussion assumed that ν_λ was known. In practice, this would usually not be a reasonable assumption, and it is, thus, desirable to treat it as an unknown parameter. In the Bayesian framework, every parameter requires a prior distribution and, at this stage, we will use the general notation $p(\nu_\lambda)$. Note that, if we do this, the prior for λ is specified in two steps, the first being (6.22), the other being $p(\nu_\lambda)$. Alternatively, the prior for λ can be written as $p(\lambda|\nu_\lambda)p(\nu_\lambda)$. Priors written in two (or more) steps in this way are referred to as hierarchical priors. Writing a prior as a hierarchical prior is often a convenient way of expressing prior information and many of the models discussed in future chapters will be written in this way. However, we do stress the convenience aspect of hierarchical priors. It is never necessary to use a hierarchical prior, since the laws of probability imply that every hierarchical prior can be written in a non-hierarchical fashion. In the present case, the result $p(\lambda) = \int p(\lambda|\nu_\lambda)p(\nu_\lambda)d\nu_\lambda$ could be used to derive the non-hierarchical version of our prior for λ.

In all of the previous empirical illustrations, we have presented posterior means as point estimates of parameters and posterior standard deviations as measures of the uncertainty associated with the point estimates. However, as mentioned in Chapter 1, means and standard deviations do not exist for all valid probability density functions. The present model is the first one we have considered where means and standard deviations do not necessarily exist. In particular, Geweke (1993) shows that if you use a common noninformative prior for β (i.e. $p(\beta) \propto 1$ on the interval $(-\infty, \infty)$), then the posterior mean does not exist, unless $p(\nu_\lambda)$ is zero on the interval $(0, 2]$. The posterior standard deviation does not exist unless $p(\nu_\lambda)$ is zero on the interval $(0, 4]$. Hence, the researcher who wants to use a noninformative prior for β should either use a prior which excludes small values for ν_λ or present posterior medians and interquartile ranges (which will exist for any valid p.d.f.). With an informative Normal prior for β like (6.9), the posterior mean and standard deviation of β will exist.

It is also risky to use a noninformative prior for ν_λ. A naive researcher who wishes to be noninformative might use an improper Uniform prior:

$$p(\nu_\lambda) \propto 1 \text{ for } \nu_\lambda \in (0, \infty)$$

thinking that it would allocate equal prior weight to every interval of equal length. But the Student-t distribution with ν_λ degrees of freedom approaches the Normal distribution as $\nu_\lambda \to \infty$. In practice, the Student-t is virtually identical to the Normal for $\nu_\lambda > 100$. Our naive 'noninformative' prior allocates virtually all its weight to this region (i.e. $\frac{p(\nu_\lambda \leq 100)}{p(\nu_\lambda > 100)} = 0$). So this prior, far from being noninformative, is extremely informative: it is saying the errors are Normally distributed! This illustrates one of the problems with trying to come up with noninformative priors. There is a large Bayesian literature on how to construct noninformative priors (Zellner, 1971, provides an introduction to this). A detailed discussion of this issue is beyond the scope of the present book (although see Chapter 12, Section 12.3). However, it is worth noting that extreme care must be taken when trying to elicit a noninformative prior.

6.4.2 Bayesian Computation

In this subsection, we develop a Gibbs sampler for posterior analysis of β, h, λ and ν_λ. The Gibbs sampler requires the derivation of the full conditional posterior distributions of these parameters. We have already derived some of these as $p(\beta|y, h, \lambda)$ and $p(h|y, \beta, \lambda)$ are given in (6.12) and (6.15), respectively.[3] Hence, we focus on $p(\lambda|y, \beta, h, \nu_\lambda)$ and $p(\nu_\lambda|y, \beta, h, \lambda)$. The former of these can be derived by plugging the prior given in (6.22) into the general form for the conditional posterior given in (6.18). An examination of the resulting density shows that the λ_is are independent of one another (conditional on the other parameters of the model) and each of the conditional posteriors for λ_i has the form of a Gamma density. Formally, we have

$$p(\lambda|y, \beta, h, \nu_\lambda) = \prod_{i=1}^{N} p(\lambda_i|y, \beta, h, \nu_\lambda) \tag{6.24}$$

and

$$p(\lambda_i|y, \beta, h, \nu_\lambda) = f_G\left(\lambda_i|\frac{\nu_\lambda + 1}{h\varepsilon_i^2 + \nu_\lambda}, \nu_\lambda + 1\right) \tag{6.25}$$

Note that, conditional on knowing β, ε_i can be calculated and, hence, the parameters of the Gamma density in (6.25) can be calculated in the Gibbs sampler.

[3]Formally, the full conditionals to be used in the Gibbs sampler should be $p(\beta|y, h, \lambda, \nu_\lambda)$ and $p(h|y, \beta, \lambda, \nu_\lambda)$. However, conditional on λ, ν_λ adds no new information and, thus, $p(\beta|y, h, \lambda, \nu_\lambda) = p(\beta|y, h, \lambda)$ and $p(h|y, \beta, \lambda, \nu_\lambda) = p(h|y, \beta, \lambda)$.

Up until now, we have said nothing about the prior for ν_λ, and its precise form has no relevance for the posterior conditional for the other parameters. However, the form of $p(\nu_\lambda)$ does, of course, affect $p(\nu_\lambda|y, \beta, h, \lambda)$ and, hence, we must specify it here. Since we must have $\nu_\lambda > 0$, we use an exponential distribution for the prior. As noted in Appendix B, Theorem B.7, the exponential density is simply the Gamma with two degrees of freedom. Hence, we write

$$p(\nu_\lambda) = f_G(\nu_\lambda|\underline{\nu}_\lambda, 2) \tag{6.26}$$

Other priors can be handled with small changes in the following posterior simulation algorithm.

$p(\nu_\lambda|y, \beta, h, \lambda)$ is relatively easy to derive, since ν_λ does not enter the likelihood and it can be confirmed that $p(\nu_\lambda|y, \beta, h, \lambda) = p(\nu_\lambda|\lambda)$. It follows from Bayes theorem that

$$p(\nu_\lambda|\lambda) \propto p(\lambda|\nu_\lambda)p(\nu_\lambda)$$

and, thus, the kernel of the posterior conditional of ν_λ is simply times (6.22) times (6.26). Thus, we obtain

$$p(\nu_\lambda|y, \beta, h, \lambda) \propto \left(\frac{\nu_\lambda}{2}\right)^{\frac{N\nu_\lambda}{2}} \Gamma\left(\frac{\nu_\lambda}{2}\right)^{-N} \exp(-\eta\nu_\lambda) \tag{6.27}$$

where

$$\eta = \frac{1}{\underline{\nu}_\lambda} + \frac{1}{2}\sum_{i=1}^{N}[\ln(\lambda_i^{-1}) + \lambda_i]$$

This density is a non-standard one. Hence, we will use a Metropolis–Hastings algorithm to take draws from (6.27). However, it should be mentioned in passing that Geweke (1993) recommends use of another useful computational technique called *acceptance sampling*. This technique is very useful when the non-standard distribution that the researcher wishes to draw from is univariate and can be bounded. We will not discuss it here, but Geweke (1993) provides more detail on acceptance sampling as it relates to the present model (see also Chapter 12, Section 12.1). Devroye (1986) offers a thorough discussion of acceptance sampling in general.

For many hypotheses (e.g. $\beta_j = 0$) the Savage–Dickey density ratio can be used for model comparison. It can be calculated as described in Chapter 4, Section 4.2.5. However, not all hypotheses are easily calculated using the Savage–Dickey ratio. For instance, in many cases you might be interested in seeing whether there is any evidence of departures from Normality. In this case, you would wish to compare M_1: $\nu_\lambda \to \infty$ to M_2: ν_λ is finite. These models do not easily fit in the nested model comparison framework for which the Savage–Dickey density ratio is suitable. However, the Bayes factor comparing these two models can be calculated using the Gelfand–Dey approach. Note that this would require a posterior simulator for each model (i.e. the posterior simulator in Chapter 4, Section 4.2 for M_1 and the one described in this section for M_2). Alternatively, posterior predictive p-values or HPDIs can be calculated to shed light on the fit and appropriateness of the model. Predictive inference in this model can be carried out using the strategy outlined in Chapter 4, Section 4.2.6.

6.4.3 Empirical Illustration: The Regression Model with Student-t Errors

We return to our familiar house price data set introduced in Chapter 3 to illustrate the use of Gibbs sampling in the linear regression model with independent Student-t errors (or, equivalently, the Normal linear regression model with heteroskedasticity of unknown form). The reader is referred to Section 3.9 for a precise description of the dependent and explanatory variables for this data set. The priors for β and h are given in (6.9) and (6.10) and we use the same values for hyperparameters as in Chapter 4, Section 4.2.7. The prior for ν_λ depends upon the hyperparameter $\underline{\nu}_\lambda$, its prior mean. We set $\underline{\nu}_\lambda = 25$, a value which allocates substantial prior weight both to very fat-tailed error distributions (e.g. $\nu_\lambda < 10$), as well as error distributions which are roughly Normal (e.g. $\nu_\lambda > 40$).

The posterior simulator is a Metropolis-within-Gibbs algorithm, with draws of β and h taken from (6.12) and (6.15), respectively. Draws from $p(\lambda|y, \beta, h, \nu_\lambda)$ are taken using (6.25). For $p(\nu_\lambda|y, \beta, h, \lambda)$, we use a Random Walk Chain Metropolis–Hastings algorithm with a Normal increment random variable (see Chapter 5, (5.10)) . Equation (6.27), evaluated at old and candidate draws, is used to calculate the acceptance probability (see Chapter 5, (5.11)). Candidate draws of ν_λ which are less than or equal to zero have the acceptance probability set to zero. The variance of the proposal density, labelled Σ in (5.12), is chosen by first setting $\Sigma = c$ and experimenting with different values of the scalar c until a value is found which yields reasonable acceptance probabilities. The posterior simulator is then run using this value to yield an estimate of the posterior variance of ν_λ, $\widehat{var(\nu_\lambda|y)}$. We then set $\Sigma = c\widehat{var(\nu_\lambda|y)}$ and experiment with different values of c until we find one which yields an average acceptance probability of roughly 0.50. Then a final long run of 30 000 replications, with 5000 burn-in replications discarded, is taken. MCMC diagnostics indicate convergence of the Metropolis-within-Gibbs algorithm and numerical standard errors indicate an approximation error which is small relative to posterior standard deviations of all parameters.

Table 6.2 contains posterior results for the key parameters and it can be seen that, although posteriors for the elements of β are qualitatively similar to those presented in Tables 4.1 and 6.1, the posterior for ν_λ indicates the errors exhibit substantial deviations from Normality. Since this crucial parameter is univariate, we also plot a histogram approximation to its posterior. Figure 6.1 indicates that $p(\nu_\lambda|y)$ has a shape which is quite skewed and confirms the finding that virtually all of the posterior probability is allocated to small values for the degrees of freedom parameter. Note, however, that there is virtually no support for extremely small values which would imply extremely fat tails. Remember that the Cauchy distribution is the Student-t with $\nu_\lambda = 1$. It has such fat tails that its mean does not exist. There is no evidence for this sort of extreme behavior in the errors for the present data set.

Table 6.2 Posterior Results for β and ν_λ

	Mean	Standard Deviation	95% HPDI
β_1	−413.15	2898.24	[−5153, 4329]
β_2	5.24	0.36	[4.65, 5.83]
β_3	2118.02	972.84	[501, 3709]
β_4	14 910.41	1665.83	[12 188, 17 631]
β_5	8108.53	955.74	[6706, 9516]
ν_λ	4.31	0.85	[3.18, 5.97]

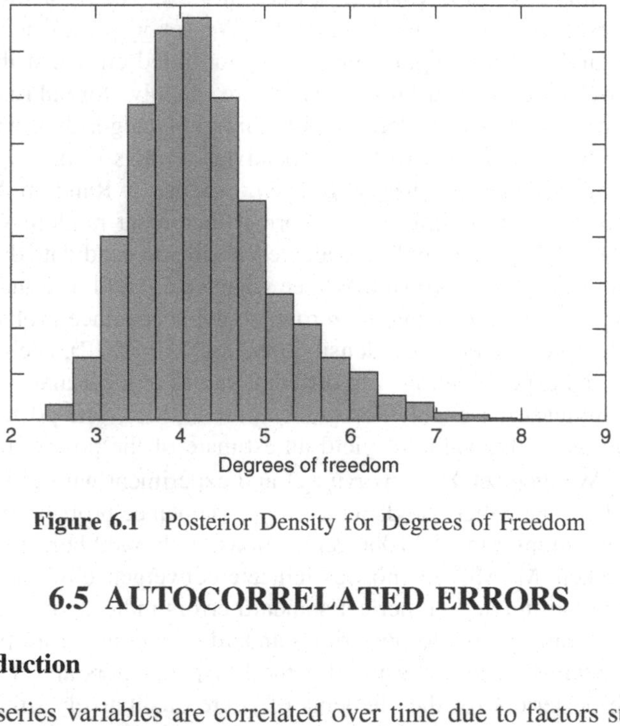

Figure 6.1 Posterior Density for Degrees of Freedom

6.5 AUTOCORRELATED ERRORS

6.5.1 Introduction

Many time series variables are correlated over time due to factors such as habit persistence or the time taken for adjustments to take place. This correlation between values of a variable at different times can spill over to the error. It is thus desirable to consider forms for the error covariance matrix which allow for this. In earlier chapters we assumed ε to be $N(0_N, h^{-1}I_N)$. In the previous sections of the present chapter, we relaxed this to allow for the error covariance matrix to be diagonal. However, so far, we have always assumed the errors to be uncorrelated with one another (i.e. $E(\varepsilon_i\varepsilon_j) = 0$ for $i \neq j$). In this section, we consider a model which relaxes this assumption.

Following common practice, we will use a subscript t to indicate time. That is, y_t for $t = 1, \ldots, T$ indicates observations on the dependent variable from period 1 through T (e.g. annual observations on GDP from 1946–2001). A simple manner of allowing for the errors to be correlated is to assume they follow an *autoregressive process of order 1* or *AR(1)* process:

$$\varepsilon_t = \rho \varepsilon_{t-1} + u_t \tag{6.28}$$

where u_t is i.i.d. $N(0, h^{-1})$. This specification allows for the error in one period to depend on that in the previous period.

The time series literature has developed a myriad of tools to aid in a more formal understanding the properties of various time series models and we digress briefly to introduce a few of them here using a general notation, z_t to indicate a time series.[4] In this section, we set $z_t = \varepsilon_t$, but the concepts are generally relevant and will be used in later chapters. It is standard to assume that the process generating the time series has been running from time period $-\infty$ and will run until period ∞. The econometrician observes this process for periods $t = 1, \ldots, T$. z_t is said to be *covariance stationary* if, for every t and s:

$$E(z_t) = E(z_{t-s}) = \mu$$

$$var(z_t) = var(z_{t-s}) = \gamma_0$$

and

$$cov(z_t, z_{t-s}) = \gamma_s$$

where μ, γ_0 and γ_s are all finite. In words, a time series is covariance stationary if it has a constant mean, variance and the covariance between any two observations depends only upon the number of periods apart they are. Many time series variables in economics do seem to be stationary or, if not, can be *differenced* to stationarity. The *first difference* of z_t is denoted by Δz_t and is defined by

$$\Delta z_t = z_t - z_{t-1}$$

In a similar fashion, we can define mth order differences for $m > 1$ as

$$\Delta^m z_t = \Delta^{m-1} z_t - \Delta^{m-1} z_{t-1}$$

To understand the economic relevance of differencing, suppose that z_t is the log of the price level, then Δz_t is (approximately) the percentage change in prices which is inflation. $\Delta^2 z_t$ would then be the percentage change in the rate of inflation. Any or all of these might be important in a macroeconomic model.

A common tool for examining the properties of stationary time series variables is γ_s which is referred to as the *autocovariance function*. Closely related is the *autocorrelation function*, which calculates correlations between observations s periods apart (i.e. it is defined as $\frac{\gamma_s}{\gamma_0}$ for $s = 0, \ldots, \infty$). These are both functions of s and it is common to plot either of them to see how they

[4] Space precludes a detailed discussion of time series methods. Bauwens, Lubrano and Richard (1999) provide an excellent Bayesian discussion of time series methods and the reader is referred to this book for more detail. Enders (1995) is a fine non-Bayesian book.

change as s increases. For instance, with macroeconomic variables we typically find autocorrelation functions decrease with s since recent happenings have more impact on current macroeconomic conditions than things that happened long ago.

Let us now return to the AR(1) process for the errors given in (6.28). To figure out its properties it is convenient to write ε_t in terms of u_{t-s} for $s = 0, \ldots, \infty$. This can be done by noting $\varepsilon_{t-1} = \rho\varepsilon_{t-2} + u_{t-1}$ and substituting this expression into (6.28), yielding

$$\varepsilon_t = \rho^2\varepsilon_{t-2} + \rho u_{t-1} + u_t$$

If we then substitute in the expression for ε_{t-2} we obtain an equation involving ε_{t-3} which we can substitute in for. Successively substituting in expressions for ε_{t-s} in this manner, (6.28) can be written as

$$\varepsilon_t = \sum_{s=0}^{\infty} \rho^s u_{t-s} \tag{6.29}$$

Written in this form, you can see that problems will occur if you try and calculate the mean, variance and covariance of ε_t since ρ^s will become infinite if $|\rho| > 1$. Even if $\rho = 1$, such calculations will involve infinite sums of finite terms. In fact, $|\rho| < 1$ is required for the time series to be stationary.

If we impose $|\rho| < 1$ it can be confirmed that $E(\varepsilon_t) = 0$

$$\gamma_0 = var(\varepsilon_t) = h^{-1}\sum_{s=0}^{\infty} \rho^{2s} = \frac{1}{h(1-\rho^2)}$$

and

$$\gamma_s = cov(\varepsilon_t, \varepsilon_{t-s}) = \frac{\rho^s}{h(1-\rho^2)}$$

Note that, since $|\rho| < 1$, the autocovariance function γ_s declines as s increases. Intuitively, with an AR(1) process, the influence of the past gradually dies away.

These results can be used to write the covariance matrix of ε as $h^{-1}\Omega$, where

$$\Omega = \frac{1}{1-\rho^2}\begin{bmatrix} 1 & \rho & \rho^2 & \cdot & \rho^{T-1} \\ \rho & 1 & \rho & \cdot & \cdot \\ \rho^2 & \rho & \cdot & \cdot & \rho^2 \\ \cdot & \cdot & \cdot & \cdot & \rho \\ \rho^{T-1} & \cdot & \rho^2 & \rho & 1 \end{bmatrix} \tag{6.30}$$

The AR(1) model can be extended to include more past time periods or *lags*. We can define the autoregressive process of order p or *AR(p) process* as

$$\varepsilon_t = \rho_1\varepsilon_{t-1} + \cdots + \rho_p\varepsilon_{t-p} + u_t \tag{6.31}$$

and methods similar to those above can be used to calculate the mean, variance and autocovariance function. As will be shown in the next section, we do not need to know the exact form of the autocovariance function in order to do

Bayesian inference with the AR(p) process. Hence, we do not write it out here. The interested reader is referred to a time series book such as Enders (1995) for detail. Suffice it to note here that the AR(p) process has similar properties to the AR(1), but is more flexible.

This is a convenient place to introduce some more time series notation. The *lag operator* is denoted by L and has the property that $L\varepsilon_t = \varepsilon_{t-1}$ or, more generally, $L^m \varepsilon_t = \varepsilon_{t-m}$. The AR(p) model can thus be written as

$$(1 - \rho_1 L - \cdots - \rho_p L^p)\,\varepsilon_t = u_t$$

or

$$\rho(L)\varepsilon_t = u_t$$

where $\rho(L) = (1 - \rho_1 L - \cdots - \rho_p L^p)$ is a polynomial of order p in the lag operator. It can be verified that an AR(p) process is stationary if the roots of the equation $\rho(z) = 0$ are all greater than one in absolute value. For future reference, define $\rho = (\rho_1, \ldots, \rho_p)'$ and let Φ denote the stationary region for this model.

6.5.2 Bayesian Computation

A posterior simulator which allows for Bayesian inference in the Normal linear regression model with AR(p) errors can be developed by adapting the formulae for the general case with Ω unspecified given in (6.12), (6.15) and (6.18). If we make one approximation, these posterior conditionals assume a simple form. This approximation involves the treatment of the initial conditions. To understand what is meant by this statement, consider how we would transform the model as in (6.2). We can do this by working out the form of Ω when AR(p) errors are present and then deriving the matrix P such that $P\Omega P' = I$. Alternatively, let us write the regression model as

$$y_t = x_t'\beta + \varepsilon_t \tag{6.32}$$

where $x_t = (1, x_{t2}, \ldots, x_{tk})'$. Multiplying both sides of (6.32) by $\rho(L)$ and defining $y_t^* = \rho(L)y_t$ and $x_t^* = \rho(L)x_t$ we obtain

$$y_t^* = x_t^{*\prime}\beta + u_t \tag{6.33}$$

We have assumed that u_t is i.i.d. $N(0, h^{-1})$ and, thus, the transformed model given in (6.33) is simply a Normal linear regression model with i.i.d. errors. Note, however, what happens to this transformation for values of $t \leq p$. y_1^*, for instance, depends upon $y_0, \ldots, y_{1-p}$. Since our data runs from $t = 1, \ldots, T$, these so-called initial conditions, $y_0, \ldots, y_{1-p}$, are not observed. The treatment of initial conditions is a subtle issue, especially if the AR process is non-stationary or nearly so. The interested reader is referred to Bauwens, Lubrano and Richard (1999) or Schotman (1994) for more detail. Here, we will assume stationarity of errors, so the treatment of initial conditions is of less importance. Accordingly, we will follow a common practice and work with the likelihood function based on data from $t = p + 1, \ldots, T$ rather than $t = 1, \ldots, T$. Provided p is small relative to T, this will result in an approximate likelihood which is quite close to the true likelihood. Since y_t^* and x_t^* for $t = p + 1, \ldots, T$ do not depend upon unobserved lagged values, the transformation given in (6.33) can be done in a straightforward fashion.

To keep the notation as simple as possible, we will not introduce a new notation for the likelihood, posterior, etc. for data from $t = p+1, \ldots, T$. Instead, for the remainder of this section, we will simply interpret y, y^*, ε and ε^* as $(T - p)$-vectors (i.e. the first p elements have been removed). X and X^* will be $(T-p)\times k$ matrices. With these changes, a Gibbs sampling algorithm can be derived in a straightforward fashion using previous results. Intuitively, $p(\beta|y, h, \rho)$ and $p(h|y, \beta, \rho)$ are given in (6.12) and (6.15). $p(\rho|y, \beta, h)$ can be derived by noting that, conditional on β and h, ε_t for $t = p+1, \ldots, T$ is known and (6.31) is simply a Normal linear regression model (with known error variance) with coefficients given by ρ. Thus, standard Bayesian results from previous chapters can be used to derive $p(\rho|y, \beta, h)$.

Formally, using the independent Normal-Gamma prior for β and h given in (6.9) and (6.10), the results of Section 6.2 can be modified to the present case as

$$\beta|y, h, \rho \sim N(\overline{\beta}, \overline{V}) \tag{6.34}$$

where

$$\overline{V} = (\underline{V}^{-1} + hX^{*\prime}X^*)^{-1} \tag{6.35}$$

and

$$\overline{\beta} = \overline{V}(\underline{V}^{-1}\underline{\beta} + hX^{*\prime}y^*) \tag{6.36}$$

The posterior for h conditional on the other parameters in the model is Gamma:

$$h|y, \beta, \rho \sim G(\overline{s}^{-2}, \overline{\nu}) \tag{6.37}$$

where

$$\overline{\nu} = T - p + \underline{\nu} \tag{6.38}$$

and

$$\overline{s}^2 = \frac{(y^* - X^*\beta)'(y^* - X^*\beta) + \underline{\nu}\,\underline{s}^2}{\overline{\nu}} \tag{6.39}$$

The posterior for ρ depends upon its prior which, of course, can be anything which reflects the researcher's non-data information. Here we assume it is multivariate Normal, truncated to the stationary region. That is,

$$p(\rho) \propto f_N(\rho|\underline{\rho}, \underline{V}_\rho)1(\rho \in \Phi) \tag{6.40}$$

where $1(\rho \in \Phi)$ is the indicator function which equals 1 for the stationary region and zero otherwise. With this prior, it is straightforward to derive

$$p(\rho|y, \beta, h) \propto f_N(\rho|\overline{\rho}, \overline{V}_\rho)1(\rho \in \Phi) \tag{6.41}$$

where

$$\overline{V}_\rho = (\underline{V}_\rho^{-1} + hE'E)^{-1} \tag{6.42}$$

$$\overline{\rho} = \overline{V}_\rho(\underline{V}_\rho^{-1}\underline{\rho} + hE'\varepsilon) \tag{6.43}$$

and E is a $(T-p)\times k$ matrix with tth row given by $(\varepsilon_{t-1},\ldots,\varepsilon_{t-p})$.

The Gibbs sampler involves sequentially drawing from (6.34), (6.37) and (6.41). The fact that (6.41) is truncated multivariate Normal, rather than simply multivariate Normal adds a slight complication. However, drawing from the truncated multivariate Normal distribution can be done by drawing from the untruncated variant and simply discarding the draws which fall outside the stationary region. Provided $\overline{\rho}$ lies within (or not too far outside) the stationary region, this strategy should work well. Alternatively, a Metropolis–Hastings algorithm can be derived or the methods of Geweke (1991) for drawing from the truncated multivariate Normal can be used. Predictive inference in this model can be carried out using the strategy outlined in Chapter 4, Section 4.2.6. Posterior predictive p-values or HPDIs can be calculated to shed light on the fit and appropriateness of the model. Bayes factors for any hypothesis of interest can be calculated using either the Savage–Dickey density ratio or the Gelfand–Dey approach. The fact that (6.41) provides only the kernel of $p(\rho|y,\beta,h)$ makes the use of the Savage–Dickey density ratio a little more complicated. Remember (see Chapter 4, Section 4.2.5) that the Savage–Dickey density ratio requires you to know the complete densities (not just the kernel), $p(\rho|y,\beta,h)$ or $p(\rho|y)$. For $p=1$, the integrating constant can be easily calculated since $p(\rho|y,\beta,h)$ is a univariate truncated Normal and the properties of this univariate density are well known (see Poirier, 1995, p. 115). However, for $p>1$ the stationary region is nonlinear and $p(\rho|y,\beta,h)$ is harder to work with analytically. Nevertheless, it is straightforward to calculate the necessary integrating constant through posterior simulation. That is, the density corresponding to (6.41) is

$$p(\rho|y,\beta,h) = \frac{f_N(\rho|\overline{\rho},\overline{V}_\rho)1(\rho\in\Phi)}{\int_\Phi f_N(\rho|\overline{\rho},\overline{V}_\rho)d\rho}$$

A common posterior simulator involves drawing from $f_N(\rho|\overline{\rho},\overline{V}_\rho)$ and discarding draws outside the stationary region. But, $\int_\Phi f_N(\rho|\overline{\rho},\overline{V}_\rho)d\rho$ is simply the proportion of draws retained. This can be estimated by, at every pass through the Gibbs sampler, calculating the number of rejected draws before an acceptable one is found. $1-\int_\Phi f_N(\rho|\overline{\rho},\overline{V}_\rho)d\rho$ is approximated by the number of rejected draws divided by the number of rejected draws plus one. As the number of Gibbs replications goes to infinity, the approximation error will go to zero. In general, the integrating constant of any truncated density can always be found by drawing from its untruncated counterpart and calculating the proportion of draws within the truncated region. Depending on the prior used, such a strategy may be necessary for calculating its integrating constant.

6.5.3 Empirical Illustration: The Normal Regression Model with Autocorrelated Errors

To illustrate Bayesian inference in the Normal regression model with autocorrelated errors, we use a data set pertaining to baseball. The dependent variable is the winning percentage of the New York Yankees baseball team every year between 1903 and 1999. Interest centers on explaining the Yankees' performance using various measures of team offensive and defensive performance. Thus

- y_t = winning percentage (PCT) in year t = wins/(wins + losses),
- x_{t2} = team on-base percentage (OBP) in year t,
- x_{t3} = team slugging average (SLG) in year t,
- x_{t4} = team earned run average (ERA) in year t.

A knowledge of baseball is not necessary to understand this empirical example. You need only note that the explanatory variables are all measures of team performance. We would expect x_{t2} and x_{t3} to be positively associated with winning percentage while x_{t4} should exhibit a negative association. Despite the prior information revealed in the previous sentence, we use a noninformative prior for β and set $\underline{V}^{-1} = 0_{k\times k}$. We also use a noninformative prior for the error precision and set $\underline{\nu} = 0$. With these choices, the values of $\underline{\beta}$ and $\underline{s}^{-2}$ are irrelevant. We use the technique described in the previous subsection to calculate the Savage–Dickey density ratios comparing models with $\rho_j = 0$ for $j = 1, \ldots, p$ to unrestricted models. This requires an informative prior for ρ and, thus, we set $\underline{\rho} = 0$ and $\underline{V}_\rho = cI_p$. Various values of c are chosen below in a prior sensitivity analysis. Throughout, we set $p = 1$. In preliminary runs with larger values of p, Bayes factors and HPDIs provided no evidence for autocorrelation of an order higher than one. To help provide intuition, note that the stationarity condition with $p = 1$ implies $|\rho_1| < 1$ and values of ρ_1 near one can be considered as implying a large degree of autocorrelation.

All results are based on 30 000 replications, with 5000 burn-in replications discarded and 25 000 replications retained. MCMC diagnostics indicate convergence of the Gibbs sampler, and numerical standard errors indicate an approximation error which is small relative to posterior standard deviations of all parameters.

Table 6.3 presents posterior results for β with $c = 0.09$, a reasonably small value reflecting a prior belief that autocorrelation in the errors is fairly small (i.e. the prior standard deviation of ρ_1 is 0.3). It can be seen that the results are as expected in that OBP and SLG are positive and ERA is negatively associated with winning.

At the beginning of the book, we emphasized the importance of doing prior sensitivity analysis. For the sake of space, our previous empirical illustrations did not include any investigation of prior sensitivity. However, we will do one here with regards to the AR(1) coefficient. Table 6.4 contains results from a prior sensitivity analysis where various values of c are used. This table reveals that prior information has little affect on the posterior, unless prior information is

Table 6.3 Posterior Results for β

	Mean	Standard Deviation	95% HPDI
β_1	0.01	0.07	[−0.11, 0.12]
β_2	1.09	0.35	[0.52, 1.66]
β_3	1.54	0.18	[1.24, 1.83]
β_4	−0.12	0.01	[−0.13, −0.10]

Table 6.4 Posterior Results for ρ_1

	Mean	Standard Deviation	95% HPDI	Bayes Factor for $\rho_1 = 0$
$c = 0.01$	0.10	0.07	[−0.02, 0.23]	0.49
$c = 0.09$	0.20	0.10	[0.03, 0.36]	0.43
$c = 0.25$	0.21	0.11	[0.04, 0.39]	0.56
$c = 1.0$	0.22	0.11	[0.05, 0.40]	0.74
$c = 100$	0.22	0.11	[0.05, 0.40]	0.84

extremely strong as in the $c = 0.01$ case. This can be seen by noting that posterior means, standard deviations and HPDIs are almost the same for all values of c between 0.09 and 100. The latter is a very large value which, to all intents and purposes, implies that the prior is flat and noninformative over the stationary region. The Bayes factors are also fairly robust to changes in the prior. As an aside, it is worth noting that this robustness of the Bayes factor is partly to do with the fact that the prior is truncated to a bounded interval – the stationary region. Don't forget the problems that can occur with Bayes factors when you use noninformative improper priors on parameters whose support is unbounded (e.g. see Chapter 3, Section 3.6.2).

6.6 THE SEEMINGLY UNRELATED REGRESSIONS MODEL

6.6.1 Introduction

The final model considered in this chapter is the Seemingly Unrelated Regressions (SUR) model. It is a multiple equation model which is both interesting in and of itself and is a component of other common models. In economics, multiple equation models arise in many contexts. For instance, in a study of consumption, the researcher may wish to estimate an equation for each category of consumption (i.e. food, consumer durables, non-durables, etc.). In a microeconomic application, the researcher may wish to estimate a factor demand equation for each factor of production.[5] In many cases, simply working with one equation at a time using the techniques of previous chapters will not lead the researcher too far wrong. However, working with all the equations together can improve estimation. This section discusses how to do so.

The SUR model can be written as

$$y_{mi} = x'_{mi}\beta_m + \varepsilon_{mi} \tag{6.44}$$

with $i = 1, \dots, N$ observations for $m = 1, \dots, M$ equations. y_{mi} is the ith observation on the dependent variable in equation m, x_{mi} is a k_m-vector containing the ith observation of the vector of explanatory variables in the mth equation and β_m is a k_m-vector of regression coefficients for the mth equation.[6] Note that this framework allows for the number of explanatory variables to differ across equations, but some or all of them may be the same in different equations.

We can put the SUR model in a familiar form. To do this we stack all equations into vectors/matrices as $y_i = (y_{1i}, \dots, y_{Mi})'$, $\varepsilon_i = (\varepsilon_{1i}, \dots, \varepsilon_{Mi})'$

$$\beta = \begin{pmatrix} \beta_1 \\ \cdot \\ \cdot \\ \beta_M \end{pmatrix}$$

$$X_i = \begin{pmatrix} x'_{1i} & 0 & \cdot & \cdot & 0 \\ 0 & x'_{2i} & 0 & \cdot & \cdot \\ \cdot & \cdot & \cdot & \cdot & \cdot \\ \cdot & \cdot & \cdot & \cdot & 0 \\ 0 & \cdot & \cdot & 0 & x'_{Mi} \end{pmatrix}$$

and define $k = \sum_{m=1}^{M} k_m$. Using this notation, it can be verified that (6.44) can be written as

$$y_i = X_i\beta + \varepsilon_i \tag{6.45}$$

We now stack all the observations together as

$$y = \begin{pmatrix} y_1 \\ \cdot \\ \cdot \\ y_N \end{pmatrix}$$

[5] For the reader with additional knowledge of econometrics, the reduced form of a simultaneous equations model is in the form of a SUR model. Similarly, a Vector Autoregression or VAR is also a SUR model (see Chapter 12, Section 12.4).

[6] Note that we have slightly changed notation from that used previously. In this section, x_{mi} is a vector and the first subscript indicates the equation number. Previously, x_{ij} was a scalar indicating the ith observation on the jth explanatory variable.

$$\varepsilon = \begin{pmatrix} \varepsilon_1 \\ \cdot \\ \cdot \\ \varepsilon_N \end{pmatrix}$$

$$X = \begin{pmatrix} X_1 \\ \cdot \\ \cdot \\ X_N \end{pmatrix}$$

and write

$$y = X\beta + \varepsilon$$

Thus, the SUR model can be written as our familiar linear regression model.

So far we have said nothing about error properties of this model. If we were to assume ε_{mi} to be i.i.d. $N(0, h^{-1})$ for all i and m, then we would simply have the Normal linear regression model of Chapters 2, 3 and 4. However, in many applications, it is common for the errors to be correlated across observations and, thus, we assume ε_i to be i.i.d. $N(0, H^{-1})$ for $i = 1, \ldots, N$ where H is an $M \times M$ error precision matrix. With this assumption it can be seen that ε is $N(0, \Omega)$ where Ω is an $NM \times NM$ block-diagonal matrix given by

$$\Omega = \begin{pmatrix} H^{-1} & 0 & \cdot & \cdot & 0 \\ 0 & H^{-1} & \cdot & \cdot & \cdot \\ \cdot & \cdot & \cdot & \cdot & \cdot \\ \cdot & \cdot & \cdot & \cdot & 0 \\ 0 & \cdot & \cdot & 0 & H^{-1} \end{pmatrix} \tag{6.46}$$

Hence, the SUR model lies in the class of models being studied in this chapter and the prior, likelihood and posterior have been discussed in Section 6.2. One minor issue you may have noticed is that there is no h in this model. This is not a substantive difference in that h was merely a scalar that was factored out for convenience in the previous sections. In this model, it is not convenient to factor out a scalar in this way (although we could have done this if we had wanted to).

6.6.2 The Prior

It is worthwhile to briefly discuss prior elicitation in the SUR model as this is a topic which has received a great deal of attention in the literature. In this section, we will use an extended version of our familiar independent Normal-Gamma prior, the independent Normal-Wishart prior:

$$p(\beta, H) = p(\beta)p(H)$$

where

$$p(\beta) = f_N(\beta|\underline{\beta}, \underline{V}) \tag{6.47}$$

and

$$p(H) = f_W(H|\underline{\nu}, \underline{H}) \tag{6.48}$$

The Wishart distribution, which is a matrix generalization of the Gamma distribution, is defined in Appendix B, Definition B.27. For prior elicitation, the most important things to note are that $E(H) = \underline{\nu H}$ and that noninformativeness is achieved by setting $\underline{\nu} = 0$ and $\underline{H}^{-1} = 0_{M \times M}$ (see Appendix B, Theorem B.16).

However, many other priors have been suggested for this model. In particular, a Normal-Wishart natural conjugate prior exists for this model analogous to that used in Chapter 3. This prior has the advantage that analytical results are available so that a posterior simulator is not required. However, the natural conjugate prior for the SUR model has been found by many to be too restrictive. For instance, it implies that the prior covariances between coefficients in each pair of equations (i.e. β_m and β_j for $j \neq m$) are all proportional to the same matrix. For this reason, only the noninformative variant of the natural conjugate prior has received much attention in empirical work. Furthermore, there have been various attempts to derive extended versions of the natural conjugate prior which are less restrictive. Readers interested in learning more about this area of the literature are referred to Dreze and Richard (1983) or Richard and Steel (1988).

6.6.3 Bayesian Computation

Bayesian computation in this model can be implemented with a Gibbs sampler using (6.12) and (6.18) based on the prior given in (6.47) and (6.48). However, both of these posterior conditionals involving inverting the $NM \times NM$ matrix Ω, which is computationally difficult. However, the block-diagonal structure of Ω allows the matrix inversion to be partly done analytically. If we do this, $p(\beta|y, H)$ and $p(H|y, \beta)$ take convenient forms. In particular,

$$\beta|y, H \sim N(\overline{\beta}, \overline{V}) \tag{6.49}$$

where

$$\overline{V} = \left(\underline{V}^{-1} + \sum_{i=1}^{N} X_i{}'HX_i\right)^{-1} \tag{6.50}$$

and

$$\overline{\beta} = \overline{V}\left(\underline{V}^{-1}\underline{\beta} + \sum_{i=1}^{N} X_i' H y_i\right) \tag{6.51}$$

The posterior for H conditional on β is Wishart:

$$H|y, \beta \sim W(\overline{\nu}, \overline{H}) \tag{6.52}$$

where

$$\overline{\nu} = N + \underline{\nu} \tag{6.53}$$

and

$$\overline{H} = \left[\underline{H}^{-1} + \sum_{i=1}^{N} (y_i - X_i\beta)(y_i - X_i\beta)' \right]^{-1} \tag{6.54}$$

Since random number generators for the Wishart distribution are available (e.g. a MATLAB variant is available in James LeSage's Econometrics Toolbox), a Gibbs sampler which successively draws from $p(\beta|y, H)$ and $p(H|y, \beta)$ can easily be developed.

Predictive inference in this model can be carried out using the strategy outlined in Chapter 4, Section 4.2.6. Posterior predictive p-values or HPDIs can be calculated to shed light on the fit and appropriateness of the model. The Savage–Dickey density ratio is particularly easy to calculate should you wish to calculate posterior odds ratios.

6.6.4 Empirical Illustration: The Seemingly Unrelated Regressions Model

To illustrate Bayesian inference in the SUR model we use an extended version of the baseball data set used in the autocorrelated errors example. In that example, we chose one baseball team, the Yankees, and investigated how team winning percentage (PCT) depended upon team on-base percentage (OBP), slugging average (SLG) and earned run average (ERA). The former two of these explanatory variables are measures of offensive performance, the last defensive performance. In the current example, we add a second equation for a second team, the Boston Red Sox (the arch-rivals of the Yankees). Hence, we have two equations, one for each team, with explanatory variables in each equation being the relevant team's OBP, SLG and ERA. Section 6.5.3 provides further detail about the data.

We use a noninformative prior for H and set $\underline{\nu} = 0$ and $\underline{H}^{-1} = 0_{2\times 2}$. For the regression coefficients, we set $\underline{\beta} = 0_k$ and $\underline{V} = 4I_k$. This prior reflects relatively noninformative prior beliefs. That is the regression coefficients are all centered over points which imply the explanatory variable has no effect on the dependent variable. But each coefficient has prior standard deviation of 2, a value which allows for the explanatory variables to have quite large impacts on the dependent variable.

Table 6.5 presents posterior results obtained from 30 000 replications from the Gibbs sampler outlined above, with 5000 burn-in replications discarded and 25 000 replications retained. MCMC diagnostics indicate convergence of the Gibbs sampler and numerical standard errors indicate an approximation error which is small relative to the posterior standard deviations of all parameters. Instead of presenting posterior results for H, which may be hard to interpret, we focus on the correlation between the errors in the two equations (i.e. corr$(\varepsilon_{1i}, \varepsilon_{2i})$ which is assumed to be the same for all $i = 1, \ldots, N$). If this correlation is equal to zero, then there is no benefit to using the SUR model over simply doing posterior inference on each equation separately. As we have emphasized throughout this book (see, e.g., Chapter 1, Section 1.2 or Chapter 3, Section 3.8), posterior

Table 6.5 Posterior Results for β and Error Correlation

	Mean	Standard Deviation	95% HPDI
		Yankees Equation	
β_1	0.03	0.06	[−0.06, 0.13]
β_2	0.92	0.30	[0.43, 1.41]
β_3	1.61	0.15	[1.36, 1.86]
β_4	−0.12	0.01	[−0.13, −0.10]
		Red Sox Equation	
β_5	−0.15	0.06	[−0.26, −0.05]
β_6	1.86	0.28	[1.41, 2.32]
β_7	1.24	0.15	[0.99, 1.50]
β_8	−0.11	0.01	[−0.12, −0.10]
		Cross-equation Error Correlation	
$\text{corr}(\varepsilon_1, \varepsilon_2)$	−0.01	0.11	[−0.18, 0.17]

simulator output can be used to do posterior inference on any function of the parameters of the model. Hence, the Gibbs draws of H can be used to derive posterior properties of $\text{corr}(\varepsilon_{1i}, \varepsilon_{2i})$. It can be seen that, with this data set, the correlation between the errors in the two equations is very near to zero. Thus, there is minimal benefit to working with the SUR model. If we had used an informative prior for H, we could have calculated a Bayes factor using the Savage–Dickey density ratio. This Bayes factor would have provided more formal evidence in favor of the hypothesis that the errors in the two equations are uncorrelated.

The regression coefficients measure the impacts of OBP, SLG and ERA on team performance. For both teams, results are sensible, indicating that higher OBP and SLG and lower ERA are associated with a higher team winning percentage. A baseball enthusiast might be interested in whether these coefficients are different in the two equations. After all, baseball wisdom has it that in some stadiums it is important to have power hitters, in others pitching is a relatively important key to success, etc. An examination of Table 6.5 indicates that, with one exception, the posterior means of comparable regression coefficients are roughly the same across equations, relative to their standard deviations. Furthermore, 95% HPDIs for comparable coefficients in different equations exhibit a large degree of overlap. The one exception is OBP where β_2 and β_6 are quite different from one another. The question of whether comparable coefficients are different in the two equations can be formally addressed by calculating Bayes factors comparing: $M_1 : \beta_j - \beta_{k_1+j} = 0$ for $j = 1, \ldots, k_1$ against M_2 where the coefficients are left unrestricted. This can be calculated using the Savage–Dickey density ratio implemented as outlined in Chapter 4, Section 4.2.5. Since the prior and conditional posterior of β are both Normal (see (6.47) and (6.49)), the prior and conditional posterior of any linear combination of the elements of β will also be Normal. This implies that the Savage–Dickey density ratio can be calculated in a simple fashion using only the Normal density. Bayes factors in favor of the models which impose equality of each pair of coefficients across equations are 2.84, 0.45, 3.05 and 255.84, respectively, for the four explanatory variables. Although these results are not overwhelming, they do provide additional support for the fact that it is only OBP which impacts differently in the two equations. Given that the coefficient on OBP is lower and that on SLG higher for the Yankees (although the latter finding is not statistically significant), a baseball researcher might conclude that the Yankees' success has been disproportionately associated with power hitting teams. A historian of baseball would remember great Yankees teams of the past and likely find this conclusion sensible.

It is worth noting that this example is purely illustrative. In a more serious application, the researcher would no doubt wish to use data on other baseball teams and have other explanatory variables. For instance, Yankees performance probably depends not only on Yankees OBP, SLG and ERA, but also on these variables for other teams they are competing against. Furthermore, the results of the previous section suggest that the model should have allowed for errors to be autocorrelated.

6.7 SUMMARY

In this chapter, we have considered Bayesian inference in the Normal linear regression model with errors distributed as $N(0, h^{-1}\Omega)$. This structure incorporates many special cases of empirical relevance. After discussing the general case, Bayesian inference in the following special cases were considered: heteroskedasticity of known and unknown forms, autocorrelation and the seemingly unrelated regressions model. Posterior computation involved either a Gibbs sampler or a Metropolis-within-Gibbs algorithm.

The Normal linear regression model with heteroskedasticity of unknown form was of particular importance, since it turned out that this model was identical to a Normal linear regression model with independent Student-t errors. This is a simple example of a mixture of Normals distribution. Such mixtures are commonly used as a way of freeing up distributional assumptions and will be discussed in more detail in Chapter 10. This model also allowed us to introduce the notion of a hierarchical prior which we will use repeatedly in subsequent chapters.

The Normal linear regression model with autocorrelated errors and the SUR model also enabled us to explain a few building blocks which we use in future chapters. For the former model, some basic time series concepts were introduced. For the latter, the multiple equation framework introduced will be used in several places in the future, including the multinomial Probit and state space models (see Chapters 8 and 9).

It is worth stressing that the modular nature of the Gibbs sampler and related algorithms makes it easy to combine and extend models. In Chapter 9, we will see how the Gibbs sampler for the SUR model forms two blocks of a more complicated Gibbs sampler for an extended variant of the SUR model called the multinomial Probit model. If you wanted to work with a linear regression model with independent Student-t errors which were autocorrelated, you could simply combine the posterior simulator for the regression model with Student-t errors with that for the model with autocorrelated errors. If you wanted a SUR model with heteroskedastic errors of known form, you could combine results from Sections 6.3 and 6.6. By incorporating a posterior simulator from Chapter 5 with ones from Sections 6.3 and 6.5 you could create a posterior simulator for a nonlinear regression model with autocorrelated Student-t errors. The possibilities are virtually endless. In general, this modularization which lies at the heart of most posterior simulators is an enormous advantage and allows for the simple extension of Bayesian methods into new areas.

6.8 EXERCISES

6.8.1 Theoretical Exercises

1. (a) Section 6.2 contains results for the Normal linear regression model with error covariance matrix Ω using the independent Normal-Gamma prior of (6.9) and (6.10). Show how posterior results would change if the natural conjugate Normal-Gamma prior of Chapter 3, Section 3.4 were used for β and h.
   (b) Using your result from part (a), show how posterior simulation in the heteroskedastic model of Section 6.3 would be altered if the natural conjugate Normal-Gamma prior were used.
   (c) Using your result from part (a), show how posterior simulation in the regression model with Student-t errors of Section 6.4 would be altered if the natural conjugate Normal-Gamma prior were used.
   (d) Using your result from part (a), show how posterior simulation in the regression model with autoregressive errors of Section 6.5 would be altered if the natural conjugate Normal-Gamma prior were used.
2. Discuss how posterior simulation could be done for all the models in this chapter if they were modified to have nonlinear regression functions as described in Chapter 5.
3. Derive a posterior simulator for the linear regression model with independent Student-t errors and AR(p) errors. (Hint: If you are having difficulty, the answer to this question is given in Chib, 1993.)

6.8.2 Computer-Based Exercises

The exercises in this chapter are closer to being small projects than standard textbook questions. Remember that some data sets and MATLAB programs are available on the website associated with this book. The house price data set is also available in the *Journal of Applied Econometrics* Data Archive listed under Anglin and Gencay (1996) (http://qed.econ.queensu.ca/jae/1996-v11.6/anglin-gencay/).

4. Use the house price data set for this question. Remember that in Chapter 4 we carried out Bayesian inference using this data and a Normal linear regression model, in the present chapter we added heteroskedasticity of a known form (see Section 6.3.2) and independent Student-t errors (see Section 6.4.3). In all cases, we used an independent Normal-Gamma prior for β and h with values for hyperparameters as in Chapter 4, Section 4.2.7. Use this prior throughout this question. For the other parameters (i.e. α for the heteroskedasticity extension and ν_λ for the Student-t errors extension) use an informative prior of your choosing.
   (a) Write programs for carrying out Bayesian inference in these three models (or obtain programs from the website associated with this book and modify as appropriate). Suppose interest centers on the marginal effect of lot size on house price (i.e. β_2). Calculate the posterior mean and standard deviation of this marginal effect using Bayesian model averaging.
   Background: Bayesian model averaging is briefly described in Chapter 2, equation 2.42 for the case of prediction. Chapter 11 will discuss it in more detail. Briefly, if $g(\beta_2)$ is a function of β_2, the rules of conditional probability imply that
$$E[g(\beta_2)|y] = \sum_{r=1}^{R} E[g(\beta_2)|y, M_r]p(M_r|y)$$
   and, thus, the Bayesian should calculate $E[g(\beta_2)|y]$ by averaging results obtained from all models where the weights in the average are given by $p(M_r|y)$.
   (b) Write a program which incorporates all of the extensions of part (a) into one model. That is, develop a posterior simulator for the linear regression model with Student-t errors which exhibit heteroskedasticity of a known form. Use this program to calculate the posterior mean and standard deviation of β_2. Compare your result with that of part (a).
5. For this exercise obtain any relevant time series data set (e.g. the Yankees data set of Section 6.5.3 or an artificially generated data set).
   (a) Write a program which carries out posterior simulation for the Normal linear regression model with AR(p) errors (or obtain from the website associated with this book the program used to do the empirical illustration in Section 6.5.3 and study and understand it).
   (b) Based on your derivations in Exercise 3, write a program which carries out posterior simulation for the independent Student-t linear regression model with AR(p) errors. Use the posterior simulator to calculate posterior means and standard deviations for all model parameters.

(c) Add to your program in part (b) code to calculate Bayes factors which can be used for choosing p, the lag length of the AR(p) model. Use the Savage–Dickey density ratio. Use this program to choose an optimal value for p for your data set.

(d) Add to your programs in parts (a) and (b) code for calculating the marginal likelihood using the Gelfand–Dey method. Use these programs to decide whether Normal or Student-t errors are preferred for your data set.

7 The Linear Regression Model with Panel Data

7.1 INTRODUCTION

In previous chapters, we have mostly considered the case where one data point was available for each observational unit. For instance, y_i was a scalar which contained a single observation of the dependent variable. However, in economics and related fields, there are many cases where several observations on each variable exist. In a microeconomic example of firm production, is it common to have several years of data on output and inputs for many firms. In the economic growth literature, data on many countries for many years is often available. In a financial example, the share price of many companies for many days might be used. In a marketing example, the purchases of many consumers on many visits to a store might be collected. All such examples are characterized by the availability of T observations for each of N individuals or firms. In econometrics, such data is referred to as *panel data*, although in the statistical literature it is referred to as *longitudinal data*. In this chapter, we discuss models and methods of Bayesian inference which are appropriate when the researcher has panel data. This chapter does not involve any new methods of computation. Instead, we combine aspects of various models and methods of posterior simulation from previous chapters. We also build upon the notion of a hierarchical prior introduced in the previous chapter. The chapter is organized according to the structure placed on the regression coefficients. We begin by assuming that the regression coefficients are the same for all individuals (*the pooled model*). We then work with a model where the regression intercepts are allowed to vary across individuals (*the individual effects model*), before proceeding to a model where all regression coefficients can vary across individuals (*the random coefficients model*). A special case of the individual effects model, called the *stochastic frontier model*, is of great empirical relevance and, hence, is discussed in detail. We also introduce a new method for calculating the marginal likelihood which we refer to as the *Chib method*, since it was introduced in Chib (1995). This method is applicable when posterior simulation is done using a Gibbs sampler and is often useful when the dimensionality of the parameter space is high.

We begin by extending the notation of previous chapters to deal with panel data (and warn the reader to exercise caution, since some notation differs slightly from previous chapters). Let y_{it} and ε_{it} denote the tth observations (for $t = 1, \ldots, T$) on the dependent variable and error, respectively, for the ith individual for $i = 1, \ldots, N$. y_i and ε_i will now denote vectors of T observations on the dependent variable and error, respectively, for the ith individual. In some of the regression models in this chapter, it is important to distinguish between the intercept and slope coefficients. Hence, we define X_i to be a $T \times k$ matrix containing the T observations on each of the k explanatory variables (including the intercept) for the ith individual. $\tilde{X}_i$ will be the $T \times (k-1)$ matrix equal to X_i with the intercept removed. Hence, $X_i = [\iota_T \ \tilde{X}_i]$. If we stack observations for all N individuals together, we obtain the TN-vectors:

$$y = \begin{bmatrix} y_1 \\ \cdot \\ \cdot \\ y_N \end{bmatrix} \quad \text{and} \quad \varepsilon = \begin{bmatrix} \varepsilon_1 \\ \cdot \\ \cdot \\ \varepsilon_N \end{bmatrix}$$

Similarly, stacking observations on all explanatory variables together yields the $TN \times K$ matrix:

$$X = \begin{bmatrix} X_1 \\ \cdot \\ \cdot \\ X_N \end{bmatrix}$$

7.2 THE POOLED MODEL

In the *pooled model*, it is assumed that the same linear regression relationship holds for every individual and, hence,

$$y_i = X_i\beta + \varepsilon_i \tag{7.1}$$

for $i = 1, \ldots, N$ where β is the k-vector of regression coefficients, including the intercept.

The form of the likelihood function depends upon assumptions made about the errors. In this chapter we will assume for $i, j = 1, \ldots, N$:

1. ε_i has a multivariate Normal distribution with mean 0_T and covariance matrix $h^{-1}I_T$.
2. ε_i and ε_j are independent of one another for $i \neq j$.

3. All elements of X_i are either fixed (i.e. not random variables) or, if they are random variables, they are independent of all elements of ε_j with a probability density function, $p(X_i|\lambda)$ where λ is a vector of parameters that does not include β and h.

These assumptions are much the same as those made in previous chapters. Note, however, that we are allowing ε_{it} and ε_{is} to be independent of one another for $t \neq s$. In words, we are allowing the T errors for a specific individual to be independent of one another. This might not be sensible if, for instance, the same individual made a consistent mistake in reporting data in every time period. If this is the case, then the researcher may wish to assume ε_i to have covariance matrix Ω. This case can be dealt with in a straightforward fashion by using ideas very similar to those used for the SUR model (see Chapter 6, Section 6.6).

The assumption that errors are independent over all individuals and time periods implies that the model simply reduces to a linear regression model of the sort discussed in Chapters 2, 3 and 4. That is, data for all individuals and time periods is simply pooled together in one big regression. Thus, we will not discuss the pooled model in any detail.

To be precise, the previous assumptions imply a likelihood function of the form

$$p(y|\beta, h) = \prod_{i=1}^{N} \frac{h^{\frac{T}{2}}}{(2\pi)^{\frac{T}{2}}} \left\{ \exp\left[-\frac{h}{2}(y_i - X_i\beta)'(y_i - X_i\beta) \right] \right\}$$

However, this likelihood can be written as

$$p(y|\beta, h) = \frac{h^{\frac{NT}{2}}}{(2\pi)^{\frac{NT}{2}}} \left\{ \exp\left[-\frac{h}{2}(y - X\beta)'(y - X\beta) \right] \right\}$$

which is the likelihood function used in Chapters 3 and 4 (e.g. see Chapter 3, (3.3)) modified to account for the fact that we now have *TN* observations. Thus, the techniques of Chapters 3 and 4 can be used directly to carry out Bayesian inference in this model.

In the empirical illustration below we will use an independent Normal Gamma prior with $\beta \sim N(\underline{\beta}, \underline{V})$ and $h \sim G(\underline{s}^{-2}, \underline{\nu})$ and, thus, methods described in Chapter 4, Section 4.2 are relevant.

7.3 INDIVIDUAL EFFECTS MODELS

The pooled model assumes the same regression relationship holds for every individual in every time period. In many empirical contexts, this assumption is unreasonable. Consider, for instance, a marketing example where y_{it} is sales of soft drink brand i in period t. The sales of a soft drink may depend on explanatory variables which are easy to observe, such as price, but also depend on more elusive quantities such as brand loyalty. Accordingly, a model of the form

$$y_{it} = \alpha_i + \beta x_{it} + \varepsilon_{it}$$

(where x_{it} is price of soft drink i in period t) might be appropriate. The fact that the intercept varies across soft drink brands (i.e. α_i has an i subscript) incorporates the effect of brand loyalty. That is, it allows for two soft drinks with an identical price to have different expected sales. It is not hard to come up with other examples which imply that the intercept should vary across individuals. We shortly study the stochastic frontier model and see how the economic theory underlying this model implies that the intercept should vary across individuals. We refer to models of this sort as *individual effect models*, with α_i being referred to as the individual effect. Similar terminology is used in the frequentist econometric literature, with *random effects* and *fixed effects* models being two popular types of individual effects models.

7.3.1 The Likelihood Function

The likelihood function for this model is based on the regression equation:

$$y_i = \alpha_i \iota_T + \tilde{X}_i \tilde{\beta} + \varepsilon_i \tag{7.2}$$

where the notation makes clear that we are using α_i to denote the intercept of ith individual's regression equation and $\tilde{\beta}$ to denote the vector of slope coefficients (which is assumed to be the same for all individuals). Equation (7.2), along with the error assumptions given after (7.1), imply a likelihood function of the form

$$p(y|\alpha, \tilde{\beta}, h) = \prod_{i=1}^{N} \frac{h^{\frac{T}{2}}}{(2\pi)^{\frac{T}{2}}} \left\{ \exp\left[-\frac{h}{2}(y_i - \alpha_i - \tilde{X}_i\tilde{\beta})'(y_i - \alpha_i - \tilde{X}_i\tilde{\beta}) \right] \right\} \tag{7.3}$$

where $\alpha = (\alpha_1, \ldots, \alpha_N)'$.

7.3.2 The Prior

In a Bayesian analysis one can, of course, use any sort of prior, including a non-informative one. Here we consider two types of priors which are computationally simple and commonly used.

A Non-hierarchical Prior

To motivate our first class of priors, note that the regression in (7.2) can be written as

$$y = X^* \beta^* + \varepsilon \tag{7.4}$$

where X^* is a $TN \times (N+k-1)$ matrix given by

$$X^* = \begin{bmatrix} \iota_T & 0_T & \cdot & \cdot & 0_T & \widetilde{X}_1 \\ 0_T & \iota_T & \cdot & \cdot & \cdot & \widetilde{X}_2 \\ \cdot & 0_T & \cdot & \cdot & \cdot & \cdot \\ \cdot & \cdot & \cdot & \cdot & 0_T & \cdot \\ 0_T & \cdot & \cdot & \cdot & \iota_T & \widetilde{X}_N \end{bmatrix}$$

and

$$\beta^* = \begin{bmatrix} \alpha_1 \\ \cdot \\ \cdot \\ \alpha_N \\ \widetilde{\beta} \end{bmatrix}$$

Written in this way, it is clear that the individual effects model can be written as a regression model of the sort used in Chapters 3 and 4 and any of the priors introduced in those chapters can be applied to (β^*, h). For those trained in frequentist econometrics, Bayesian analysis with such a non-hierarchical prior leads to a model which is analogous to the fixed effects model. To see this, note that X^* is the matrix, which includes the explanatory variables attached to a matrix containing a dummy variable for each individual.

In this chapter, we will use the independent Normal-Gamma prior of Chapter 4, Section 4.2 and, thus, assume β^* and h are *a priori* independent of each other with

$$\beta^* \sim N(\underline{\beta}^*, \underline{V}) \tag{7.5}$$

and

$$h \sim G(\underline{s}^{-2}, \underline{\nu}) \tag{7.6}$$

A Hierarchical Prior

In modern statistics, interest increasingly centers on models where the parameter vector is of high dimension. As discussed in Chapter 6, Section 6.4, Bayesian methods are gaining in popularity due to the fact that hierarchical priors can surmount some of the problems caused by high dimensional parameter spaces. The individual effects model is a model with a parameter space which contains $N + k$ parameters (i.e. N intercepts in α, $k-1$ slope coefficients in $\widetilde{\beta}$ plus the error precision, h). If T is small relative to N, the number of parameters is quite large relative to sample size.[1] This suggests that a hierarchical prior might be appropriate and it is indeed the case that such priors are commonly used.

[1] Panel data methods are often used with survey data collected by the government. These typically involve questioning a large number of people (e.g. $N = 10\,000$) every few years (e.g. $T = 5$) on various topics (e.g. their expenditure or work histories). In such cases N is invariably huge relative to T.

A convenient hierarchical prior assumes that, for $i = 1, \ldots, N$,

$$\alpha_i \sim N(\mu_\alpha, V_\alpha) \tag{7.7}$$

with α_i and α_j being independent of one another for $i \neq j$. The hierarchical structure of the prior arises if we treat μ_α and V_α as unknown parameters which require their own prior. We assume μ_α and V_α to be independent of one another with

$$\mu_\alpha \sim N(\underline{\mu}_\alpha, \underline{\sigma}^2_\alpha) \tag{7.8}$$

and

$$V_\alpha^{-1} \sim G(\underline{V}_\alpha^{-1}, \underline{\nu}_\alpha) \tag{7.9}$$

Both the hierarchical and non-hierarchical priors allow for every individual to have a different intercept. However, the hierarchical prior places more structure in that it assumes all intercepts are drawn from the same distribution. This extra structure (if consistent with patterns in the data) allows for more accurate estimation.

For the remaining parameters, we assume a non-hierarchical prior of the independent Normal-Gamma variety. Thus,

$$\tilde{\beta} \sim N(\underline{\beta}, \underline{V}_\beta) \tag{7.10}$$

and

$$h \sim G(\underline{s}^{-2}, \underline{\nu}) \tag{7.11}$$

For those trained in frequentist econometrics, Bayesian analysis with such a hierarchical prior leads to a model which is analogous to the frequentist random effects model.

7.3.3 Bayesian Computation

Posterior Inference under the Non-hierarchical Prior

Under the non-hierarchical prior given in (7.5) and (7.6), we have a linear regression model with independent Normal-Gamma prior. Hence, posterior inference can be carried out using the methods described in Chapter 4. In particular, a Gibbs sampler can be set up which takes sequential draws from

$$\beta^*|y, h \sim N(\overline{\beta^*}, \overline{V}) \tag{7.12}$$

and

$$h|y, \beta^* \sim G(\overline{s}^{-2}, \overline{\nu}) \tag{7.13}$$

where

$$\overline{V} = (\underline{V}^{-1} + hX^{*\prime}X^*)^{-1}$$

$$\overline{\beta^*} = \overline{V}(\underline{V}^{-1}\underline{\beta^*} + hX^{*\prime}y)$$

$$\overline{\nu} = TN + \underline{\nu}$$

and

$$\overline{s}^2 = \frac{\sum_{i-1}^{N}(y_i - \alpha_i \iota_T - \tilde{X}_i\tilde{\beta})'(y_i - \alpha_i \iota_T - \tilde{X}_i\tilde{\beta}) + \underline{\nu}\underline{s}^2}{\overline{\nu}}$$

The convergence and degree of approximation implicit in the Gibbs sampler can be calculated using the MCMC diagnostics described in Chapter 4, Section 4.2.4. Predictive inference in this model can be carried out using the strategy outlined in Chapter 4, Section 4.2.6 and model comparison can be done using any of the methods described in previous chapters.

Numerical problems can arise if N is very large since $\overline{V}$ is an $(N+k-1)\times(N+k-1)$ matrix which must be inverted. Computer algorithms for matrix inversion can become unreliable if the dimension of the matrix is large. In such cases, theorems for the inverse of a partitioned matrix (e.g. Appendix A, Theorem A.9) can be used to reduce the dimension of the matrices for which inversion is required.

Posterior Inference under the Hierarchical Prior

The derivation of the posterior under the hierarchical prior given in (7.7) through (7.11) is straightforward and details will not be provided here. The derivation involves multiplying prior times likelihood and then examining the result for each of $\tilde{\beta}$, h, α, μ_α and V_α to find the kernels of each conditional posterior distribution. A Gibbs sampler which sequentially draws from these posterior conditionals can then be used for posterior simulation.

The relevant posterior distributions for $\tilde{\beta}$ and h, conditional on α, are derived in the same fashion as those for the linear regression model with independent Normal-Gamma prior and, hence,[2]

$$\tilde{\beta}|y, h, \alpha, \mu_\alpha, V_\alpha \sim N(\overline{\beta}, \overline{V}_\beta) \tag{7.14}$$

and

$$h|y, \tilde{\beta}, \alpha, \mu_\alpha, V_\alpha \sim G(\overline{s}^{-2}, \overline{\nu}) \tag{7.15}$$

where

$$\overline{V}_\beta = \left(\underline{V}_\beta^{-1} + h\sum_{i=1}^{N}\tilde{X}_i'\tilde{X}_i\right)^{-1}$$

$$\overline{\beta} = \overline{V}\left(\underline{V}_\beta^{-1}\underline{\beta} + h\sum_{i=1}^{N}\tilde{X}_i'[y_i - \alpha_i\iota_T]\right)$$

$$\overline{\nu} = TN + \underline{\nu}$$

[2]Note that in the following equations $p(\tilde{\beta}|y, h, \alpha, \mu_\alpha, V_\alpha)$ and $p(h|y, \tilde{\beta}, \alpha, \mu_\alpha, V_\alpha)$ do not depend upon μ_α, and V_α, and thus are equivalent to $p(\tilde{\beta}|y, h, \alpha)$ and $p(h|y, \tilde{\beta}, \alpha)$. We adopt the more complete notation here and in later sections to emphasize that Gibbs sampling involves drawing from the full posterior conditional distributions (see Chapter 4, Section 4.2.3).

and

$$\overline{s}^2 = \frac{\sum_{i-1}^{N}(y_i - \alpha_i \iota_T - \tilde{X}_i \tilde{\beta})'(y_i - \alpha_i \iota_T - \tilde{X}_i \tilde{\beta}) + \underline{\nu s}^2}{\overline{\nu}}$$

The conditional posterior for each α_i is independent of α_j for $i \neq j$ and is given by

$$\alpha_i | y, \tilde{\beta}, h, \mu_\alpha, V_\alpha \sim N(\overline{\alpha}_i, \overline{V}_i) \tag{7.16}$$

where

$$\overline{V}_i = \frac{V_\alpha h^{-1}}{T V_\alpha + h^{-1}}$$

and

$$\overline{\alpha}_i = \frac{V_\alpha (y_i - \tilde{X}_i \tilde{\beta})' \iota_T + h^{-1} \mu_\alpha}{(T V_\alpha + h^{-1})}$$

Finally, the conditional posteriors for the hierarchical parameters, μ_α and V_α, are

$$\mu_\alpha | y, \tilde{\beta}, h, \alpha, V_\alpha \sim N(\overline{\mu}_\alpha, \overline{\sigma}^2_\alpha) \tag{7.17}$$

and

$$V_\alpha^{-1} | y, \tilde{\beta}, h, \alpha, \mu_\alpha \sim G(\overline{V}_\alpha^{-1}, \overline{\nu}_\alpha) \tag{7.18}$$

where

$$\overline{\sigma}^2_\alpha = \frac{V_\alpha \underline{\sigma}^2_\alpha}{V_\alpha + N \underline{\sigma}^2_\alpha}$$

$$\overline{\mu}_\alpha = \frac{V_\alpha \underline{\mu}_\alpha + \underline{\sigma}^2_\alpha \sum_{i=1}^{N} \alpha_i}{V_\alpha + N \underline{\sigma}^2_\alpha}$$

$$\overline{\nu}_\alpha = \underline{\nu}_\alpha + N$$

and

$$\overline{V}_\alpha = \frac{\sum_{i=1}^{N}(\alpha_i - \mu_\alpha)^2 + \underline{V}_\alpha \underline{\nu}_\alpha}{\overline{\nu}_\alpha}$$

Note that the Gibbs sampler, involving (7.14)–(7.18), requires only random number generation from the Normal and Gamma distributions. Thus, although the formulae look a bit messy, it is not hard to write a program which carries out posterior simulation in this model. Predictive analysis can be done using the methods described in Chapter 4 and MCMC diagnostics can be used to verify convergence of the Gibbs sampler.

7.4 THE RANDOM COEFFICIENTS MODEL

The pooled model assumed that the same regression line was appropriate for all individuals, while the individual effects models assumed all individuals had regression lines with the same slopes, but possibly different intercepts. In some cases, it is desirable to free up the common slope assumption of the latter class of models and work with

$$y_i = X_i\beta_i + \varepsilon_i \tag{7.19}$$

where β_i is a k-vector of regression coefficients including intercept as well as slopes. Equation (7.19) holds for $i = 1, \ldots, N$ and, thus, the entire model contains $Nk + 1$ parameters (i.e. k regression coefficients for each of N individuals plus the error precision, h). Unless T is large relative to N it is very difficult to estimate all the parameters in this model with any degree of precision. Hence, it is common to use a hierarchical prior for the regression coefficients. Such a model is referred to as a *random coefficients* model.

To motivate this model, let us return to the marketing example, where the dependent variable is sales of a particular soft drink and X_i contains an intercept and the price of the soft drink. Since β_i differs over soft drink brands, the model allows for two brands with identical prices to have different expected sales (i.e. due to different intercepts). Furthermore, it allows for the marginal effect of price on sales to vary across brands (i.e. due to different slope coefficients). If brand loyalty is important, then such cross-brand differences might occur. The use of a hierarchical prior places some structure on these differences by modeling them as coming from a common distribution. Informally speaking, the random coefficients model allows for every individual brand to be different, but the hierarchical prior means they are not too different. Such a model might be reasonable in many applications.

7.4.1 The Likelihood Function

Our error assumptions (see the discussion after (7.1)) plus (7.19) yields a likelihood function of the form

$$p(y|\beta, h) = \prod_{i=1}^{N} \frac{h^{\frac{T}{2}}}{(2\pi)^{\frac{T}{2}}} \left\{ \exp\left[-\frac{h}{2}(y_i - X_i\beta_i)'(y_i - X_i\beta_i) \right] \right\} \tag{7.20}$$

where we will let $\beta = (\beta_1', \ldots, \beta_N')'$ denote all the regression coefficients for all individuals stacked together.

7.4.2 A Hierarchical Prior for the Random Coefficients Model

A convenient hierarchical prior is one which assumes that β_i for $i = 1, \ldots, N$ are independent draws from a Normal distribution and, thus, we assume

$$\beta_i \sim N(\mu_\beta, V_\beta) \tag{7.21}$$

The second stage of the hierarchical prior is given by

$$\mu_\beta \sim N(\underline{\mu}_\beta, \underline{\Sigma}_\beta) \tag{7.22}$$

and

$$V_\beta^{-1} \sim W(\underline{\nu}_\beta, \underline{V}_\beta^{-1}) \tag{7.23}$$

Remember that the Wishart distribution, which was first introduced in Chapter 6, Section 6.6, in the context of the SUR model, is a matrix generalization of the Gamma distribution (see Appendix B, Definition B.27). The distribution in (7.23) is parameterized so that $E(V_\beta^{-1}) = \underline{\nu}_\beta \underline{V}_\beta^{-1}$. Remember also that noninformativeness is achieved by setting $\underline{\nu} = 0$.

For the error precision, we use our familiar Gamma prior:

$$h \sim G(\underline{s}^{-2}, \underline{\nu}) \tag{7.24}$$

7.4.3 Bayesian Computation

As in the individual effects models, posterior inference can be carried out by setting up a Gibbs sampler. Since the Gibbs sampler requires only the full conditional posterior distributions we proceed directly to these. The derivation of these is relatively straightforward and can be done by multiplying the likelihood in (7.20) by the prior given in (7.21)–(7.24). An examination of the resulting expression reveals the kernels of all the relevant posterior conditional distributions. The conditional posteriors for the β_i's are independent of one another, for $i = 1, \ldots, N$, with

$$\beta_i | y, h, \mu_\beta, V_\beta \sim N(\overline{\beta}_i, \overline{V}_i) \tag{7.25}$$

where

$$\overline{V}_i = (hX_i'X_i + V_\beta^{-1})^{-1}$$

and

$$\overline{\beta}_i = \overline{V}_i(hX_i'y_i + V_\beta^{-1}\mu_\beta)$$

For the hierarchical parameters, μ_β and V_β, the relevant posterior conditionals are

$$\mu_\beta | y, \beta, h, V_\beta \sim N(\overline{\mu}_\beta, \overline{\Sigma}_\beta) \tag{7.26}$$

and

$$V_\beta^{-1} | y, \beta, h, \mu_\beta \sim W(\overline{\nu}_\beta, [\overline{\nu}_\beta \overline{V}_\beta]^{-1}) \tag{7.27}$$

where

$$\overline{\Sigma}_\beta = \left(NV_\beta^{-1} + \underline{\Sigma}_\beta^{-1}\right)^{-1}$$

$$\overline{\mu}_\beta = \overline{\Sigma}_\beta \left(V_\beta^{-1} \sum_{i=1}^{N} \beta_i + \underline{\Sigma}_\beta^{-1} \underline{\mu}_\beta\right)$$

$$\overline{\nu}_\beta = N + \underline{\nu}_\beta$$

$$\overline{V}_\beta = \sum_{i=1}^{N} (\beta_i - \mu_\beta)(\beta_i - \mu_\beta)' + \underline{V}_\beta$$

and $\sum_{i=1}^{N} \beta_i$ should be understood to be the k-vector containing the sums of the elements of β_i.

The posterior conditional for the error precision has the familiar form:

$$h|y, \beta, \mu_\beta, V_\beta \sim G(\overline{s}^{-2}, \overline{\nu}) \tag{7.28}$$

where

$$\overline{\nu} = TN + \underline{\nu}$$

and

$$\overline{s}^2 = \frac{\sum_{i-1}^{N} (y_i - X_i\beta_i)'(y_i - X_i\beta_i) + \underline{\nu}\underline{s}^2}{\overline{\nu}}$$

The Gibbs sampler, involving (7.25)–(7.28), requires only random number generation from the Normal, Gamma and Wishart distributions. Thus, it is not hard to write a program which carries out posterior simulation in this model. Predictive analysis can be done using the methods described in previous chapters and MCMC diagnostics can be used to verify convergence of the Gibbs sampler.

7.5 MODEL COMPARISON: THE CHIB METHOD OF MARGINAL LIKELIHOOD CALCULATION

You may have noticed that little has been said so far about model comparison in this class of models. For many types of models one might wish to compare, the methods used in the previous chapter can be used. For instance, investigating exact restrictions involving $\tilde{\beta}$ in the individual effects models can be done using the Savage–Dickey density ratio as in Chapter 4, Section 4.2.5. For the researcher who does not wish to calculate posterior odds (e.g. if she is using a noninformative prior), HPDIs and posterior predictive p-values can be calculated. Some types of model comparison are, however, difficult to carry out in the models introduced in this chapter. Suppose, for instance, that you wish to compare the individual effects model with hierarchical prior to the pooled model. It can be seen that the latter model is simply the former with $V_\alpha = 0$ imposed. The fact that one model is nested within the other suggests that the Savage–Dickey density ratio might be an appropriate tool for Bayes factor calculation. An examination of (7.18), however, shows the problem with this approach. That is, one would have to set $V_\alpha^{-1} = \infty$ to calculate the Savage–Dickey density ratio. One might handle this problem by setting V_α^{-1} to a large finite value, but this would provide at best a rough approximation of the true Bayes factor. Hence, the Savage–Dickey density ratio would likely be unsatisfactory. The researcher might then consider using the Gelfand–Dey approach. In theory, this should be an acceptable method of calculating the marginal likelihood. However, in high-dimensional problems the Gelfand–Dey method can be inaccurate due, especially, to the difficulties in making a good choice for the function we called $f(\theta)$ (see Chapter 5, (5.21) and (5.22)). In general, marginal likelihood calculation can be hard when the dimensionality of the parameter space is high and Raftery (1996) (which is Chapter 10 in Gilks *et al.*, 1996, pp. 173–176) discusses various approaches to this problem. In this section, we describe one approach which, in many cases, allows for efficient calculation of the marginal likelihood in cases where the dimensionality of the parameter space is high.

As we have done previously when introducing a new concept, we adopt our general notation where θ is a vector of parameters and $p(y|\theta)$, $p(\theta)$ and $p(\theta|y)$ are the likelihood, prior and posterior, respectively. Chib's method for marginal likelihood calculation, outlined in Chib (1995), begins with a very simple observation. Bayes rule says

$$p(\theta|y) = \frac{p(y|\theta)p(\theta)}{p(y)}$$

This equation can be rearranged to give an expression for the marginal likelihood, $p(y)$:

$$p(y) = \frac{p(y|\theta)p(\theta)}{p(\theta|y)}$$

But $p(y)$ does not depend upon θ, so that the right-hand side of the previous equation can be evaluated at any point, θ^*, and the result will be the marginal likelihood. Thus, for any point θ^*, we obtain

$$p(y) = \frac{p(y|\theta^*)p(\theta^*)}{p(\theta^*|y)} \tag{7.29}$$

which Chib refers to as the *basic marginal likelihood identity*. Note that all the densities on the right-hand side of (7.29) are evaluated at a point. For instance, $p(\theta^*)$ is short-hand notation for $p(\theta = \theta^*)$. Thus, if we know the exact forms of the likelihood function, prior and posterior (i.e. not just their kernels, but the exact p.d.f.s), we can calculate the marginal likelihood by simply evaluating them at any point and using (7.29). In most cases, we do know the exact form of the likelihood and prior, but we do not know the exact form of the posterior. Thus, to implement Chib's method, we need to figure out how to evaluate the posterior at a point (i.e. calculate $p(\theta^*|y)$), and Chib's paper describes methods for doing this in various cases. Here we describe some setups of particular relevance for the models introduced in the present chapter.

Many models used by econometricians have a structure where the parameter vector breaks down into a low-dimensional vector, θ, and a high-dimensional vector, z. In many cases, z can be interpreted as *latent data* and, thus, a posterior simulation procedure which sequentially draws from $p(\theta|y, z)$ and $p(z|y, \theta)$ is sometimes referred to as *Gibbs sampling with data augmentation*. In future chapters, we will come across cases where z does have such a latent data interpretation, but in the present chapter think of z as being the individual effects (i.e. $z = \alpha$) or random coefficients (i.e. $z = \beta$ in the random coefficients model). In such cases, we can use (7.29) directly to calculate the marginal likelihood. That is, we can integrate out the high-dimensional parameter vector z and work only with the low-dimensional vector, θ. To do this, note that the rules of probability imply that

$$p(\theta^*|y) = \int p(\theta^*|y, z)p(z|y)dz \tag{7.30}$$

Equation (7.30) can be evaluated in the context of the Gibbs sampler by simply calculating $p(\theta^*|y, z^{(s)})$ for each draw (i.e. for $s = 1, \ldots, S$) and averaging the result. To be precise, the weak law of large numbers which is used to justify the Gibbs sampler (see Chapter 4, Section 4.2.3) implies that, if $z^{(s)}$ for $s = 1, \ldots, S$ are draws from the Gibbs sampler, then

$$\widehat{p(\theta^*|y)} = \frac{1}{S}\sum_{s=1}^{S} p(\theta^*|y, z^{(s)}) \tag{7.31}$$

converges to $p(\theta^*|y)$ as S goes to infinity.

Thus, if $p(y|\theta^*)$, $p(\theta^*)$ and $p(\theta^*|y, z)$ can all be calculated (i.e. not just the kernels, but the exact p.d.f.s are known), then output from a Gibbs sampler with data augmentation can be used as described in (7.31) to obtain the marginal likelihood. In theory, any point, θ^*, can be chosen. However, in practice, the Chib method works much better if θ^* is chosen to be in an area of appreciable posterior probability. Setting θ^* to be the posterior mean, based on an initial run with the Gibbs sampler, is a common practice.

Unfortunately, the Chib method as described above does not work with the our panel data models since $p(\theta^*|y, z)$ is not known. However, it can be generalized to deal with our case. Suppose the parameter vector is broken up into two blocks, θ_1 and θ_2 (i.e. $\theta = (\theta_1', \theta_2')'$) and that a Gibbs sampler is available which sequentially draws from $p(\theta_1|y, z, \theta_2)$, $p(\theta_2|y, z, \theta_1)$ and $p(z|y, \theta_1, \theta_2)$. Thus, we have posterior simulator output, $\theta_1^{(s)}$, $\theta_2^{(s)}$ and $z^{(s)}$ for $s = 1, \ldots, S$. To use the Chib method, we must calculate $p(\theta_1^*, \theta_2^*|y)$ where θ_1^* and θ_2^* are any points. The rules of probability imply that

$$p(\theta_1^*, \theta_2^*|y) = p(\theta_1^*|y)p(\theta_2^*|y, \theta_1^*) \tag{7.32}$$

$$p(\theta_1^*|y) = \int\int p(\theta_1^*|y, \theta_2, z)p(\theta_2, z|y)\, d\theta_2 dz \tag{7.33}$$

and

$$p(\theta_2^*|y, \theta_1^*) = \int p(\theta_2^*|y, \theta_1^*, z)p(z|y, \theta_1^*)\, dz \tag{7.34}$$

In the same way as (7.31) provides an estimate for $p(\theta^*|y)$, we can use the Gibbs sampler output to provide an estimate of $p(\theta_1^*|y)$. That is, (7.33) and the weak law of large numbers implies

$$\widehat{p(\theta_1^*|y)} = \frac{1}{S}\sum_{s=1}^{S} p(\theta_1^*|y, z^{(s)}, \theta_2^{(s)}) \tag{7.35}$$

converges to $p(\theta_1^*|y)$ as S goes to infinity.

To use (7.32), we now need to figure out how to calculate $p(\theta_2^*|y, \theta_1^*)$. This can be done separately through Gibbs sampling. That is, if we set up a second Gibbs sampler which sequentially draws from $p(\theta_2|y, z, \theta_1^*)$ and $p(z|y, \theta_1^*, \theta_2)$ we can use this posterior simulator output to obtain an estimate of $p(\theta_2^*|y, \theta_1^*)$. That is, if $\theta_2^{(s^*)}$ and $z^{(s^*)}$, for $s^* = 1, \ldots, S^*$, is posterior simulator output from the second Gibbs sampler, then (7.34) and the weak law of large numbers imply that

$$\widehat{p(\theta_2^*|y, \theta_1^*)} = \frac{1}{S^*}\sum_{s^*=1}^{S^*} p(\theta_2^*|y, z^{(s^*)}, \theta_1^*), \tag{7.36}$$

converges to $p(\theta_2^*|y, \theta_1^*)$ as S^* goes to infinity. $\widehat{p(\theta_2^*|y, \theta_1^*)}$ and $\widehat{p(\theta_1^*|y)}$ can be multiplied together to provide the estimate of $p(\theta_1^*, \theta_2^*|y)$ needed to calculate (7.29).

Note that the disadvantage of the Chib method in this case is that it requires two Gibbs samplers to be run and, thus, increases the computational burden. However, the actual programming costs are quite low. Once you have written computer code for the basic Gibbs sampler which sequentially draws from $p(\theta_1|y, z, \theta_2)$, $p(\theta_2|y, z, \theta_1)$ and $p(z|y, \theta_1, \theta_2)$, little additional coding is required to do the Chib method. That is, (7.35) and (7.36) each only require the evaluation of a p.d.f. which typically requires one line of code. The second Gibbs sampler is virtually identical to the first. All that is required is to remove the code which draws from $p(\theta_1|y, z, \theta_2)$ and replace it with a line which sets $\theta_1 = \theta_1^*$. So although the Chib method looks a bit complicated, it is in fact reasonably easy to implement in practice.

The ideas outlined above can be extended to the case where the Gibbs sampler involves blocking θ into B segments (i.e. $\theta = (\theta_1', \ldots, \theta_B')'$). That is, we can use the fact that

$$p(\theta_1^*, \theta_2^*, \ldots, \theta_B^*|y) = p(\theta_1^*|y)p(\theta_2^*|y, \theta_1^*)\ldots p(\theta_B^*|y, \theta_1^*, \ldots, \theta_{B-1}^*)$$

and use the original Gibbs sampler to calculate $p(\theta_1^*|y)$, a second Gibbs sampler to calculate $p(\theta_2^*|y, \theta_1^*)$, all the way up to a Bth Gibbs sampler to calculate $p(\theta_B^*|y, \theta_1^*, \dots, \theta_{B-1}^*)$. Once again, the Chib method can be computationally demanding, but writing code is quite easy since all the B Gibbs samplers have essentially the same structure.

In the individual effects model with hierarchical prior, we have $B = 4$ (i.e. $\theta_1 = \tilde{\beta}, \theta_2 = h, \theta_3 = \mu_\alpha, \theta_4 = V_\alpha^{-1}$) and $z = \alpha$. In the random coefficients model, we have $B = 3$ (i.e. $\theta_1 = h, \theta_2 = \mu_\beta, \theta_3 = V_\beta^{-1}$) and $z = \beta$. Since all the conditional posteriors are of known form (see Sections 7.3 and 7.4), estimates of $p(\theta^*|y)$ can be calculated as outlined above. To use the Chib method to calculate the marginal likelihood for these models, we must also evaluate $p(\theta^*)$ and $p(y|\theta^*)$. The prior can be calculated directly using (7.8)–(7.11) for the individual effects and (7.22)–(7.24) for the random coefficients model. Evaluation of the likelihood function is a little harder since the expressions given in (7.3) and (7.20) give, in terms of our general notation, $p(y|\theta, z)$ rather than the $p(y|\theta)$ which is required for the Chib method. However, the integration necessary to move from $p(y|\theta, z)$ to $p(y|\theta)$ can be done analytically using the properties of the multivariate Normal distribution. For the individual effects model with hierarchical prior, we have $p(y|\tilde{\beta}, h, \mu_\alpha, V_\alpha^{-1}) = \prod_{i=1}^N p(y_i|\tilde{\beta}, h, \mu_\alpha, V_\alpha^{-1})$, where

$$y_i|\tilde{\beta}, h, \mu_\alpha, V_\alpha^{-1} \sim N(\mu_i, V_i)$$

where

$$\mu_i = \mu_\alpha \iota_T + \tilde{X}_i \tilde{\beta}$$

and

$$V_i = V_\alpha \iota_T \iota_T' + h^{-1} I_T$$

For the random coefficients model, a similar derivation implies $p(y|h, \mu_\beta, V_\beta^{-1}) = \prod_{i=1}^N p(y_i|h, \mu_\beta, V_\beta^{-1})$, where

$$y_i|h, \mu_\beta, V_\beta^{-1} \sim N(\mu_i, V_i) \tag{7.37}$$

where

$$\mu_i = X_i \mu_\beta$$

and

$$V_i = X_i V_\beta X_i' + h^{-1} I_T$$

Thus, the Chib method can be used to calculate the marginal likelihood of any of the models described in this chapter. Chib and Jeliazkov (2001) shows how this method of marginal likelihood calculation can be extended to work with output from a Metropolis–Hastings algorithm. However, if the added complication required to use this method are not warranted, the researcher may wish to use other methods of model comparison. Methods of model selection and comparison in models with high-dimensional parameter spaces (e.g. caused by the presence of latent data) is a topic of current research interest in the statistics literature.

The interested reader is referred to Carlin and Louis (2000, Section 6.5) for more detail.

7.6 EMPIRICAL ILLUSTRATION

In this section, we illustrate empirical inference in the pooled, individual effects with non-hierarchical and hierarchical prior and random coefficients models, which we label $M_1, \ldots, M_4$, respectively, using several different artificial data sets. The use of artificial data means that we know what process has generated the data and, thus, can see how well our Bayesian analysis works in a controlled setting. Gibbs samplers to carry out posterior inference in all models are described above and the method of Chib is used to calculate the marginal likelihood. Posterior results for all models are based on 30 000 replications, with 5000 burn-in replications discarded and 25 000 replications retained. MCMC diagnostics indicate convergence of all the Gibbs samplers and numerical standard errors indicate an approximation error which is small relative to posterior standard deviations of all parameters.

We use two artificial data sets generated from the individual effects model with an intercept and one other explanatory variable (i.e. $k = 2$). In both we set $N = 100$, $T = 5$, $\tilde{\beta} = 2$ and $h = 25$. In the first data set we draw the intercepts independently from the Normal distribution with

$$\alpha_i \sim N(0, 0.25)$$

a large degree of intercept variance relative to the error variance and magnitude of the coefficients. The second data set is generated with $\alpha_i = -1$ with 25% probability and $\alpha_i = 1$ with 75% probability. Such a specification might be reasonable in a labor economics application where there are two groups of individuals and the difference between groups manifests itself through the intercept.

With regards to the prior, for the error precision we use the same prior hyperparameter values for all four models, $\underline{s}^{-2} = 25$ and $\underline{\nu} = 1$. For M_1 we set the remaining prior hyperparameters to $\underline{\beta} = 0_2$ and $\underline{V} = I_2$. For M_2 we set the remaining prior hyperparameters to $\underline{\beta}^* = 0_{N+1}$ and $\underline{V} = I_{N+1}$. For M_3 we set the remaining prior hyperparameters to $\underline{\beta} = 0$, $\underline{V} = 1$, $\underline{\mu}_\alpha = 0$, $\underline{\sigma}_\alpha^2 = 1$, $\underline{V}_\alpha^{-1} = 1$ and $\underline{\nu}_\alpha = 2$. For M_4 we set the remaining prior hyperparameters to $\underline{\mu}_\beta = 0_2$, $\underline{\Sigma}_\beta = I_2$, $\underline{V}_\beta^{-1} = I_2$ and $\underline{\nu}_\beta = 2$. An examination of these choices reveals that they are all relatively noninformative (i.e. degrees of freedom parameters are all set to very small values relative to sample size and prior variances and covariance matrices are relatively large), and are roughly consistent with the true data generating processes.

These models have many parameters and, for the sake of brevity, we present results only for a few key ones. Note that the slope coefficient is common to M_1, M_2 and M_3 and we simply label it $\tilde{\beta}$ in the tables below. Under M_4, μ_β is the parameter most comparable to $\tilde{\beta}$, and thus we present results for these two parameters in the same row.

Table 7.1 Posterior Results for First Artificial Data Set

	M_1	M_2	M_3	M_4
$E(\tilde{\beta}\|y)$ or $E(\mu_\beta\|y)$	2.04	2.04	2.04	2.04
$\sqrt{var(\tilde{\beta}\|y)}$ or $\sqrt{var(\mu_\beta\|y)}$	0.08	0.03	0.03	0.04
$E(h\|y)$	3.61	28.41	27.86	30.20
$\sqrt{var(h\|y)}$	0.23	1.98	2.07	2.25
$\log[p(y)]$	−407	-3×10^4	−66	−91

Table 7.2 Posterior Results for Second Artificial Data Set

	M_1	M_2	M_3	M_4
$E(\tilde{\beta}\|y)$ or $E(\mu_\beta\|y)$	1.88	2.03	2.01	2.00
$\sqrt{var(\tilde{\beta}\|y)}$ or $\sqrt{var(\tilde{\beta}\|y)}$	0.14	0.04	0.04	0.05
$E(h\|y)$	1.25	24.12	23.75	27.17
$\sqrt{var(h\|y)}$	0.08	1.67	1.73	2.08
$\log[p(y)]$	−669	-2×10^4	−152	−168

Table 7.1, which presents results for the first artificial data set, indicates that all four models provide good estimates of the slope coefficient (remember that the slope coefficient was set to 2 when the data was generated). For the error precision, M_2, M_3 and M_4 all provide good estimates (remember that this parameter was set to 25 when the data was generated), although the estimate under M_1 is much too small. Since M_1 does not allow for any variation in the intercept, all such variation is assigned to the error variance. Thus, under M_1, the estimate of the error variance is much too large (and, thus, the error precision is much too small). These findings are mostly repeated in Table 7.2, which provides results for the second data set. Note, though, that under M_1 the posterior mean of the slope coefficient is too low. This illustrates how ignoring coefficient variation can lead to misleading inferences. That is, a researcher examining Table 7.1 might say "If I am only interested in the slope coefficient, it does not matter whether I properly allow for variation in the intercept", but Table 7.2 indicates that this is not the case. In fact, even for parameters which are not of key interest to the researcher, it is important to model variation correctly in order to avoid misleading inferences with regard to other coefficients.

Figures 7.1 and 7.2 present information about the intercepts using the first artificial data set. In particular, for M_2 and M_3, these figures take $E(\alpha_i|y)$ for $i = 1, \ldots, N$ and plot a histogram.

These figures look very similar to one another, indicating that the hierarchical and non-hierarchical priors are providing basically the same estimates of the

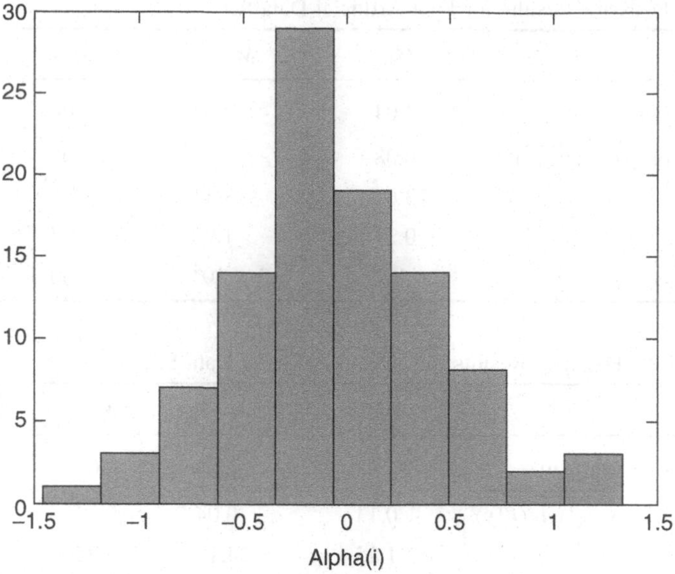

Figure 7.1 Histogram of Posterior Means of Alpha(i)s, Non-hier. Prior

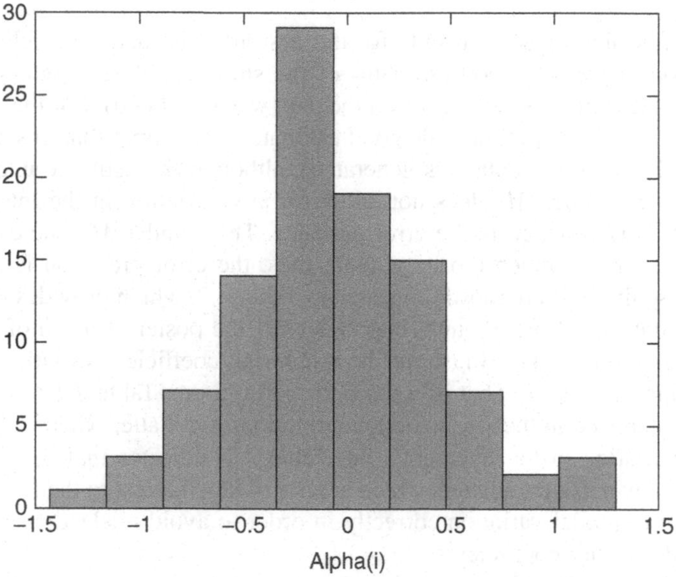

Figure 7.2 Histogram of Posterior Means of Alpha(i)s, Hierarchical Prior

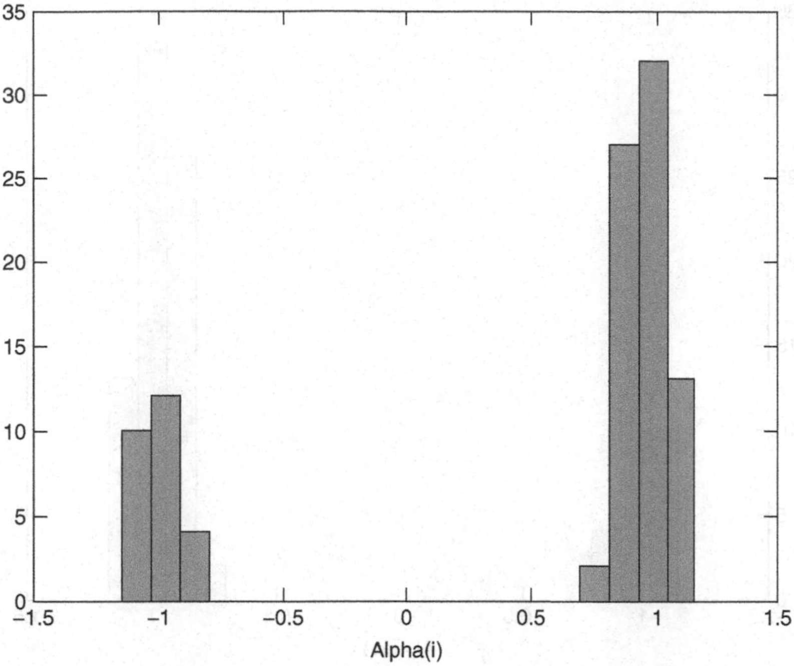

Figure 7.3 Histogram of Posterior Means of Alpha(i)s, Non-hier. Prior

intercepts. Furthermore, the data generating process used to produce the first data set had $\alpha_i \sim N(0, 0.25)$ and the figures above look to be replicating this pattern quite accurately.

Figures 7.3 and 7.4 repeat Figures 7.1 and 7.2 using posterior means calculated using the second artificial data set. These figures look quite similar to one another and are replicating the data generating process quite well. Remember that, in this second data set, intercepts are generated with $\alpha_i = -1$ with 25% probability and $\alpha_i = 1$ with 75% probability. Figure 7.4 is particularly interesting in that the hierarchical prior assumes that intercept variation is Normally distributed. But, given the relatively noninformative prior used for M_3, this model is able to pick up the very non-Normal intercept variation pattern in the data.

An examination of all of these figures indicates that both of our individual effects models are doing a very good job of picking out the variation in the intercepts. Results for the random coefficients model (not presented here), indicate a similarly fine performance.

So far our Bayesian methods seem to be doing a good job at finding complicated patterns in the data. However, an oddity arises when we turn to the issue of model comparison. The logs of the marginal likelihoods for each model are presented in the last row of Tables 7.1 and 7.2. For both data sets, the individual effects model with hierarchical prior is favored with the highest marginal

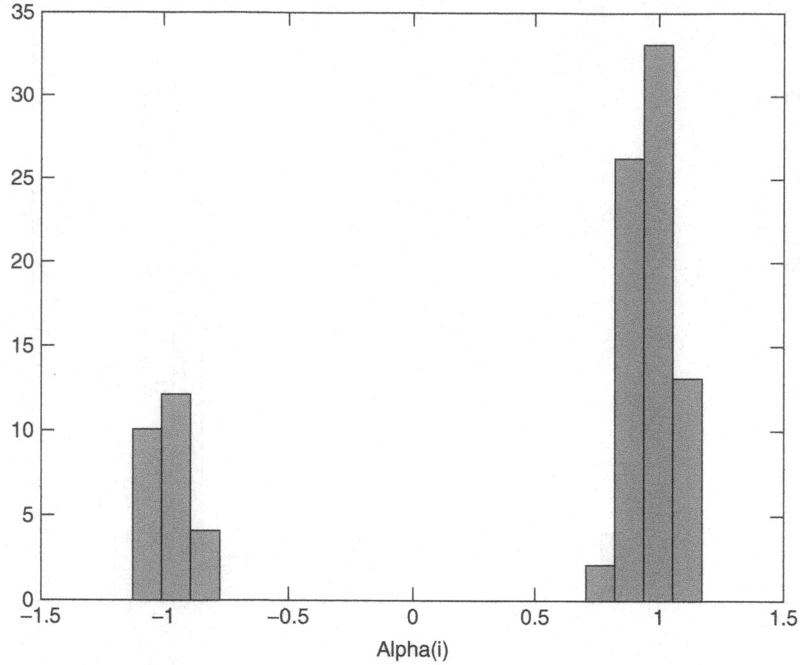

Figure 7.4 Histogram of Posterior Means of Alpha(i)s, Hierarchical Prior

likelihood, with the random coefficients model being the second choice. This is sensible in that both data sets have been generated so as to have substantial variation in the intercept, but no variation in the slope coefficient. Thus, M_1, which allows for no variation in the intercept does poorly, while M_4, which allows for additional (unnecessary) variation in the slope is unable to beat M_3. The oddity in both tables is why M_2, which looks similar to M_3 except that it does not have a hierarchical prior, does so poorly. An examination of the Tables 7.1 and 7.2 as well as Figures 7.1 and 7.2 indicate that these two individual effects models are yielding very similar posterior results in every regard except for the marginal likelihood. Furthermore, M_3 assumes the intercepts to be Normally distributed whereas, for the second data set, this assumption is incorrect. Surely, M_2, which does not use the Normality assumption, should be preferred to M_3? But, in fact, M_3 has a higher marginal likelihood, even for this second data set.

The explanation of this oddity relates to the dimensionality of the parameter space and the problems associated with the use of Bayes factors with noninformative priors. Remember that, in Chapter 3, Section 3.6.2, we showed how when comparing two models, one of which was nested in the other, great care had to be taken in prior elicitation. For instance, if M_A was an unrestricted model and M_B was equal to M_A except that the restriction $\theta_1 = 0$ was imposed, then a noninformative prior for θ_1 could lead to nonsensical results (e.g. M_B would always be preferred, regardless of what the data was). As a concrete example, let us suppose that the true value is $\theta_1 = 100$ and, under M_A, a prior for θ_1 is used which is Uniform over $(-\infty, \infty)$. Some intuition is provided by remembering that, for the Bayesian, the model involves the likelihood function and the prior. Under M_A, the prior allocates virtually all prior weight to ridiculous values for θ_1. For instance, the prior implies $\frac{p(|\theta_1|\geq 1\,000\,000|M_A)}{p(|\theta_1|<1\,000\,000|M_A)} \approx 1$ so that almost all prior probability is located in the region $|\theta_1| \geq 1\,000\,000$. Very loosely speaking, the Bayesian approach looks at M_A and says "the data says θ_1 should be around 100, but the prior is saying $|\theta_1| \geq 1\,000\,000$, this model is silly so it cannot be correct". The Bayesian approach then looks at M_B and says "this model says $\theta_1 = 0$. This is at odds which the data, but even so the model is not as silly as M_A, hence M_B should be supported." This example considers a case where the prior is completely noninformative, but similar considerations hold less strongly in cases where priors are relatively noninformative.

To return to our panel data models, note that the use of hierarchical priors effectively means that the parameter spaces for M_1, M_3 and M_4 are $k+1$, $k+2$ and $\frac{k(k+1)}{2} + k + 1$, respectively. However, the parameter space for M_2 is $N + k$ dimensional. In our examples (and in most applications), N is much larger than k so that the dimensionality of M_2 is much higher than the other models. Here we are using a relatively noninformative 102 dimensional prior in M_2, whereas in M_3 we are using a 4-dimensional prior. A careful consideration of this issue raised in the previous paragraph shows that the problem caused by the use of relatively noninformative priors will increase with the dimensionality of the prior. When comparing models which are of similar dimension, this problem can often be ignored. However, M_2 is of such higher dimensionality that it cannot be ignored here.

The discussion of the previous two paragraphs was intuitive and informal. Even if the reader is having difficulty understanding the points made, the following rule of thumb, which builds on the one in Chapter 3, should be remembered. *When comparing models using posterior odds ratios, it is acceptable to use noninformative priors over parameters which are common to all models. However, informative, proper priors should be used over all other parameters. If the two models being compared have a similar number of parameters, then it is often acceptable for these proper priors to be weakly informative priors of convenience. However, the use of weakly informative priors of convenience should be avoided if the dimensionality of the two models is very different.*

In our empirical example, the reason M_2 has such a low marginal likelihood is the prior was $\beta^* \sim N(\underline{\beta}^*, \underline{V})$, and we chose $\underline{V} = I_{N+1}$. In many contexts, it is hard to imagine the researcher having stronger prior beliefs. In such contexts, it is probably best to avoid the use of posterior odds ratios altogether (at least for model comparisons involving M_2) and rely on alternative model comparison techniques. For instance, for the second data set the posterior predictive p-value approach would clearly indicate that M_3 (with a Normal hierarchical prior) is not modeling the variation in intercepts in an appropriate way (see Figure 7.4).

In contrast, this approach would indicate that M_2 (which does not assume the variation in intercepts follows a Normal distribution) is appropriate. Even more informally, an examination of histograms like Figures 7.1–7.4 would show that M_3 is suitable for the first data set, but M_2 would be preferred for the second. Alternatively, methods of model selection and comparison in models with high-dimensional parameter spaces (e.g. caused by the presence of latent data) is a topic of current research interest in the statistics literature and Carlin and Louis (2000, Section 6.5) suggest some interesting approaches.

7.7 EFFICIENCY ANALYSIS AND THE STOCHASTIC FRONTIER MODEL

In this section, we draw on some economic theory to develop *the stochastic frontier model*, which falls into the class of individual effects models, but with a different hierarchical prior than the one in Section 7.3.2. This model is important in its own right, as it is used in studies where interest centers on the efficiency or productivity of a firm or individual. Furthermore, the derivation of the stochastic frontier model is a good illustration of how the applied economist can take economic theory and use it to construct an econometric model.

7.7.1 Introduction to the Stochastic Frontier Model

The ideas underlying this class of models can be demonstrated using an economic model of production where output of firm i at time t, Y_{it}, is produced using a vector of inputs, X_{it}^*, $(i = 1, \ldots, N, t = 1, \ldots, T)$. Firms have access to a common best-practice technology for turning inputs into output. This technology depends upon a vector of unknown parameters, β, and is given by:

$$Y_{it} = f(X_{it}^*; \beta) \tag{7.38}$$

This so-called *production frontier* measures the maximum amount of output that can be obtained from a given level of inputs. In practice, actual output of a firm may fall below the maximum possible. The deviation of actual from maximum feasible output is a measure of inefficiency and is the focus of interest in many applications. Formally, (7.38) can be extended to

$$Y_{it} = f(X_{it}^*; \beta)\tau_i \tag{7.39}$$

where $0 < \tau_i \leq 1$ is a measure of firm-specific efficiency and $\tau_i = 1$ indicates firm i is fully efficient. A value of, say, $\tau_i = 0.75$ means that firm i is producing only 75% of the output it could have if it were operating according to best-practice technology. In this specification, we have assumed each firm has a particular efficiency level which is constant over time. This assumption can be relaxed, and the reader is referred to Koop and Steel (2001) for details.

Following standard econometric practice, we allow for a random error in the model, ζ_{it}, to capture measurement (or specification) error, resulting in

$$Y_{it} = f(X_{it}^*; \beta)\tau_i \zeta_{it} \tag{7.40}$$

The inclusion of measurement error makes the frontier stochastic, hence the name *stochastic frontier* model. If the production frontier, $f()$, is log-linear (e.g. Cobb–Douglas or translog) we can take logs and write (7.40) as

$$y_{it} = X_{it}\beta + \varepsilon_{it} - z_i \tag{7.41}$$

where $\beta = (\beta_1, \ldots, \beta_k)'$, $y_{it} = ln(Y_{it})$, $\varepsilon_{it} = ln(\zeta_{it})$, $z_i = -ln(\tau_i)$ and X_{it} is the counterpart of X_{it}^* with the inputs transformed to logarithms. z_i is referred to as inefficiency and, since $0 < \tau_i \leq 1$, it is a non-negative random variable. X_{it} is assumed to contain an intercept and β_1 is its coefficient. Note that this model is of the form of an individual effects model. That is, $\beta_1 - z_i$ plays the same role that α_i did in Section 7.3.2. However, in the stochastic frontier model, economic theory gives us some guidance in selecting a hierarchical prior.

It is worth noting in passing that, if the production function is not log-linear (e.g. the constant elasticity of scale production function), then Bayesian inference can be done by combining the techniques developed in Chapter 5 with those presented here.

Equation (7.41) can be written as

$$y_i = X_i\beta + \varepsilon_i - z_i \iota_T \tag{7.42}$$

if we stack all variables into matrices as described at the beginning of this chapter (see Section 7.1). Remember that ι_T is our notation for a T-vector of ones.

7.7.2 The Likelihood Function

The form of the likelihood function depends upon assumptions made about the errors. In addition to the standard error assumptions set out at the beginning of the chapter (see Section 7.1), we also assume that z_i and ε_j are independent of one another for all i and j. The resulting likelihood function is

$$p(y|\beta, h, z) = \prod_{i=1}^{N} \frac{h^{\frac{T}{2}}}{(2\pi)^{\frac{T}{2}}} \left\{ \exp\left[-\frac{h}{2}(y_i - X_i\beta + z_i\iota_T)'(y_i - X_i\beta_i + z_i\iota_T) \right] \right\} \tag{7.43}$$

where $z = (z_1, \ldots, z_N)'$.

In this specification, we are treating z as a vector of unknown parameters which enter the likelihood function. In frequentist analyses, the likelihood would be defined as $p(y|\beta, h, \theta) = \int p(y|\beta, h, z)p(z|\theta)\,dz$, where $p(z|\theta)$ is a distributional assumption for the inefficiencies, which depends upon a vector of unknown parameters, θ. But such a procedure is mathematically equivalent to the Bayesian procedure of using $p(z|\theta)$ as a hierarchical prior. In other words, in models such as this one, the choice as to what we label the 'likelihood function' and what we label the 'hierarchical prior' is a purely semantic one which has no implications for statistical inference. The reader interested in more discussion of this point is referred to Bayarri, DeGroot and Kadane (1988).

7.7.3 A Hierarchical Prior for the Stochastic Frontier Model

For the coefficients in the production frontier and the error precision, we use our familiar independent Normal-Gamma prior:

$$\beta \sim N(\underline{\beta}, \underline{V}) \tag{7.44}$$

and

$$h \sim G(\underline{s}^{-2}, \underline{\nu}) \tag{7.45}$$

For the inefficiencies, we use a hierarchical prior. Since $z_i > 0$, the Normal hierarchical prior of Section 7.3.2 is not suitable. In the literature, common choices for this prior include the truncated-Normal and members of the family of Gamma distributions. Here we illustrate Bayesian inference in stochastic frontier models using the exponential distribution, which is the Gamma with two degrees of freedom (see Definition B.22 and Theorem B.7). Thus, we assume that z_i and z_j are *a priori* independent for $i \neq j$ with

$$z_i \sim G(\mu_z, 2) \tag{7.46}$$

The hierarchical nature of the prior means that we treat the mean of the inefficiency distribution as a parameter which requires its own prior. Since $z_i > 0$, it follows that $\mu_z > 0$. In the same way that working with the error precision (h), instead of the error variance (σ^2) allows us to stay in the familiar class of error distributions, it proves easier to work with μ_z^{-1} instead of μ_z. Hence, we take a prior of the form

$$\mu_z^{-1} \sim G(\underline{\mu}_z^{-1}, \underline{\nu}_z) \tag{7.47}$$

The hyperparameters $\underline{\mu}_z^{-1}$ and $\underline{\nu}_z$ can often be elicited through consideration of the efficiency distribution. That is, researchers may often have prior information about the location of the efficiency distribution. Let τ^* denote the prior median of this distribution. If the researcher expects the firms in her sample to be quite efficient, she may set τ^* to a high value (e.g. 0.95). If she expects many firms to be inefficient she may set it to a lower value. As shown in van den Broeck *et al.* (1994), setting $\underline{\nu}_z = 2$ implies a relatively noninformative prior, and setting $\underline{\mu}_z = -ln(\tau^*)$ implies the median of the prior efficiency distribution is τ^*.

This illustrates a common strategy for prior elicitation. The researcher elicits a prior in terms of hyperparameters which are easy to interpret in terms of the underlying economic theory (i.e. in this case, τ^*), then transforms back to find values for the hyperparameters used in the model (i.e. in this case, $\underline{z}$ and $\underline{\nu}_z$).

Another point worth stressing is that economic theory often provides restrictions that can be imposed through the prior. For instance, the researcher might wish to impose the restriction that the production frontier is monotonically increasing in inputs. In other variants of the stochastic frontier model, it may be desirable to impose the restriction that a cost function is concave or that technological regress is impossible. All of these are inequality restrictions on the parameters which can be imposed using the methods of Chapter 4, Section 4.3.

7.7.4 Bayesian Computation

As in the individual effects models, posterior inference can be carried out by setting up a Gibbs sampler. Since the Gibbs sampler requires only the full conditional posterior distributions, we do not present the posterior itself. Rather we present only the relevant conditional distributions which, except for the ones relating to z and μ_z, are the same as those of the individual effects model with hierarchical prior.

For the parameters in the production frontier, we obtain

$$\beta|y, h, z, \mu_z \sim N(\overline{\beta}, \overline{V}) \tag{7.48}$$

where

$$\overline{V} = \left(\underline{V}^{-1} + h\sum_{i=1}^{N} X_i' X_i\right)^{-1}$$

and

$$\overline{\beta} = \overline{V}\left(\underline{V}^{-1}\underline{\beta} + h\sum_{i=1}^{N} X_i'[y_i + z_i \iota_T]\right)$$

For the error precision, we have the standard result:

$$h|y, \beta, z, \mu_z \sim G(\overline{s}^{-2}, \overline{\nu}) \tag{7.49}$$

where

$$\overline{\nu} = TN + \underline{\nu}$$

and

$$\overline{s}^2 = \frac{\sum_{i=1}^{N}(y_i + z_i \iota_T - X_i\beta)'(y_i + z_i \iota_T - X_i\beta) + \underline{\nu}\underline{s}^2}{\overline{\nu}}$$

The posterior conditionals for the inefficiencies are independent of one another (i.e. z_i and z_j are independent for $i \neq j$) and are each Normal, truncated to be positive with p.d.f. given by

$$p(z_i|y_i, X_i, \beta, h, \mu_z) \propto f_N(z_i|\overline{X}_i\beta - \overline{y}_i - (Th\mu_z)^{-1}, (Th)^{-1})1(z_i \geq 0) \tag{7.50}$$

where $\overline{y}_i = \frac{\sum_{t=1}^{T} y_{it}}{T}$ and $\overline{X}_i$ is a $(1 \times k)$ matrix containing the average value of each explanatory variable for individual i. Remember that $1(z_i \geq 0)$ is the indicator function which equals 1 if $z_i \geq 0$ and is otherwise equal to zero.

The posterior conditional for μ_z^{-1} given by

$$\mu_z^{-1}|y, \beta, h, z \sim G(\overline{\mu}_z, \overline{\nu}_z) \tag{7.51}$$
$$\overline{\nu} = 2N + \underline{\nu}_z$$

and

$$\overline{\mu}_z = \frac{N + \frac{\underline{\nu}_z}{2}}{\sum_{i=1}^{N} z_i + \underline{\mu}_z}$$

A Gibbs sampler which involves sequentially drawing from (7.48) through (7.51) can be used to carry out Bayesian inference in the stochastic frontier model. Note that drawing from the truncated Normal distribution (see (7.50)) can either be done by drawing from the corresponding Normal and simply discarding draws for which $z_i < 0$. Alternatively, algorithms for drawing from this distribution are available. One such algorithm, programmed in MATLAB is available on the website associated with this book. Predictive analysis can be done using the methods described in Chapter 4, Section 4.2.6 and MCMC diagnostics can be used to verify convergence of the Gibbs sampler. Model comparison can be done using the methods described in previous sections and chapters. For instance, the Chib method can be used to calculate the marginal likelihood.

It is worthwhile noting in passing that the cross-sectional version of this model (i.e. with $T = 1$) is often used and all the methods described above are still relevant. However, in the $T = 1$ case it is not acceptable to use certain improper noninformative priors as this leads to an improper posterior. Intuitively, if $T = 1$ the number of parameters in the entire model is larger than sample size (i.e. z, μ_z, β and h together contain $N + k + 2$ parameters and we have only N observations) which precludes meaningful posterior inference in the absence of prior information. Issues relating to the use of noninformative priors in stochastic frontier models are discussed in Fernandez, Osiewalski and Steel (1997).

7.7.5 Empirical Illustration: Efficiency Analysis with Stochastic Frontier Models

To illustrate Bayesian inference in the stochastic frontier model, artificial data was generated from

$$y_{it} = 1.0 + 0.75x_{2,it} + 0.25x_{3,it} - z_i + \varepsilon_{it}$$

for $i = 1, \ldots, 100$ and $t = 1, \ldots, 5$. We assume $\varepsilon_{it} \sim N(0, 0.04)$, $z_i \sim G(-ln[0.85], 2)$, $x_{2,it} \sim U(0, 1)$ and $x_{2,it} \sim U(0, 1)$ where all the random variables are independent of one another and independent over all i and t. Note that in a production frontier example, $x_{2,it}$ and $x_{3,it}$ would be inputs. This motivates the selection of 0.75 and 0.25 for these coefficients as these values imply constant returns to scale. The inefficiency distribution is selected so as to imply the median of the efficiency distribution is 0.85.

The required priors are given (7.44), (7.45) and (7.47). We choose

$$\underline{\beta} = \begin{bmatrix} 0.0 \\ 0.5 \\ 0.5 \end{bmatrix}$$

and

$$\underline{V} = \begin{bmatrix} 100.0 & 0.0 & 0.0 \\ 0.0 & 0.25^2 & 0.0 \\ 0.0 & 0.0 & 0.25^2 \end{bmatrix}$$

These values are chosen to be relatively noninformative about the intercept, but are slightly more informative about the slope coefficients. In particular, they reflect a belief that large deviations from constant returns to scale are unlikely. For the error precision, we make the noninformative choice of $\underline{\nu} = 0$, which makes the choice of $\underline{s}^2$ irrelevant. As described above, we make the relatively noninformative choice of $\underline{\nu}_z = 2$ with $\underline{\mu}_z = -ln(0.85)$ which implies the median of the prior efficiency distribution is 0.85.

Posterior results for all models are based a Gibbs sampler using (7.48) through (7.51). We take 30 000 replications, with 5000 burn-in replications discarded and 25 000 replications retained. MCMC diagnostics indicate convergence of all the Gibbs samplers and numerical standard errors indicate an approximation error which is small relative to posterior standard deviations of all parameters. To focus on the efficiencies, we do not carry out a model comparison exercise in this section.

Table 7.3 contains posterior means and standard deviations for the parameters of the stochastic frontier model. With stochastic frontier models, interest often centers on the firm-specific efficiencies, τ_i for $i = 1, \ldots, N$. Since $\tau_i = \exp(-z_i)$, and the Gibbs sampler yields draws of z_i, we can simply transform them and average to obtain estimates of $E(\tau_i|y)$ in the usual way (see Chapter 4, Section 4.2.3). For the sake of brevity, we do not present results for all $N = 100$ efficiencies. Rather we select the firms which have the minimum, median and maximum values for $E(\tau_i|y)$. These are labelled $\tau_{\min}$, τ_{med} and $\tau_{\max}$ in Table 7.3.

It can be seen that the posterior means of all parameters are quite close to the true values used to generate the data set (note that $-ln(0.85) = 0.16$) and they are all accurately estimated. The posterior means of the efficiencies are reasonable, although their posterior standard deviations are relatively large (a point which we will return to shortly). In a policy study, the researcher might report the posterior means of the efficiencies (e.g. "we estimate that the least efficient firm in the sample is producing only 56% of what it could if it was on the production frontier"). A histogram such as Figure 7.5, which uses the posterior means of the efficiencies of all 100 firms, might be presented to give a rough idea of how efficiencies are distributed across firms.

Table 7.3 Posterior Results for Artificial Data Set from Stochastic Frontier Model

	Mean	Standard Deviation
β_1	0.98	0.03
β_2	0.74	0.03
β_3	0.27	0.03
h	26.69	1.86
μ_z	0.15	0.02
$\tau_{\min}$	0.56	0.05
τ_{med}	0.89	0.06
$\tau_{\max}$	0.97	0.03

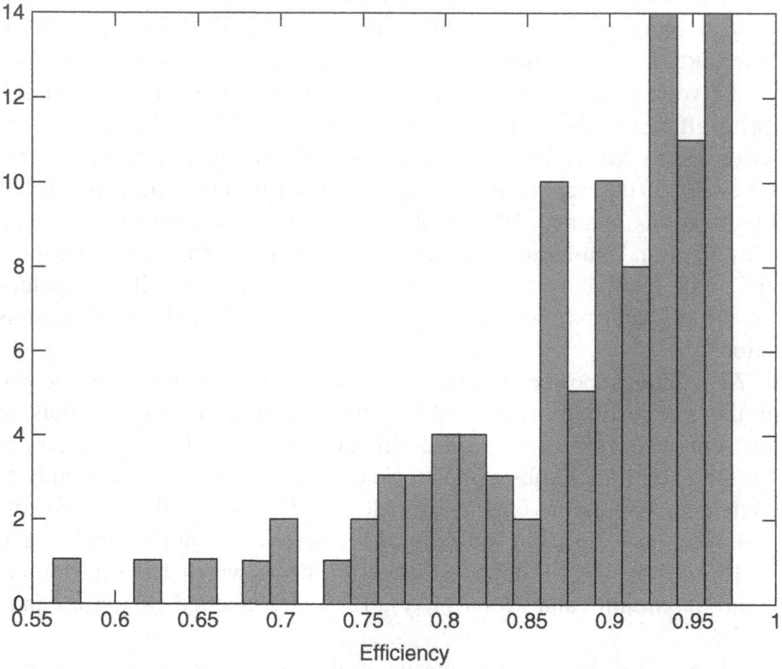

Figure 7.5 Histogram of Posterior Means of Efficiencies

An important issue in efficiency analysis is whether point estimates can be treated as a reliable guide to the ranking of firms. After all, important policy recommendations may hang on a finding that firm A is less efficient than firm B. However, Table 7.3 indicates that efficiency point estimates can have large standard deviations associated with them. This is common in empirical studies involving stochastic frontier models. Simply relying on point estimates which indicate that firm A is less efficient than firm B may lead to inappropriate policy advice. Gibbs sampler output can be used in a straightforward manner to shed light on this issue. For instance, $p(\tau_A < \tau_B|y)$ is the probability firm A is less efficient than firm B. This is a function of the models parameters and can be calculated in the standard way. To be precise, $p(\tau_A < \tau_B|y)$ can be written as $E[g(\tau)|y]$ for a particular choice of $g()$ and the Gibbs sampler is designed to calculate such quantities (see Chapter 4, Section 4.2.3).

In our data set, we find $p(\tau_{\max} > \tau_{med}|y) = 0.89$, $p(\tau_{\max} > \tau_{\min}|y) = 1.00$ and $p(\tau_{med} > \tau_{\min}|y) = 1.00$. In words, we are 100% sure that the firm which appears least efficient truly is less efficient than the median or most efficient firms. We are 89% sure that the firm which appears most efficient truly is more efficient than the median efficient firm. Thus, we can conclude that firms which are ranked far apart in terms of their efficiency estimates do truly differ in efficiency. However, it is likely the case that, for example, the researcher would be very uncertain about saying the 12th ranked firm is more efficient than the 13th ranked. Figure 7.6 plots the full posteriors for $\tau_{\min}$, τ_{med} and $\tau_{\max}$. It can be seen that these posteriors are fairly spread out.

The fact that efficiencies (or, more generally, individual effects) are hard to estimate precisely is a common finding in empirical studies. Stochastic frontier models are sometimes referred to as *composed error models* as they can be

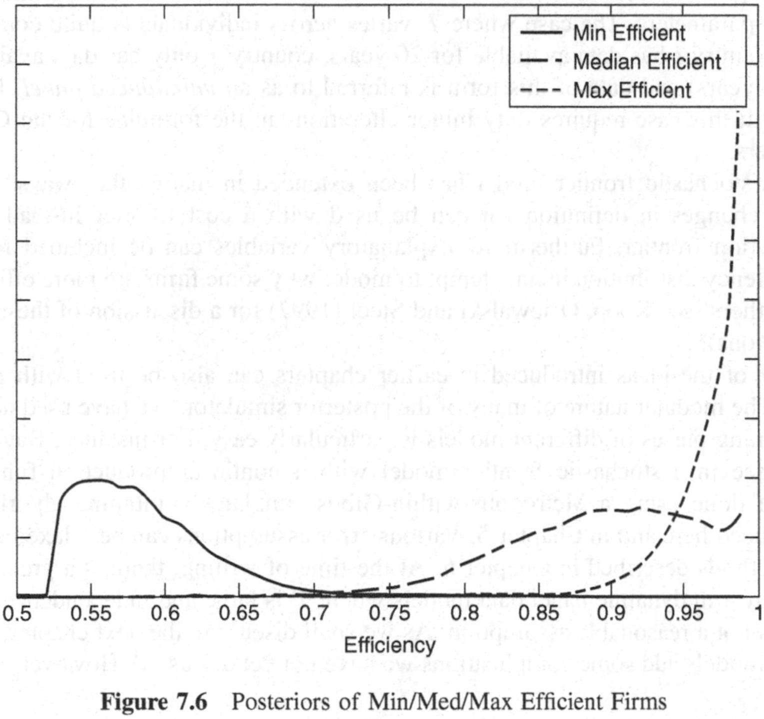

Figure 7.6 Posteriors of Min/Med/Max Efficient Firms

interpreted as having an error composed of two parts: ε_{it} and z_i. Decomposing a single error into two parts is hard to do statistically, particularly if T is small. Nevertheless, stochastic frontier models are quite popular and, if used with care, can be used to shed light on issues relating to efficiency.

7.8 EXTENSIONS

The panel data models introduced in this chapter are useful for modeling heterogeneity of various sorts. In many fields of economics, there is a growing interest in unobserved heterogeneity of the individual under study. For instance, in labor economics, individuals may vary in many ways that cannot be directly observed by the econometrician (e.g. they may differ in their returns to schooling, their value of leisure, their productivity, etc.). This motivates a growing interest in the models introduced in this section, which can be extended in many directions in a straightforward manner. With the exception of the pooled model, all of the models involve heterogeneity across individuals or firms (i.e. variation is over the i subscript). But similar methods can be used to model heterogeneity across time (i.e. variation over the t subscript is easy to handle). For instance, in the stochastic frontier model the researcher may want to have the production frontier shifting over time to represent technical progress. Koop, Osiewalski and Steel (2000) presents such a model which allows for variation over both i and t in model parameters. The case where T varies across individuals is quite common (e.g. country i has data available for 20 years, country j only has data available for 10 years) and data of this form is referred to as an *unbalanced panel*. Dealing with this case requires only minor alterations in the formulae for the Gibbs samplers.

The stochastic frontier model has been extended in many other ways. With slight changes in definitions, it can be used with a cost frontier instead of a production frontier. Furthermore, explanatory variables can be included in the inefficiency distribution in an attempt to model why some firms are more efficient than others (see Koop, Osiewalski and Steel (1997) for a discussion of these two extensions).

Any of the ideas introduced in earlier chapters can also be used with panel data. The modular nature of many of the posterior simulators we have used means combining pieces of different models is particularly easy. For instance, Bayesian inference in a stochastic frontier model with a nonlinear production function can be done using a Metropolis-within-Gibbs simulator combining algorithms introduced here and in Chapter 5. Various error assumptions can be relaxed using the methods described in Chapter 6. At the time of writing, there is a great deal of interest in dynamic panel data models where T is large and independence over time is not a reasonable assumption. As we shall discuss in the next chapter, time series models add some complications we have not yet discussed. However, crude versions of dynamic panel data models can be easily handled either by including lagged dependent variables in X_i or by allowing for autocorrelated errors. The first of these extensions is trivial, the second can be done by combining aspects of the Gibbs sampler introduced in Chapter 6, Section 6.5.2 with the Gibbs sampler introduced in the present chapter.

7.9 SUMMARY

In this chapter, we have introduced several models which are used with panel data. These models differed in the degree of variability they allowed in the parameters across individuals. The first model, the pooled model, allowed no such variation and was equivalent to the Normal linear regression model discussed in previous chapters. The second class comprised individual effects models. This class allowed for the intercept to vary across individuals. Two variants were considered, one of which had a hierarchical and the other a non-hierarchical prior. The distinction between hierarchical and non-hierarchical priors is analogous to the frequentist econometric distinction between random and fixed effects panel data models. The third model was the random coefficient model which allowed for both intercept and slope coefficients to vary across individuals. A fourth model, the stochastic frontier model, was also introduced, although it really belongs in the class of individual effects models. Stochastic frontier models are used for efficiency analysis and the 'efficiency of firm i' is essentially equivalent to 'the individual effect of firm i'.

Posterior inference in all of these models can be done using Gibbs sampling and, thus, no new methods of posterior simulation were introduced in this chapter. However, a new method of marginal likelihood calculation was introduced: the Chib method. This method can be computationally intensive, but it does work well in some cases where the dimensionality of the parameter space is very high and, thus, the Gelfand–Dey method is hard to implement.

This chapter also stressed how many extensions can be handled in straightforward fashion by adding posterior simulation ideas from earlier chapters to the Gibbs samplers introduced for panel data models.

7.10 EXERCISES

7.10.1 Theoretical Exercises

1. *Some general results for Normal linear hierarchical regression models* (see Lindley and Smith (1972) for more details).

   Let y be an N-vector and θ_1, θ_2 and θ_3 be parameter vectors of length k_1, k_2 and k_3, respectively. Let X, W and Z be known $N \times k_1$, $k_1 \times k_2$ and $k_2 \times k_3$

matrices and C_1, C_2 and C_3 be $k_1 \times k_1$, $k_2 \times k_2$ and $k_3 \times k_3$ known positive definite matrices. Assume

$$y|\theta_1, \theta_2, \theta_3 \sim N(X\theta_1, C_1)$$
$$\theta_1|\theta_2, \theta_3 \sim N(W\theta_2, C_2)$$

and

$$\theta_2|\theta_3 \sim N(Z\theta_3, C_3)$$

Throughout this exercise, we will treat θ_3 as known (e.g. as a vector of prior hyperparameters selected by the researcher).

(a) Show that

$$y|\theta_2, \theta_3 \sim N(XW\theta_2, C_1 + XC_2X')$$

and, thus, that this Normal linear regression model with hierarchical prior can be written as a different Normal linear regression model with non-hierarchical prior.

(b) Derive $p(\theta_1|y, \theta_3)$. What happens to this density as $C_3^{-1} \to 0_{k_3 \times k_3}$?

(c) Derive $p(\theta_2|y, \theta_3)$.

2. Consider a model which combines some elements of the individual effects model and some elements of the pooled model. The intercept and some of the regression coefficients vary across individuals, but other regression coefficients do not. That is,

$$y_i = X_i\beta_i + W_i\gamma + \varepsilon_i$$

where the explanatory variables with varying coefficients are located in X_i and the explanatory variables with constant coefficients are located in W_i. Write out a Gibbs sampler for carrying out Bayesian inference in this model. Discuss how you might test which explanatory variables have constant coefficients.

3. Suppose you have an unbalanced panel. That is, for each individual data for T_i periods is available.
   (a) Describe how the posterior simulators for the individual effects models (see Section 7.3.3) would be altered by the unbalanced panel.
   (b) Describe how the posterior simulators for the random coefficients model (see Section 7.4.3) would be altered by the unbalanced panel.
4. Panel data with $T = 1$ reduces to the cross-sectional data sets used in previous chapters. The pooled model is identical to the Normal linear regression model of Chapters 3 and 4. But the other models in this chapter differ from those in earlier chapters even if $T = 1$. Discuss Bayesian inference in both individual effects models (i.e. with hierarchical and non-hierarchical priors) when $T = 1$.
5. Consider the individual effects model with hierarchical prior with likelihood given in (7.3), prior given in (7.7)–(7.11) and posterior conditionals given in (7.14) through (7.18). Extend these results to allow for independent Student-t errors (see Chapter 6, Section 6.4).

6. The posterior simulator for the stochastic frontier model (see Section 7.7.4) used an exponential inefficiency distribution (i.e. $z_i \sim G(\mu_z, 2)$). Derive a posterior simulator for the case where $z_i \sim G(\mu_z, 4)$ and $z_i \sim G(\mu_z, 6)$.

Note: The Gamma distribution with (known) degrees of freedom equal to an even integer (i.e. $\nu = 2, 4, 6, \ldots$) is referred to as the *Erlang distribution*. Bayesian inference in stochastic frontier models with Erlang inefficiency distribution using importance sampling is described in van den Broeck, Koop, Osiewalski and Steel (1994). Inference using Gibbs sampling is described in Koop, Steel and Osiewalski (1995). Bayesian inference in the unrestricted Gamma case ($z_i \sim G(\mu_z, \nu)$ where ν is an unknown parameter) is described in Tsionas (2000). If you are having difficulty with this question you may wish to take a look at some of these references.

7.10.2 Computer-Based Exercises

The exercises in this chapter are closer to being small projects than standard textbook questions. Remember that some data sets and MATLAB programs are available on the website associated with this book.

7. (a) Write a computer program which uses your answer to Exercise 5 to carry out posterior simulation in the individual effects model with hierarchical prior and independent Student-t errors.
   (b) Using artificial data sets simulated as in the empirical illustration of Section 7.6, test your program of part (a).
   (c) Add computer code which does an informal test for Normality of errors by calculating $p(\nu > 30|y)$ and use this on the artificial data set.
   (d) Repeat parts (b) and (c) using different artificial data sets, including some with Student-t errors.
8. The empirical illustration for the stochastic frontier model (see Section 7.7.5) used an exponential inefficiency distribution (i.e. $z_i \sim G(\mu_z, 2)$).
   (a) Repeat the empirical illustration assuming $z_i \sim G(\mu_z, 4)$ and $z_i \sim G(\mu_z, 6)$ using your answer to Exercise 6.
   (b) Write up computer code which uses the Chib method to calculate the marginal likelihood for three models which differ in their inefficiency distributions with M_1: $z_i \sim G(\mu_z, 2)$, M_2: $z_i \sim G(\mu_z, 4)$ and M_3: $z_i \sim G(\mu_z, 6)$. Use this code to compare the three models.
   (c) Repeat parts (a) and (b) using various artificial data sets generated using different inefficiency distributions.

8 Introduction to Time Series: State Space Models

8.1 INTRODUCTION

Time series data is, as its name suggests, ordered by time. It commonly arises in the fields of macroeconomics (e.g. one might have data on the unemployment rate observed every quarter since 1960) and finance (e.g. one might have data on the price of a particular stock every day for a year). There is a huge literature on econometric methods for time series, and it is impossible to do it justice in just one chapter. In this chapter we offer an introduction to one class of models referred to as *state space models* which are commonly-used with time series data. We make this choice for three reasons. First, as we shall see, state space models are hierarchical in nature. As was stressed in the previous chapter, Bayesian methods with hierarchical priors are particularly attractive. Secondly, Bayesian analysis of the main alternative approach[1] to time series econometrics has already been covered in detail in a recent textbook: Bauwens, Lubrano and Richard (1999). To avoid overlap, the present book offers a different way of looking at time series issues. Thirdly, state space models are not so much a different class of models than is used in Bauwens, Lubrano and Richard (1999), but rather offer a different way of writing the same models.[2] Hence, by using state space models, we can address all the same issues as Bauwens, Lubrano and Richard (1999), but stay in a hierarchical framework which is both familiar and computationally convenient.

We have already introduced many time series concepts in Chapter 6, Section 6.5, which discussed the linear regression model with autocorrelated errors. You may wish to review this material to remind yourself of basic concepts and notation. For instance, with time series we use t and T instead of i and N, so that y_t for $t = 1, \dots, T$ indicates observations on the dependent variable from

[1]For readers with some knowledge of time series methods, note that this alternative approach includes autoregressive moving average (ARMA) models and extensions to dynamic regression models which allow for the discussion of issues like unit roots and cointegration.

[2]For instance, there is a state space representation for any ARMA model.

period 1 through T. Before discussing state space models, it is worth briefly mentioning that the techniques discussed in Chapter 6 (Section 6.5) can take you quite far in practice. For instance, the linear regression model with autocorrelated errors is a time series model which may be appropriate in many cases. This model has the errors, ε_t, following an AR(p) process. A common univariate time series model (i.e. a model for investigating the behavior of the single time series y) has y_t following an AR(p) process:

$$(1 - \rho_1 L - \cdots - \rho_p L^p) y_t = u_t \tag{8.1}$$

Computational methods for Bayesian analysis of this model are a straightforward simplification of those presented previously. In fact, (8.1) is really just a linear regression model where the explanatory variables are lags of the dependent variable

$$y_t = \rho_1 y_{t-1} + \cdots + \rho_p y_{t-p} + u_t \tag{8.2}$$

Thus, all the basic regression techniques discussed in previous chapters are relevant. Equation (8.2) can even be extended to include other explanatory variables (and their lags) while still remaining within the regression framework:

$$y_t = \rho_1 y_{t-1} + \cdots + \rho_p y_{t-p} + \beta_0 x_t + \beta_1 x_{t-1} + \cdots + \beta_q x_{t-q} + u_t \tag{8.3}$$

However, several complications arise in this regression-based approach. Loosely speaking, a good deal of the time series literature relates to placing restrictions on (or otherwise transforming) the coefficients of (8.3). There are also some important issues relating to prior elicitation which do not arise in cross-sectional contexts.[3]

Even if we stay within the class of state space models, we cannot possibly offer more than a superficial coverage of a few key issues in a single chapter. Accordingly, we will begin with the simplest univariate state space model called the *local level* model. Most of the basic issues involving prior elicitation and computation can be discussed in the context of this model. We then proceed to a more general state space model. For readers interested in more detail West and Harrison (1997) is a popular Bayesian textbook reference in this field.[4] Kim and Nelson (1999) is another Bayesian book which introduces and extends state space models.

In this chapter, we also use state space models to introduce empirical Bayes methods. These methods are increasingly popular with hierarchical models of all sorts. They provide a data-based method for eliciting prior hyperparameters. For the researcher who does not wish to subjectively elicit informative priors and

---

[3]In addition to Bauwens, Lubrano and Richard (1999), the reader interested in more detail is referred to the papers in themed issues of *Econometric Theory* (volume 10, August/October, 1994) and the *Journal of Applied Econometrics* (volume 6, October/December, 1991).

[4]A few other recent journal articles on Bayesian analysis of state space models are Carlin, Polson and Stoffer (1992), Carter and Kohn (1994), de Jong and Shephard (1995), Fruhwirth-Schnatter (1995), Koop and van Dijk (2000) and Shively and Kohn (1997). Durbin and Koopman (2001) is a good textbook source which has some Bayesian content.

does not wish to use a noninformative prior (e.g. since Bayes factors are hard to interpret with improper priors), empirical Bayesian methods offer an attractive alternative.[5]

8.2 THE LOCAL LEVEL MODEL

The local level model is given by

$$y_t = \alpha_t + \varepsilon_t \tag{8.4}$$

where ε_t is i.i.d. $N(0, h^{-1})$. The unique aspect of this model is the term α_t which is not observed and is assumed to follow a *random walk*

$$\alpha_{t+1} = \alpha_t + u_t \tag{8.5}$$

where u_t is i.i.d. $N(0, \eta h^{-1})$ and ε_t and u_s are independent of one another for all s and t. In (8.4) t runs from 1 through T while in (8.5) it runs from 1 through $T-1$. Equation (8.5) does not explicitly provide an expression for α_1, which is referred to as an *initial condition*. Equation (8.4) is referred to as the *observation* (or measurement) *equation*, while (8.5) is referred to as the *state equation*.

Before discussing Bayesian inference in the local level model, it is worthwhile to spend some time motivating this model. In Chapter 6, Section 6.5, we discussed the AR(1) model, and noted that if the coefficient on the lagged dependent variable, ρ, equalled one then the time series was nonstationary. Here it can be seen that (8.5) implies that α_t is nonstationary. In particular, it implies that α_t has a *stochastic trend*. The term *stochastic trend* arises from the fact that models such as (8.5) imply that a series can wander widely (i.e. trend) over time, but that an element of randomness enters the trend behavior. That is, in contrast to a *deterministic trend* such as

$$\alpha_t = \alpha + \beta t$$

where the variable is an exact function of time, a stochastic trend involves a random error, u_t. The fact that (8.5) implies that α_t exhibits trend behavior can be seen by noting that (8.5) can be written as

$$\alpha_t = \alpha_1 + \sum_{j=1}^{t-1} u_j \tag{8.6}$$

and, thus (ignoring the initial condition) $var(\alpha_t) = (t-1)\eta h^{-1}$. In addition, α_t and α_{t-1} tend to be close to one another (i.e. $E(\alpha_t|\alpha_{t-1}) = 0$). In words, the stochastic trend term has variance which is increasing with time (and thus can wander over an increasing wide range), but α_t changes only gradually over

[5]Carlin and Louis (2000) provides an excellent introduction to empirical Bayesian methods, although it is a statistics as opposed to econometrics textbook.

time. This is consistent with the intuitive concept of a trend as something which increases (or decreases) gradually over time.

To return to the local level model, we can see that (8.4) decomposes the observed series, y_t, into a trend component, α_t, and an error or irregular component, u_t.[6] In general, state space models can be interpreted as decomposing an observed time series into various parts. In the local level model, there are two components, a trend and an error. In more complicated state space models, the observed series can be decomposed into more components (e.g. trend, error and seasonal components).

It is worth mentioning that the local level model has been used for measuring the relative sizes of the trend and irregular components. This motivates the way that we have written the variances of the two errors (i.e. error variances are written as h^{-1} and ηh^{-1}). In this manner, η is directly interpreted as the size of the random walk relative to the error variance in the measurement equation. That is, it can be seen that if $\eta \to 0$, then the error drops out of (8.5) and $\alpha_t = \alpha_1$ for all t and (8.4) becomes $y_t = \alpha_1 + \varepsilon_t$. In this case, y_t exhibits random fluctuations around a constant level, α_1, and is not trending at all. However, as η becomes larger (i.e. the variance of u_t becomes larger), then the stochastic trend term plays a bigger role. Examining η is, thus, a nice way of measuring the importance of trend behavior in an economic time series. For the reader with previous knowledge of time series econometrics, note that the test of whether $\eta = 0$ is one way of testing for a *unit root*. We will not discuss unit root testing in any detail here. Suffice it to note that, unit root testing has played an important role in modern empirical macroeconomics, and that state space models allow for this to be done in an intuitive and straightforward manner.

Another way of interpreting (8.4) and (8.5) is by noting that α_t is the mean (or level) of y_t. Since this mean is varying over time, the terminology *local level model* is used. Interpreting α_t in this way, as a parameter, is natural in a Bayesian setup. That is, (8.4) can be interpreted as a very simple example of a linear regression model involving only an intercept. The innovative thing is that the intercept varies over time. Thus, the local level model is a simple example of a *time varying parameter model*. More sophisticated state space models can allow for time varying regression coefficients or time varying error variances. If $\alpha = (\alpha_1, \ldots, \alpha_T)'$ is interpreted as a vector of parameters then, as Bayesians, we must elicit a prior for it. But (8.5) provides us with such a prior. That is, (8.5) can be interpreted as defining a hierarchical prior for α. Note that, with such an interpretation, the local level model is very similar to the individual effects panel data model of Chapter 7 (Section 7.3) with $T = 1$. Of course, the individual effects model has an intercept which varies across individuals, while the local level model has an intercept which varies across time, but the basic structure of

[6]For the macroeconomist, some imperfect intuition for this would be that the trend term captures the long run trend growth of the economy (e.g. due to growth of the labor force, building up of capital stock and gradual technical improvements), whereas the irregular component reflects the random short term shocks hitting the economy (e.g. business cycle effects).

the two models is the same. Thus, the basic tools developed in Chapter 7 using an independent Normal-Gamma prior can be used here with some modifications. For this reason, in this section we do something new. We use a natural conjugate prior and introduce a new type of prior elicitation procedure.

That is, Bayesian methods using an independent Normal-Gamma prior are very similar to those described in Chapter 7, so we do not repeat them here. In particular, a Gibbs sampler with data augmentation can be developed as in Chapter 7. In Section 8.3 we develop such an algorithm in the context of a more general state space model. This can be used for the local level model and the reader interested in using the independent Normal-Gamma prior is referred to Section 8.3. In the present section, we will use a natural conjugate framework to introduce empirical Bayesian methods.

8.2.1 The Likelihood Function and Prior

If we define $y = (y_1, \ldots, y_T)'$ and $\varepsilon = (\varepsilon_1, \ldots, \varepsilon_T)'$, then we can write the local level model in matrix notation as

$$y = I_T \alpha + \varepsilon \tag{8.7}$$

If we make the standard error assumptions, that ε has a multivariate Normal distribution with mean 0_T and covariance matrix $h^{-1} I_T$, then this model is simply a Normal linear regression model where the matrix of explanatory variables is the identity matrix (i.e. $X = I_T$) and α is the T-vector of regression coefficients. Thus, the likelihood function has the standard form for the Normal linear regression model (e.g. see Chapter 3, (3.3)).

Of course, as in any Bayesian exercise, we can use any prior we wish. However, the state equation given in (8.5) suggests a hierarchical prior. We use one involving natural conjugate form. To draw out the similarities with results in Chapter 3 for the Normal linear regression model with natural conjugate prior, it is convenient to write this model in a slightly different way. To do this we begin by defining the $(T-1) \times T$ *first difference matrix*:

$$D = \begin{bmatrix} -1 & 1 & 0 & 0 & \cdots & \cdots & 0 \\ 0 & -1 & 1 & 0 & \cdots & \cdots & 0 \\ \cdots & \cdots & \cdots & \cdots & \cdots & \cdots & \cdots \\ 0 & \cdots & \cdots & 0 & 0 & -1 & 1 \end{bmatrix} \tag{8.8}$$

To draw out the connections with the state space model, note that

$$D\alpha = \begin{pmatrix} \alpha_2 - \alpha_1 \\ \cdot \\ \cdot \\ \alpha_T - \alpha_{T-1} \end{pmatrix}$$

and thus the state equation given in (8.5) can be written as:

$$D\alpha = u$$

where $u = (u_1, \ldots, u_{T-1})'$. The assumption that u is Normal can thus be interpreted as saying that the state equation is defining a Normal hierarchical prior for $D\alpha$.

To specify a complete prior for all the parameters in the model, we also need to specify a prior for h and α_1. To do this, we first write (8.7) as

$$y = W\theta + \varepsilon \tag{8.9}$$

where

$$\theta = \begin{pmatrix} \alpha_1 \\ \alpha_2 - \alpha_1 \\ \cdot \\ \cdot \\ \alpha_T - \alpha_{T-1} \end{pmatrix}$$

and

$$W = \begin{pmatrix} 1 & 0'_{T-1} \\ \iota_{T-1} & C \end{pmatrix}$$

where ι_{T-1} is an $(T-1)$-vector of ones. Direct matrix multiplication can be used to verify that (8.9) is exactly equivalent to (8.7). Direct matrix inversion can be used to show that C is a $(T-1) \times (T-1)$ lower triangular matrix with all non-zero elements equalling one (it is the inverse of D with its first column removed). That is, C has all elements on or below the diagonal equalling 1, and all elements above the diagonal equalling 0.

We begin by eliciting a natural conjugate prior for θ and h:

$$\theta, h \sim NG(\underline{\theta}, \underline{V}, \underline{s}^{-2}, \underline{\nu}) \tag{8.10}$$

The reader is referred to Chapter 3 for a reminder of notation and properties of this Normal-Gamma prior.

We consider a particular structure for $\underline{\theta}$ and $\underline{V}$ which embodies the prior information contained in the state equation:

$$\underline{\theta} = \begin{pmatrix} \underline{\theta}_1 \\ 0 \\ \cdot \\ \cdot \\ 0 \end{pmatrix} \tag{8.11}$$

$$\underline{V} = \begin{pmatrix} \underline{V}_{11} & 0'_{T-1} \\ 0_{T-1} & \eta I_{T-1} \end{pmatrix} \tag{8.12}$$

Note that this prior implies $\alpha_{t+1} - \alpha_t$ is $N(0, \eta h^{-1})$, which is exactly what we assumed at the beginning of this section. The fact that this prior depends upon the parameter η makes it hierarchical. In addition, we have provided a prior for the initial condition, α_1, as being $N(\underline{\theta}_1, h^{-1}\underline{V}_{11})$.

At this point it is worth summarizing what we have done. We have written the local level model as a familiar Normal linear regression model with natural conjugate prior. The fact that this is a time series problem involving a state space model manifests itself solely through the prior we choose. In a Bayesian paradigm, the interpretation of the state equation as being a prior is natural and attractive. However, it is worth mentioning that the non-Bayesian econometrician would interpret our hierarchical prior as part of a likelihood function. As stressed in the previous chapter, in many models there is a degree of arbitrariness as to what part of a model is labelled the 'likelihood function' and what part is labelled the 'prior'.

8.2.2 The Posterior

Using standard results for the Normal linear regression model with natural conjugate prior (see Chapter 3), it follows that the posterior for θ and h, denoted by $p(\theta, h|y)$ is $NG(\overline{\theta}, \overline{V}, \overline{s}^{-2}, \overline{\nu})$ where

$$\overline{\theta} = \overline{V}(\underline{V}^{-1}\underline{\theta} + W'y) \tag{8.13}$$

$$\overline{V} = (\underline{V}^{-1} + W'W)^{-1} \tag{8.14}$$

$$\overline{\nu} = \underline{\nu} + T \tag{8.15}$$

and

$$\overline{\nu}\overline{s}^2 = \underline{\nu}\underline{s}^2 + (y - W\overline{\theta})'(y - W\overline{\theta}) + (\overline{\theta} - \underline{\theta})'\underline{V}^{-1}(\overline{\theta} - \underline{\theta}) \tag{8.16}$$

The properties of the Normal-Gamma distribution imply that it is easy to transform back from the parameterization in (8.9) to the original parameterization given in (8.7). That is, $p(\theta|h, y)$ is Normal and we know linear combinations of Normal are Normal (see Appendix B, Theorem B.10). Thus, if the posterior for (θ, h) is $NG(\overline{\theta}, \overline{V}, \overline{s}^{-2}, \overline{\nu})$ then the posterior for (α, h) is $NG(\overline{\alpha}, \overline{V}_\alpha, \overline{s}^{-2}, \overline{\nu})$ where

$$\overline{\alpha} = W\overline{\theta} \tag{8.17}$$

and

$$\overline{V}_\alpha = W\overline{V}W' \tag{8.18}$$

Since we have used a natural conjugate prior, analytical posterior results are available and there is no need for a posterior simulator. It is also interesting to note that the local level model is a regression model where the number of regression coefficients is equal to the number of observations. In a regression analysis, it is usually the case that the number of regression coefficients is much less than the number of observations (i.e. in the notation of previous chapters $k \ll N$). However, the local level model shows that prior information can, in many cases, be used to provide valid posterior inferences even in models with a huge number of parameters. Expressed in another way, the question arises as to why we don't just obtain a degenerate posterior distribution at the point $\alpha = y$. After all, setting $\alpha_t = y_t$ for all t would yield a perfectly fitting model in the sense that $\varepsilon_t = 0$ for all t. It can be verified that the likelihood is infinite at this point. However, the Bayesian posterior is not located at this point of infinite likelihood because of prior information. The state equation says that α_{t+1} and α_t are close to one another, which pulls the posterior away from the point of perfect fit. In the state space literature, this is referred to as smoothing the state vector.

Since the model considered here is simply a Normal linear regression model with natural conjugate prior, model comparison and prediction can be done using methods outlined in Chapter 3.

8.2.3 Empirical Bayesian Methods

In previous chapters, we have either elicited priors subjectively or used noninformative priors. In the present context, this would mean choosing values for $\underline{\theta}$, $\underline{V}$, $\underline{s}^{-2}$, $\underline{\nu}$ or setting them to their noninformative values (see Chapter 3, Section 3.5) of $\underline{\nu} = 0$ and $\underline{V}^{-1} = 0_{T \times T}$.[7] However, both of these approaches had potential drawbacks. Subjective elicitation of priors may be difficult to do, or it may be subject to criticism by other researchers with different priors. Noninformative priors often make it difficult to do Bayesian model comparison since the resulting marginal likelihood may be undefined. Accordingly, some Bayesians use so-called empirical Bayes methods which surmount these two problems. The local level model is a convenient place to introduce empirical Bayes methods because some interesting issues arise in its application. However, empirical Bayes methods can be used with any model and are particularly popular with hierarchical prior models such as those of Chapter 7 and the present chapter. It should be noted, however, that empirical Bayesian methods have been criticized for implicitly double-counting the data. That is, the data is first used to select prior hyperparameter values. Once these values are selected, the data are used a second time in a standard Bayesian analysis.

Empirical Bayesian methods involve estimating prior hyperparameters from the data, rather than subjectively choosing values for them or setting them to noninformative values. The marginal likelihood is the preferred tool for this. In particular, for any choice of prior hyperparameters a marginal likelihood can be calculated. The values of the prior hyperparameters which yield the largest marginal likelihood are those used in an empirical Bayes analysis. However, searching over all possible prior hyperparameters can be a very difficult thing to do. Accordingly, empirical Bayes methods are often used on one or two key prior hyperparameters. Here we show how this might be done for the local level model.

The prior for the local level model specified in (8.10), (8.11) and (8.12) depends upon four hyperparameters η, $\underline{\theta}_1$, $\underline{V}_{11}$, $\underline{s}^{-2}$ and $\underline{\nu}$. Of these, η is almost invariably the most important and seems a candidate for the empirical Bayes approach. After

[7]Remember that, with these noninformative choices, the values of $\underline{\theta}$ and $\underline{s}^{-2}$ are irrelevant.

all, it can be interpreted as relating to the size of the random walk component in the state space model and it may be hard to elicit subjectively a value for it. Furthermore, setting it to an apparently 'noninformative' limiting value, $\eta \to \infty$, makes little sense since this implies the stochastic trend term completely dominates the irregular component. This is not 'noninformative', but rather quite informative. Accordingly, we focus on η. We will begin by assuming the researcher is able to subjectively elicit values for $\underline{\theta}_1$, $\underline{V}_{11}$, $\underline{s}^{-2}$ and $\underline{\nu}$.

The results of Chapter 3 (see (3.34)) imply that the marginal likelihood for the present model takes the form

$$p(y|\eta) = c\left(\frac{|\overline{V}|}{|\underline{V}|}\right)^{\frac{1}{2}}(\overline{\nu}\overline{s}^2)^{-\frac{\overline{\nu}}{2}} \tag{8.19}$$

where

$$c = \frac{\Gamma\left(\frac{\overline{\nu}}{2}\right)(\underline{\nu}\underline{s}^2)^{\frac{\underline{\nu}}{2}}}{\Gamma\left(\frac{\underline{\nu}}{2}\right)\pi^{\frac{T}{2}}} \tag{8.20}$$

The notation in (8.19) makes clear that we are treating the marginal likelihood as a function of η (i.e. in previous chapters we used notation $p(y)$ or $p(y|M_j)$ to denote the marginal likelihood, but here we make explicit the dependence on η). The standard way of carrying out an empirical Bayes analysis would be to choose, $\widehat{\eta}$, the value of η which maximizes $p(y|\eta)$ in (8.19). $\widehat{\eta}$ would then be plugged in (8.12), and posterior analysis could then be done in the standard way using (8.13)–(8.18). In the present model, $\widehat{\eta}$ could be found by using grid search methods. That is, the researcher could simply try every value for η in some appropriate grid and choose $\widehat{\eta}$ as being the value which maximizes $p(y|\eta)$.

A more formal way of carrying out empirical Bayesian estimation would involve explicitly treating η as a parameter and using the laws of conditional probability to carry out Bayesian inference. If η is treated as an unknown parameter, then Bayes theorem implies $p(\eta|y) \propto p(y|\eta)p(\eta)$ where $p(\eta)$ is a prior and we can write

$$p(\eta|y) \propto c\left(\frac{|\overline{V}|}{|\underline{V}|}\right)^{\frac{1}{2}}(\overline{\nu}\overline{s}^2)^{-\frac{\overline{\nu}}{2}}p(\eta) \tag{8.21}$$

This posterior can be used to make inferences about η. If interest centers on the other parameters in the model, then we can use the fact that

$$p(\theta, h, \eta|y) = p(\theta, h|y, \eta)p(\eta|y)$$

Since $p(\theta, h|y, \eta)$ is Normal-Gamma (i.e. conditional on a specific value for η the posterior results in (8.13)–(8.18) hold) and $p(\eta|y)$ is one-dimensional, Monte Carlo integration can be used to carry out posterior inference in this model. That is, drawing from $p(\eta|y) \propto p(y|\eta)p(\eta)$ and, conditional upon this draw, drawing from $p(\theta, h|y, \eta)$ yields a draw from the joint posterior. As an aside, just how one draws from $p(\eta|y)$ depends on the exact form of $p(\eta)$. However, a simple way of drawing from any univariate distribution involves approximating it by a discrete alternative. That is, evaluating $p(\eta|y)$ at B different points on a grid, $\eta_1, \ldots, \eta_B$, will yield $p(\eta_1|y), \ldots, p(\eta_B|y)$. Draws of η taken from the resulting discrete distribution (i.e. the distribution defined by $p(\eta = \eta_i) = p(\eta_i|y)$ for $i = 1, \ldots, B$), will be approximately equal to draws from $p(\eta|y)$. As B increases, the quality of the approximation will get better. In the empirical illustration below, we use this crude but effective strategy for carrying out Bayesian inference in the local level model.

The empirical Bayes methods for the local level model as described so far requires the researcher to choose $\underline{\theta}_1, \underline{V}_{11}, \underline{s}^{-2}$ and $\underline{\nu}$ (and $p(\eta)$ for the second approach outlined in the preceding paragraph). It is common to make noninformative choices for such prior hyperparameters and, for most models with hierarchical priors (e.g. the panel data models of Chapter 7), such a strategy works well. However, with the local level model, such a strategy does not work. It is worthwhile to discuss in detail why this is so, as it illustrates a problem which can occur in Bayesian inference in models with large numbers of parameters.

Consider first what happens when we set $\underline{\nu}$ and $\underline{V}_{11}^{-1}$ to their limiting values $\underline{\nu} = \underline{V}_{11}^{-1} = 0$. With these choices, the values of $\underline{s}^2$ and $\underline{\theta}_1$ are irrelevant. For these noninformative choices, it can be directly verified that $p(\theta, \sigma^{-2}|y, \eta)$ is a well-defined posterior. However, with regards to the marginal likelihood, two problems arise. First, the integrating constant in (8.20) is indeterminate. This is the standard problem we have discussed previously (e.g. see Chapter 2, Section 2.5). Insofar as interest centers on η, or the marginal likelihood is used for comparing the present model to another with the same noninformative prior for the error variance, this first problem is not a serious one. The constant c either does not enter or cancels out of any derivation (e.g. a Bayes factor) and can be ignored. Secondly, the term $\overline{\nu s}^2$ goes to zero as $\eta \to \infty$. To see this, note that with all the hyperparameters set to noninformative values $\overline{\theta} = (W'W)^{-1}W'y$ and $y - W\overline{\theta} = 0_T$. We will not provide a formal proof, but it is the case that this degeneracy is enough to imply that the marginal likelihood in (8.10) becomes infinite as $\eta \to \infty$. Hence, an empirical Bayes analysis will set $\widehat{\eta} \to \infty$ for any data set. It can be shown that this implies $E(\alpha|y) = y$ and no smoothing of the state vector occurs. Thus, empirical Bayes methods fail in the local level model when we set $\underline{\nu}$ and $\underline{V}_{11}^{-1}$ to noninformative values. This problem (which does not arise in most models) occurs because the number of explanatory variables in the linear regression model given in (8.7) is equal to the number of observations and, thus, it is possible for the regression line to fit perfectly. The general point to note here is that, in models with a large number of parameters, the researcher must be very careful when working with improper noninformative priors.

In the local level model, we have seen that we cannot use empirical Bayes methods with $\underline{\nu} = \underline{V}_{11}^{-1} = 0$. However, it can be verified that if we set either $\underline{\nu} > 0$ or $\underline{V}_{11}^{-1} > 0$ (and make an appropriate choice for $\underline{s}^2$ or $\underline{\theta}_1$), then we can use empirical Bayes methods. Intuitively, either of these will stop $\overline{\nu s}^2$ in (8.16) from going to zero as $\eta \to \infty$. It is worth stressing that empirical Bayes methods work in the present model if *either* $\underline{\nu} > 0$ or $\underline{V}_{11}^{-1} > 0$, it is not necessary to have an informative prior for both h and θ_1.

In the alternative approach which involves treating η as a parameter (see (8.21)), a similar pathology occurs if we set $\underline{\nu} = \underline{V}_{11}^{-1} = 0$ and use an improper prior for η. For instance, if we set $\underline{\nu} = \underline{V}_{11}^{-1} = 0$ and choose $p(\eta)$ to be an improper Uniform distribution over the interval $(0, \infty)$ then it turns out that $p(\eta|y)$ is not a valid probability density function (i.e. it is improper). However, if we either set $\underline{\nu} > 0$ or $\underline{V}_{11}^{-1} > 0$ or choose $p(\eta)$ to be a proper p.d.f. then $p(\eta|y)$ is a valid posterior density. Thus, if we treat η as an unknown parameter, Bayesian inference can be carried out if prior informative about η or h or θ_1 is available.

8.2.4 Empirical Illustration: The Local Level Model

To illustrate empirical Bayesian inference in the local level model, we artificially generated data from the model given in (8.4) and (8.5) with $\eta = 1$, $h = 1$ and $\theta_1 \equiv \alpha_1 = 1$. For a prior we use $\theta, h \sim NG(\underline{\theta}, \underline{V}, \underline{s}^{-2}, \underline{\nu})$ with $\underline{\theta}$ and $\underline{V}$ as described in (8.11) and (8.12). We begin by considering four priors. The first of these is weakly informative for all parameters and sets $\underline{\nu} = 0.01$, $\underline{s}^{-2} = 1$, $\underline{\theta}_1 = 1$ and $\underline{V}_{11} = 100$. Note that this prior is centered over the values used to generate the data (i.e. $\underline{s}^{-2} = 1$ and $\underline{\theta}_1 = 1$), but expresses extreme uncertainty about these values. That is, the prior for h contains as much information as 0.01 of an observation and the prior variance for the initial condition is 100. The second prior is the same as the first, except that it is completely noninformative for h (i.e. $\underline{\nu} = 0$). The third prior is the same as the first, except that it is completely noninformative for θ_1 (i.e. $\underline{V}_{11}^{-1} = 0$). The fourth prior is completely noninformative for both parameters (i.e. $\underline{\nu} = \underline{V}_{11}^{-1} = 0$). Of course, the preceding discussion implies that empirical Bayesian methods should fail for this last prior.

Figure 8.1 plots the marginal likelihoods for a grid of values of η between 0 and 10. The plots corresponding to the four priors are very similar to one another. For the first three priors, we find empirical Bayes estimates of η being $\widehat{\eta} = 0.828$, $\widehat{\eta} = 0.828$ and $\widehat{\eta} = 0.823$, respectively. In fact, even the completely noninformative case (which has $\widehat{\eta} \to \infty$) would yield $\widehat{\eta} = 0.829$ if we limit consideration to the interval $(0, 10)$. The pathology noted with the use of a completely noninformative prior only occurs for extremely large values of η. Equation (8.21) can be used to derive $p(\eta|y)$ and, since we have not specified $p(\eta)$, our empirical illustration implicitly holds for the case where $p(\eta)$ is an improper Uniform prior over the interval $(0, \infty)$. Interpreted in this manner, our empirical illustration shows that if we use a completely noninformative prior for all parameters, $p(\eta|y)$ is a skewed (improper) distribution. It has a mode at the point $\eta = 0.829$, but then gradually increases to infinity as $\eta \to \infty$. Using results solely based on Figure 8.1 is equivalent to using a Uniform prior over

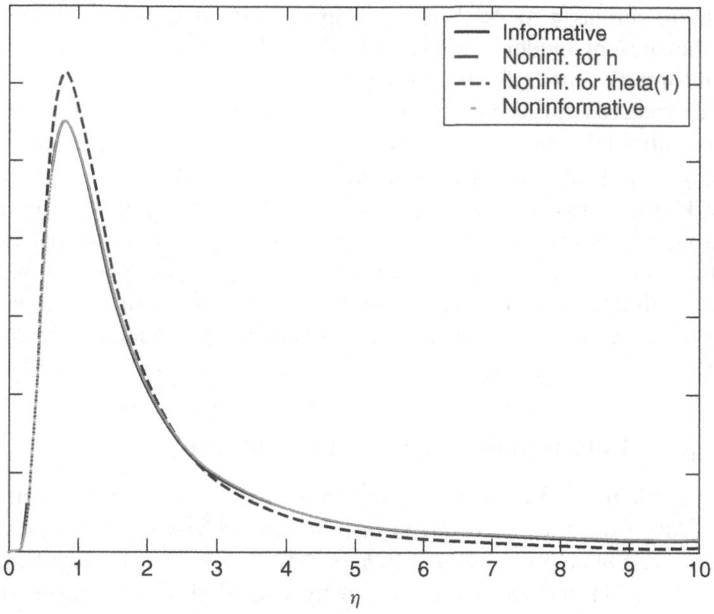

Figure 8.1 Marginal Likelihoods for Four Different Priors

the interval (0, 10) for η. Using such a prior for η is enough to ensure sensible empirical Bayes results.

In summary, users of empirical Bayes methods are often interested in focusing on one parameter and using noninformative priors over the rest. In the local level model with natural conjugate prior, this amounts to setting $\underline{\nu} = \underline{V}_{11}^{-1} = 0$ and using empirical Bayesian methods to estimate η. In the previous subsection we have shown that, in theory, this is not possible to do since we will always obtain $\widehat{\eta} \to \infty$. However, in practice, the empirical illustration shows that this pathology is probably not an important problem. That is, only a minuscule amount of prior information about either h or the initial condition or η is required to ensure empirical Bayesian methods will work.

So far we have focused exclusively on η, however it is often the case that interest centers on the state equation and, in particular, estimating the stochastic trend in the model. To investigate how well empirical Bayes methods work in this regard, we focus on the second prior of the previous section which uses a minuscule amount of prior information about h (i.e. $\underline{\nu} = 0.01$, $\underline{s}^{-2} = 1$), but is noninformative in all other respects. The other priors yield virtually identical results. We simulate four artificial data sets all of which have $h = 1$ and $\theta_1 = 1$ but have $\eta = 0$, 0.1, 1 and 100, respectively.

Figures 8.2a–8.2d plot the four data sets along with $E(\alpha|y)$ obtained using (8.17) for the value of η chosen using empirical Bayes methods. $E(\alpha|y)$ is referred to as the 'Fitted Trend' in the figures. Remember that α can be interpreted as the stochastic trend in the time series, and is often of interest in a time series analysis. Before discussing the stochastic trend it is worthwhile to discuss the data itself. A wide variety of values for η have been chosen to show its role in determining the properties of the data. In Figure 8.2a we see how time series with no stochastic trend ($\eta = 0$) exhibit random fluctuations about a mean. However, as η increases, the trend behavior becomes more and more apparent. As η becomes very large (see Figure 8.2d), the stochastic trend becomes predominant and the series wanders smoothly over a wide range of values.

The estimates of η selected by empirical Bayes are similar to those used to generate the artificial data and the resulting fitted trends are quite sensible. In Figure 8.2a, where there is no trend, the fitted stochastic trend is almost nonexistent (i.e. it is close to simply being a horizontal line). In Figure 8.2d, where the trend predominates, the fitted stochastic trend matches the data very closely (indeed it is hard to see the difference between the two lines in Figure 8.2d). Figures 8.2b and 8.2c present intermediate cases.

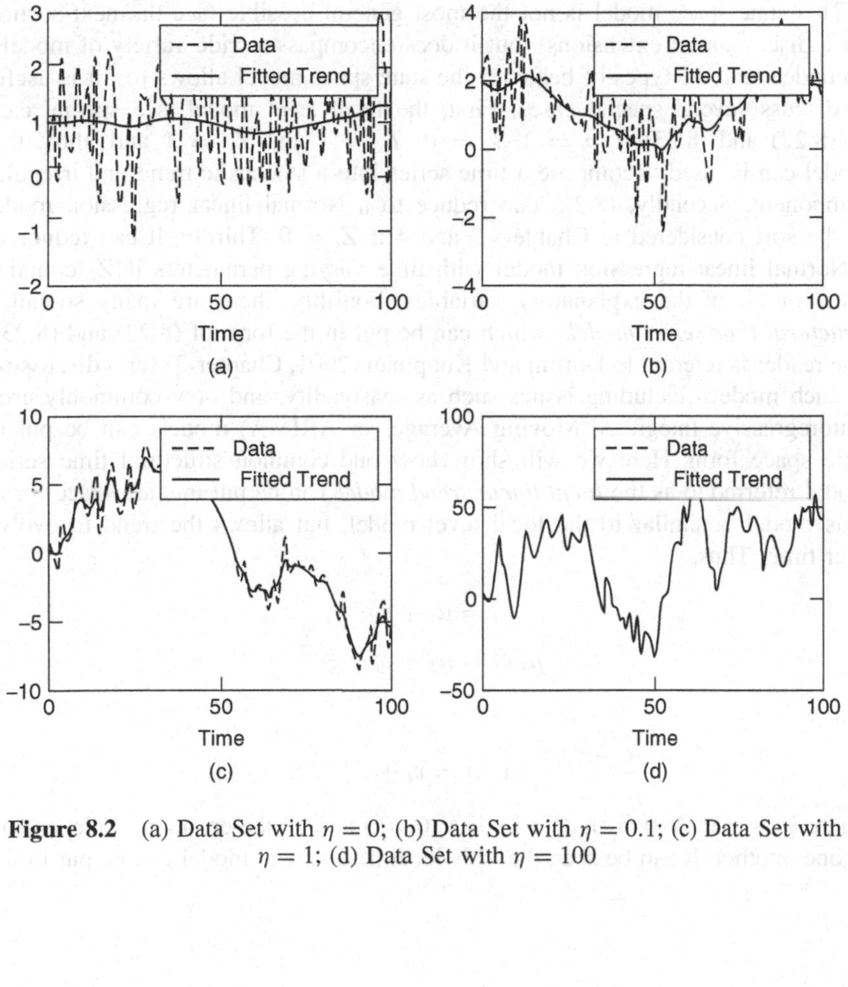

Figure 8.2 (a) Data Set with $\eta = 0$; (b) Data Set with $\eta = 0.1$; (c) Data Set with $\eta = 1$; (d) Data Set with $\eta = 100$

8.3 A GENERAL STATE SPACE MODEL

In this section, we discuss a more general state space model which we will simply refer to it as the *state space model* and write as

$$y_t = X_t\beta + Z_t\alpha_t + \varepsilon_t \tag{8.22}$$

and

$$\alpha_{t+1} = T_t\alpha_t + u_t \tag{8.23}$$

This model uses slightly different notation from the local level model, in that we allow α_t to be a $p \times 1$ vector containing p state equations. We assume ε_t to be i.i.d. $N(0, h^{-1})$, but u_t is now a $p \times 1$ vector which is i.i.d. $N(0, H^{-1})$ and ε_t and u_s are independent of one another for all s and t. X_t and Z_t are $1 \times k$ and $1 \times p$ vectors, respectively, containing explanatory variables and/or known constants. T_t is a $p \times p$ matrix of known constants. The case where T_t contains unknown parameters can be handled in a straightforward fashion, as noted below.

This state space model is not the most general possible (see the next section for a discussion of extensions), but it does encompass a wide variety of models. To understand the types of behavior the state space model allows for, it is useful to discuss several special cases. First, the local level model is a special case of (8.22) and (8.23) if $p = 1$, $k = 0$, $T_t = 1$ and $Z_t = 1$ and, thus, this model can be used decompose a time series into a stochastic trend and irregular component. Secondly, (8.22) can reduce to a Normal linear regression model of the sort considered in Chapters 3 and 4 if $Z_t = 0$. Thirdly, it can reduce to a Normal linear regression model with time varying parameters if Z_t contains some or all of the explanatory variables. Fourthly, there are many so-called *structural time series models* which can be put in the form of (8.22) and (8.23). The reader is referred to Durbin and Koopman (2001, Chapter 3) for a discussion of such models, including issues such as seasonality, and how commonly-used Autoregressive Integrated Moving Average (or ARIMA) models can be put in state space form. Here we will show how one common structural time series model referred to as the *local linear trend model* can be put in state space form. This model is similar to the local level model, but allows the trend to evolve over time. Thus,

$$y_t = \mu_t + \varepsilon_t$$

$$\mu_{t+1} = \mu_t + v_t + \xi_t$$

and

$$v_{t+1} = v_t + \zeta_t$$

where ξ_t is i.i.d. $N(0, \sigma_\xi^2)$, ζ_t is i.i.d. $N(0, \sigma_\zeta^2)$ and all the errors are independent of one another. It can be seen that this local linear trend model can be put in the form of the state space model by setting

$$\alpha_t = \begin{pmatrix} \mu_t \\ v_t \end{pmatrix}$$

$$u_t = \begin{pmatrix} \xi_t \\ \zeta_t \end{pmatrix}$$

$$T_t = \begin{pmatrix} 1 & 1 \\ 0 & 1 \end{pmatrix}$$

$$Z_t = \begin{pmatrix} 1 & 0 \end{pmatrix}$$

$$H = \begin{pmatrix} \sigma_\xi^2 & 0 \\ 0 & \sigma_\zeta^2 \end{pmatrix}$$

and $\beta = 0$. In short, a wide variety of useful regression and time series models can be written as state space models.

8.3.1 Bayesian Computation in the State Space Model

We have stressed throughout this book that an advantage of Bayesian inference is that it is often modular in nature. Methods for posterior computation in many complicated models can be developed by simply combining results from simpler models. The state space model is a good example of how this can be done. Hence, rather than go through the steps of writing out the likelihood, prior and posterior, we jump straight to the issue of Bayesian computation, and show how we can draw on results from earlier chapters to carry out Bayesian inference in this model. As we shall see, a complication for posterior simulation arises since the posterior conditional for α analogous to Chapter 7 (7.17) will not be independent across time (i.e. (8.23) implies that α_t and α_{t-1} will not be independent of one another). Thus, we cannot easily draw from the α_ts one at a time and a direct implementation of the Gibbs sampler would involve drawing from a T-dimensional Normal distribution. In general, this can be a bit slow, but De Jong and Shephard (1995) describe an efficient method for Gibbs sampling in this class of models.

An examination of (8.22) reveals that, if α_t for $t = 1, \ldots, T$ were known (as opposed to being unobserved), then the state space model would reduce to a Normal linear regression model:

$$y_t^* = X_t\beta + \varepsilon_t$$

where $y_t^* = y_t - Z_t\alpha_t$. Thus, all the results of previous chapters for the Normal linear regression model could be used, except the dependent variable would be y_t^* instead of y. This suggests that a Gibbs sampler with data augmentation can be set up for the state space model. That is, depending on the prior chosen, $p(\beta, h|y, \alpha_1, \ldots, \alpha_T)$ will have one of the simple forms given in Chapters 3 or 4. Similarly, if α_t for $t = 1, \ldots, T$ were known then the state equations given in

(8.23) are a simple variant of the Seemingly Unrelated Regression (SUR) linear regression model discussed in Chapter 6 (Section 6.6) and $p(H|y, \alpha_1, \ldots, \alpha_T)$ has a familiar form.[8] Thus, if we can derive a method for taking random draws from $p(\alpha_1, \ldots, \alpha_T|y, \beta, h, H)$ then we have completely specified a Gibbs sampler with data augmentation which allows for Bayesian inference in the state space model. In the following material, we develop such a Gibbs sampler for a particular prior choice, but we stress that other priors can be used with minor modifications.

Here we will use an independent Normal-Gamma prior for β and h, a Wishart prior for H and the prior implied by the state equation for $\alpha_1, \ldots, \alpha_T$. In particular, we assume a prior of the form

$$p(\beta, h, H, \alpha_1, \ldots, \alpha_T) = p(\beta)p(h)p(H)p(\alpha_1, \ldots, \alpha_T|H)$$

where

$$p(\beta) = f_N(\beta|\underline{\beta}, \underline{V}) \tag{8.24}$$

$$p(h) = f_G(h|\underline{s}^{-2}, \underline{\nu}) \tag{8.25}$$

and

$$p(H) = f_W(H|\underline{\nu}_H, \underline{H}) \tag{8.26}$$

For the elements of the state vector we treat (8.23) as a hierarchical prior. If we treat the time index for (8.23) as beginning at 0 (i.e. $t = 0, 1, \ldots, T$) and assume $\alpha_0 = 0$, then the state equation even provides a prior for the initial condition. Formally, this amounts to writing this part of the prior as

$$p(\alpha_1, \ldots, \alpha_T|H) = p(\alpha_1|H)p(\alpha_2|\alpha_1, H) \ldots p(\alpha_T|\alpha_{T-1}, H)$$

where, for $t = 1, \ldots, T - 1$

$$p(\alpha_{t+1}|\alpha_t, H) = f_N(\alpha_{t+1}|T_t\alpha_t, H) \tag{8.27}$$

and

$$p(\alpha_1|H) = f_N(\alpha_1|0, H) \tag{8.28}$$

Note that H is playing a similar role to η in the local level model. However, H is a $p \times p$ matrix, so it would be difficult to use empirical Bayes methods with this high dimensional model. Furthermore, we are no longer using a natural conjugate prior so that the analytical results of Section 8.2 no longer hold.

The reasoning above suggests that our end goal is a Gibbs sampler with data augmentation which sequentially draws from $p(\beta|y, \alpha_1, \ldots, \alpha_T)$, $p(h|y, \alpha_1, \ldots, \alpha_T)$, $p(H|y, \alpha_1, \ldots, \alpha_T)$ and $p(\alpha_1, \ldots, \alpha_T|y, \beta, h, H)$. The first three of these posterior conditional distributions can be dealt with by using

[8]The case where T_t contains unknown parameters would involve drawing from $p(H, T_1, \ldots, T_T|y, \alpha_1, \ldots, \alpha_T)$ which can usually be done fairly easily. In the common time-invariant case where $T_1 = \cdots = T_T$, $p(H, T_1, \ldots, T_T|y, \alpha_1, \ldots, \alpha_T)$ will have precisely the form of a SUR model.

results from previous chapters. In particular, from Chapter 4 (Section 4.2.2) we find

$$\beta|y, h, \alpha_1, \ldots, \alpha_T \sim N(\overline{\beta}, \overline{V}) \tag{8.29}$$

and

$$h|y, \beta, \alpha_1, \ldots, \alpha_T \sim G(\overline{s}^{-2}, \overline{\nu}) \tag{8.30}$$

where

$$\overline{V} = \left(\underline{V}^{-1} + h \sum_{t=1}^{T} X_t' X_t \right)^{-1} \tag{8.31}$$

$$\overline{\beta} = \overline{V} \left(\underline{V}^{-1} \underline{\beta} + h \sum_{t=1}^{T} X_t'(y_t - Z_t \alpha_t) \right) \tag{8.32}$$

$$\overline{\nu} = T + \underline{\nu} \tag{8.33}$$

and

$$\overline{s}^2 = \frac{\sum_{t=1}^{T} (y_t - X_t \beta - Z_t \alpha_t)^2 + \underline{\nu}\underline{s}^2}{\overline{\nu}} \tag{8.34}$$

Using results for the SUR model (with no explanatory variables) from Chapter 6 (Section 6.6.3) we obtain

$$H|y, \alpha_1, \ldots, \alpha_T \sim W(\overline{\nu}_H, \overline{H}) \tag{8.35}$$

where

$$\overline{\nu}_H = T + \underline{\nu}_H \tag{8.36}$$

and

$$\overline{H} = \left[\underline{H}^{-1} + \sum_{t=0}^{T-1} (\alpha_{t+1} - T_t \alpha_t)(\alpha_{t+1} - T_t \alpha_t)' \right]^{-1} \tag{8.37}$$

To complete our Gibbs sampler, we need to derive $p(\alpha_1, \ldots, \alpha_T|y, \beta, h, H)$ and a means of drawing from it. Although it is not hard to write out this multivariate Normal distribution, it can be hard to draw from it in practice since it is T-dimensional, and its elements can be highly correlated with one another. Accordingly, there have been many statistical papers which seek to find efficient ways of drawing from this distribution (Carter and Kohn (1994) and DeJong and Shephard (1995) are two prominent contributions to this literature). Here we present the method described in DeJong and Shephard (1995), which has been found to work very well in many applications. The reader interested in proofs and derivations can look at this paper. DeJong and Shephard (1995) works with a slightly more general version of our state space model, written as

$$y_t = X_t\beta + Z_t\alpha_t + G_t v_t \tag{8.38}$$

and

$$\alpha_{t+1} = T_t\alpha_t + J_t v_t \tag{8.39}$$

for $t = 1, \ldots, T$ in (8.38) and $t = 0, \ldots, T$ in (8.39) and $\alpha_0 = 0$. v_t is i.i.d. $N(0, h^{-1}I_{p+1})$. Other variables and parameters are as defined for our state space model. It can be see that our state space model is equivalent to the one given in (8.38) and (8.39) if we set

$$v_t = \begin{pmatrix} \varepsilon_t \\ u_t \end{pmatrix}$$

G_t to be a $(p + 1)$ row vector given by

$$G_t = (\ 1 \quad 0 \quad . \quad . \quad 0\ )$$

and J_t to be a $p \times (p + 1)$ matrix given by

$$J_t[\ 0_p \quad A\ ]$$

where A is a $p \times p$ matrix implicitly defined by

$$H^{-1} = \frac{1}{h}AA'$$

Since the Gibbs sampler involves drawing from $p(\alpha_1, \ldots, \alpha_T|y, \beta, h, H)$, everything in (8.38) and (8.39) except for α_t and v_t can be treated as known. The contribution of DeJong and Shephard (1995)[9] was to develop an efficient algorithm for drawing from $\eta_t = F_t v_t$ for various choices of F_t. Draws from η_t can then be transformed into draws from α_t. We set out their algorithm for arbitrary F_t, but note that the usual choice is to set $F_t = J_t$, as this yields draws from the state equation errors which can be directly transformed into the required draws from α_t.

DeJong and Shephard (1995) refer to their algorithm as the *simulation smoother*. The simulation smoother begins by setting $a_1 = 0$, $P_1 = J_0J_0'$ and calculating for $t = 1, \ldots, T$ the quantities:[10]

$$e_t = y_t - X_t\beta - Z_t a_t \tag{8.40}$$

$$D_t = Z_t P_t Z_t' + G_t G_t' \tag{8.41}$$

$$K_t = (T_t P_t Z_t' + J_t G_t')D_t^{-1}, \tag{8.42}$$

$$a_{t+1} = T_t a_t + K_t e_t \tag{8.43}$$

[9]There are other advantages of the algorithm proposed by DeJong and Shephard (1995) involving computer storage requirements and avoiding certain degeneracies which will not be discussed here.

[10]For readers with some knowledge of the state space literature, these calculations are referred to as running the *Kalman filter*.

and

$$P_{t+1} = T_t P_t (T_t - K_t Z_t)' + J_t (J_t - K_t G_t)' \tag{8.44}$$

and storing the quantities e_t, D_t and K_t. Then a new set of quantities are calculated in reverse time order (i.e. $t = T, T-1, \ldots, 1$). These begin by setting $r_T = 0$ and $U_T = 0$, and then calculating

$$C_t = F_t(I - G_t' D_t^{-1} G_t - [J_t - K_t G_t]' U_t [J_t - K_t G_t]) F_t' \tag{8.45}$$

$$\xi_t \sim N(0, h^{-1} C_t) \tag{8.46}$$

$$V_t = F_t(G_t' D_t^{-1} Z_t + [J_t - K_t G_t]' U_t [T_t - K_t Z_t]), \tag{8.47}$$

$$r_{t-1} = Z_t' D_t^{-1} e_t + (T_t - K_t Z_t)' r_t - V_t' C_t^{-1} \xi_t \tag{8.48}$$

$$U_{t-1} = Z_t' D_t^{-1} Z_t + (T_t - K_t Z_t)' U_t (T_t - K_t Z_t) + V_t' C_t^{-1} V_t \tag{8.49}$$

and

$$\eta_t = F_t(G_t' D_t^{-1} e_t + [J_t - K_t G_t]' r_t) + \xi_t \tag{8.50}$$

where $G_0 = 0$. This algorithm will yield $\eta = (\eta_0, \ldots, \eta_T)'$, and it can be proved that this is a random draw from $p(\eta|y, \beta, h, H)$. Depending on the form for F_t, this can be transformed into the required random draw of α_t to $t = 1, \ldots, T$. For the common choice of $F_t = J_t$, this algorithm provides draws from the errors in the state equation (i.e. $\eta_t = J_t v_t$) which can be transformed into draws from α_t using (8.39) and the fact that $\alpha_0 = 0$.

These formulae may look complicated. However, the algorithm is simply a series of a calculations involving matrices that are of low dimension plus random draws from the Normal distribution to get ξ_t. This greatly speeds up computation since manipulating high dimensional (e.g. $T \times T$) matrices is very slow indeed. Furthermore, for most applications the matrices F_t, G_t, J_t and T_t will have simple forms, and thus the previous equations will simplify. Thus, with a bit of care, programming up this component of the Gibbs sampler is a straightforward task.

In summary, a Gibbs sampler with data augmentation which sequentially draws from $p(\beta|y, \alpha_1, \ldots, \alpha_T)$, $p(h|y, \alpha_1, \ldots, \alpha_T)$, $p(H|y, \alpha_1, \ldots, \alpha_T)$ and $p(\alpha_1, \ldots, \alpha_T|y, \beta, h, H)$ has been derived using results from previous chapters along with an algorithm developed in DeJong and Shephard (1995). Given output from such a posterior simulator, posterior inference can be carried out as in previous chapters (see Chapter 4, Sections 4.2.3 and 4.2.4). Predictive inference in this model can be carried out using the strategy outlined in Chapter 4, Section 4.2.6. Posterior predictive p-values or HPDIs can be calculated to shed light on the fit and appropriateness of the model. The marginal likelihood for the state space model can be calculated using the method of Chib (see Chapter 7, Section 7.5). The implementation of the Chib method is similar to that described for the individual effects model of Chapter 7 with $\alpha_1, \ldots, \alpha_T$ being treated as latent data.

8.3.2 Empirical Illustration: The State Space Model

To illustrate Bayesian methods in the state space model, we use one of the data sets and some of the models analyzed in Koop and Potter (2001). The data set has been used by economic historians interested in epochs such as the industrial revolution and the Great Depression (e.g. see Greasley and Oxley, 1994). It consists of the annual percentage change in UK industrial production from 1701 to 1992. There are many questions of interest which can be investigated with this data set. In this empirical illustration, we will focus on the question of whether the basic structure of the time series model driving the growth in industrial production is changing over time. To this end we consider an AR(p) model with time varying coefficients:

$$y_t = \alpha_{0t} + \alpha_{1t} y_{t-1} + \cdots + \alpha_{pt} y_{t-p} + \varepsilon_t \tag{8.51}$$

where for $i = 0, \ldots, p$

$$\alpha_{it+1} = \alpha_{i,t} + u_{it} \tag{8.52}$$

We assume ε_t to be i.i.d. $N(0, h^{-1})$ and u_{it} to i.i.d. $N(0, \lambda_i h^{-1})$ where ε_t, u_{is} and u_{jr} are independent of one another for all s, t, r, i and j. In words, this is an autoregressive model, but the autoregressive coefficients (and the intercept) may be gradually evolving over time. It can be seen that this model is a special case of the state space model in (8.22) and (8.23) if we exclude X_t, and define

$$\alpha_t = \begin{pmatrix} \alpha_{0t} \\ \alpha_{1t} \\ \cdot \\ \cdot \\ \alpha_{pt} \end{pmatrix}$$

$$u_t = \begin{pmatrix} u_{0t} \\ u_{1t} \\ \cdot \\ \cdot \\ u_{pt} \end{pmatrix}$$

$$Z_t = \begin{pmatrix} 1 & y_{t-1} & \cdot & \cdot & y_{t-p} \end{pmatrix}$$

and set $T_t = I_{p+1}$ and

$$H^{-1} = h^{-1} \begin{bmatrix} \lambda_0 & 0 & 0 & . & 0 \\ 0 & \lambda_1 & . & . & . \\ . & 0 & . & . & . \\ . & . & . & . & 0 \\ 0 & . & . & 0 & \lambda_p \end{bmatrix}$$

We choose $p = 1$ to illustrate results for this model, although our previous work with this data set indicates larger values of p should be used in a serious piece of empirical work. To simplify issues relating to initial conditions, for our dependent variable we use data beginning in 1705. This means that the y_{t-4} term in (8.51) will always be observed.

We use an informative prior for the parameters h and λ_i for $i = 0, \ldots, p$.[11] For h we use the Gamma prior from (8.25) with $\underline{\nu} = 1$ and $\underline{s}^{-2} = 1$. Since the data is measured as a percentage change, the prior for h is centered over a value which implies over 95% of the errors are less than 2%. However it is relatively noninformative, since the prior contains the same information as one data point (i.e. $\underline{\nu} = 1$). Note that if H were not a diagonal matrix we would probably want to use a Wishart prior for it, here we have assumed the state equations to have errors which are uncorrelated with one another and, hence, we only need elicit $p+1$ univariate priors for the λ_is. Thus, the Wishart prior for H given in (8.36) simplifies here to

$$p(\lambda_i^{-1}) = f_G(\lambda_i^{-1}|\underline{\lambda}_i^{-1}, \underline{\nu}_i)$$

for $i = 0, \ldots, p$. We choose the relatively noninformative values of $\underline{\nu}_i = 1$ for all i, but center the prior for λ_i over 1 by setting $\underline{\lambda}_i = 1$. Since $AR(p)$ coefficients tend to be quite small (e.g. in the stationary AR(1) case the coefficient is less than one in absolute value), this prior allows for fairly substantive changes in the coefficients over time. With this prior, the conditional posterior for H given in (8.35) simplifies to

$$p(\lambda_i^{-1}|y, \alpha_1, \ldots, \alpha_T) = f_G(\lambda_i^{-1}|\overline{\lambda}_i^{-1}, \overline{\nu}_i)$$

for $i = 0, \ldots, p$, where

$$\overline{\nu}_i = T + \underline{\nu}_i$$

and

$$\overline{\lambda}_i = \frac{h\sum_{t=0}^{T-1}(\alpha_{i,t+1} - \alpha_{it})(\alpha_{i,t+1} - \alpha_t)' + \underline{\nu}_i\underline{\lambda}_i}{\overline{\nu}_i}$$

Table 8.1 contains posterior results for the state space model using the data and prior discussed above. The Gibbs sampler was run for 21 000 replications, with 1000 burn-in replications discarded and 20 000 replications retained. The last column of Table 8.1 presents Geweke's convergence diagnostic which indicates that convergence of the posterior simulator has been achieved. Posterior means and standard deviations for λ_0 and λ_1 indicate that a substantial amount of parameter variation occurs both in the intercept and the AR(1) coefficient. Thus, in addition to there being a stochastic trend in the growth in industrial

[11]Note that we are not using the natural conjugate prior and, hence, the results relating to noninformative prior which we derived for the local level model do not apply here. The results of Fernandez, Osiewalski and Steel (1997) are relevant, and imply that a proper prior is required for these parameters if we are to obtain a proper posterior.

Table 8.1 Posterior Results for State Space Model

	Mean	Stand. Dev.	Geweke's CD
h	0.17	0.04	0.99
λ_0	0.93	0.16	0.28
λ_1	0.61	0.11	0.64

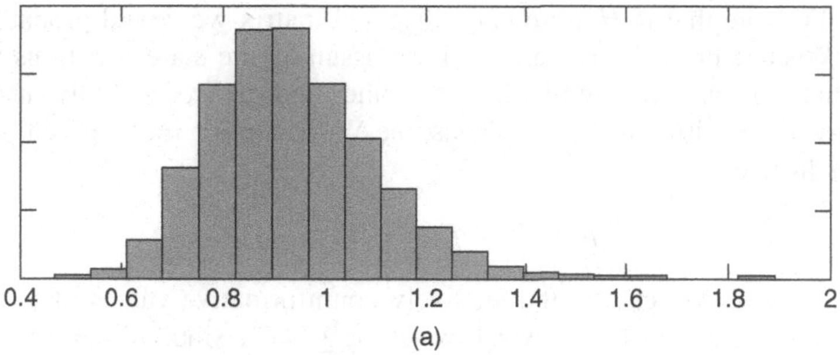

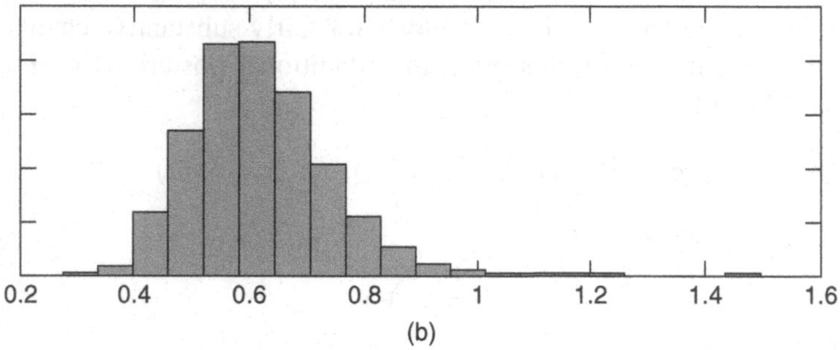

Figure 8.3 (a) Posterior Density for λ_0; (b) Posterior Density for λ_1

production, the AR process itself is changing over time. These findings are supported by an examination of Figures 8.3a and 8.3b which plot the entire posterior densities of each of these parameters.[12]

8.4 EXTENSIONS

The state space model introduced in (8.22) and (8.23) covers a wide range of interesting time series models (e.g. the local linear trend model, time varying parameter models, models with seasonality, etc.). However, there are numerous

[12]We stress that this is only an illustration of Bayesian methods in the state space models and should not necessarily be taken to imply a particular model for industrial production. A serious piece of empirical work involving this time series would involve a consideration of other models (e.g. a model with a structural break).

extensions of this model that can be made in a straightforward manner. Some of these extensions can be handled using methods outlined in this book. For instance, we have discussed the *Normal* state space model, but extending this to the *Student-t* state space model can be done by adding one block to the Gibbs sampler. That is, the Gibbs sampler of Chapter 6, Section 6.4 for handling the linear regression model with Student-t errors can be combined with the Gibbs sampler developed in this chapter.

Other important extensions cannot be directly handled using the methods outlined in this book, but a small amount of additional reading or thought would suffice to create methods for Bayesian inference in these models. Examples of this sort include various nonlinear, regime shifting or structural break models (see Kim and Nelson (1999), Bauwens, Lubrano and Richard (1999) or Chapter 12, Section 12.4.1 for a discussion of some of these models). In finance, a particularly important model, known as the *stochastic volatility model*, is a modest extension of that discussed in this chapter. The stochastic volatility model has time varying error variances of the sort that seem to occur in stock returns and many other financial time series (see Jacquier, Polson and Rossi (1994) for the first Bayesian work on this model). It can be written as

$$y_t = \varepsilon_t$$

where ε_t is i.i.d. $N(0, \sigma_t^2)$

$$\log(\sigma_t^2) = \log(\sigma^2) + \beta \log(\sigma_{t-1}^2) + u_t$$

where u_t is i.i.d. $N(0, 1)$. If we define $\alpha_t \equiv \log(\sigma_t^2)$, then it can be seen that this is a state space model where the state equation relates to the conditional variance of the error as opposed to the conditional mean. A Gibbs sampler which can be used to carry out Bayesian inference in this model is very similar to the one we have developed in this chapter. In fact, the algorithm of DeJong and Shephard (1995) can be directly used to draw from the posterior of the states, α_t, conditional on the other parameters in the model. Hence, all that is required is an algorithm to draw from the other parameters conditional on the states. But this is relatively straightforward (see DeJong and Shephard, 1995, for details).

Perhaps the most important extension of the state space model discussed in this chapter is to allow y_t to be a vector instead of a scalar. After all, economists are usually interested in multivariate relationships between time series. This extension is very easy since (8.22) can be re-interpreted with y_t being a q-vector containing q different time series and very little will change in the development of the posterior simulator. In fact, the DeJong and Shephard (1995) model and algorithm in (8.38)–(8.50) has been deliberately written so as to hold in the multivariate case. Thus, these equations can be used to draw from $p(\alpha_1, \ldots, \alpha_T | y, \beta, h, H)$ in a posterior simulator. $p(H | y, \alpha_1, \ldots, \alpha_T)$ will be similarly unaffected by the move from a univariate to a multivariate state space model. A Gibbs sampler for the multivariate model can be completed by drawing on methods for the SUR model (see Chapter 6, Section 6.6.3) to derive $p(\beta | y, \alpha_1, \ldots, \alpha_T)$ and

$p(h|y, \alpha_1, \ldots, \alpha_T)$. Thus, Bayesian inference in the multivariate state space model involves only minor alterations of the methods of Section 8.3.

As an aside, we should mention that multivariate state space models can be used to investigate the presence of *cointegration* in time series. Cointegration is an important concept in empirical macroeconomics and relates to the number of stochastic trends which are present in a set of time series. To motivate this concept further, consider the following multivariate state space model:

$$y_t = Z_t\alpha_t + \varepsilon_t \tag{8.53}$$

and

$$\alpha_{t+1} = \alpha_t + u_t \tag{8.54}$$

where y_t is a q-vector of q time series and α_t is a p-vector of state equations and H is a diagonal matrix. If $p = q$ and $Z_t = I_q$, then this becomes a multivariate local level model. In this case, each of the q time series contains a stochastic trend. That is, we can write

$$y_{it} = \alpha_{it} + \varepsilon_{it}$$

and

$$\alpha_{i,t+1} = \alpha_{it} + u_{it}$$

for $i = 1, \ldots, q$ and each individual time series follows a local level model. There are q independent stochastic trends driving the q time series.

Consider what happens, however, if $p < q$ and Z_t is a $q \times p$ matrix of constants. In this case, there are p stochastic trends driving the q time series. Since there are fewer stochastic trends than time series, some of the trends must be common to more than one time series. For this reason, if $p < q$ this model is referred to as a *common trends model*. Other ways of expressing this common trend behavior is to say the q time series are trending together or *co-trending* or *cointegrated*.

The macroeconometric literature on cointegration is very voluminous; suffice it to note here that cointegration is thought to hold in many economic time series. That is, many economic time series seem to exhibit stochastic trend behavior. However, economic theory also suggests many time series variables should be related through equilibrium concepts. In practice, these two considerations suggest cointegration should occur. Suppose, for instance, that y_{1t} and y_{2t} are two time series which should be equal to one another in equilibrium. In reality, we expect perturbations and random shocks to imply that equilibrium is rarely, if ever, perfectly achieved. A bivariate local level model with a single stochastic trend would fit with this theoretical expectation. That is, both time series would exhibit stochastic trend behavior. Furthermore, (8.53) and (8.54) with $q = 2$, $p = 1$ and $Z_t = 1$ can be written as

$$y_{1t} = y_{2t} + (\varepsilon_{1t} - \varepsilon_{2t})$$

Hence, $y_{1t} = y_{2t}$ apart from a random equilibrium error $(\varepsilon_{1t} - \varepsilon_{2t})$. Thus, it can be argued that cointegration is how macroeconomic equilibrium concepts should manifest themselves empirically. Economic theories used to justify cointegration include purchasing power parity, the permanent income hypothesis and various theories of money demand and asset pricing.

Cointegration is, thus, a potentially important thing to look for in a wide variety of applications in macroeconomics and finance. In multivariate state space models the number of common trends can be directly investigated by comparing models with different numbers of state equations. For instance, a researcher could calculate the marginal likelihood for the model described in (8.53) and (8.54) for various values of p. If substantial posterior probability is assigned to models with $p < q$, then the researcher could conclude that evidence exists for cointegration.

8.5 SUMMARY

In this chapter, we have introduced a fairly general state space model along with an interesting special case referred to as the local level model. State space models are commonly used when working with time series data and are suitable for modeling a wide variety of behaviors (e.g. trending, cycling or seasonal). State space models are especially attractive for the Bayesian since they can be interpreted as flexible models with hierarchical priors. Thus, the interpretation and computational methods are similar to those for other models such as the individual effects or random coefficients panel data models.

For the local level model, we used a natural conjugate prior and showed how this allowed for analytical results. We introduced a new method for prior elicitation referred to as empirical Bayes. This method, which is especially popular in models involving hierarchical priors, chooses as values for prior hyperparameters those which maximize the marginal likelihood. Such an approach allows the researcher to avoid subjective elicitation of prior hyperparameters or using a noninformative prior. An empirical illustration involving artificial data showed how the empirical Bayes approach could be implemented in practice.

For the more general state space model, we used an independent Normal-Gamma prior and showed how this, along with a hierarchical prior defined by the state equation, meant that a posterior simulator was required. Such a posterior simulator was developed by combining results for the Normal linear regression model, and the SUR model along with a method developed in DeJong and Shephard (1995) for drawing from the states (i.e. α_t for $t = 1, \ldots, T$). The state space model thus provides a good illustration of the modular nature of Bayesian computation where model extensions often simply involve adding a new block to a Gibbs sampler. An application of interest to economic historians, involving a long time series of industrial production and an AR(p) model with time varying coefficients, was used to illustrate Bayesian inference in the state space model.

The chapter ended with a discussion of several possible extensions to the state space models considered in this chapter. Of particular importance were the stochastic volatility model and multivariate state space models. The former of these extensions is commonly used with financial time series while the latter can be used in macroeconomic applications to investigate cointegration and related issues. We stressed how Bayesian analysis of these extensions can be implemented through minor modifications of the posterior simulator described in Section 8.3.

Time series econometrics is such a huge field that a single chapter such as the present one necessarily skips over many important issues. Chapter 12 (Section 12.4.1) provides a brief discussion of some additional time series topics.

8.6 EXERCISES

The exercises in this chapter are closer to being small projects than standard textbook questions. Remember that some data sets and MATLAB programs are available on the website associated with this book.

1. (a) Use the derivations in Section 8.2 and Chapter 6, Section 6.4 to obtain a posterior simulator for the local level model with independent Student-t errors. You may use whatever prior you wish for the model parameters (although the natural conjugate one of Section 6.4 will be the easiest).
   (b) Write a program which uses your result from part (a) and test the program on either real data (e.g. the industrial production data from the empirical illustration in Section 8.3.2) or artificial data generated according to various data generating processes.
   (c) Add to your program code for calculating the marginal likelihood for the local level model with independent Student-t errors. For your data set(s) calculate the Bayes factor comparing the local level model with Normal errors to the local level model with Student-t errors.
2. *Unit root testing with the local level model.* To do this exercise, use either real data (e.g. the industrial production data from the empirical illustration in Section 8.3.2) or artificial data generated according to various data generating processes of your choice. Use the local level model with natural conjugate Normal-Gamma prior described in Section 8.2 with the second variant on the empirical Bayesian methodology described in Section 8.2.3. That is, treat η as an unknown parameter, choose a prior for η of your choice (e.g. a Gamma or Uniform prior) and obtain $p(\eta|y)$. Remember that a unit root is present in the model $M_1 : \eta > 0$, but is not present in the model $M_2 : \eta = 0$. You want to calculate the Bayes factor comparing M_1 to M_2.
   (a) Derive the formula for the marginal likelihood of M_2.
   (b) Using your result for part (a), write a program for calculating the required Bayes factor and test the program using your data set(s).

(c) Consider an approximate strategy where you calculate the Bayes factor comparing M_1 to $M_2^* : \eta = a$ where a is a very small number. Using the Savage–Dickey density ratio, derive a formula for calculating the Bayes factor comparing M_1 to M_2^*.

(d) For your data set(s) compare the approximate Bayes factor of part (c) to that obtained in part (b) for various values of a (e.g. $a = 0.01$, 0.0001, 0.0000001, etc.). How well does the approximate strategy work?

3. Use the methods of Section 8.3 for the general state space model to answer this question. Use either real data (e.g. the industrial production data from the empirical illustration in Section 8.3.2) or artificial data generated according to various data generating processes of your choice.
   (a) Write a program which carries out posterior simulation in the local level model described in Section 8.3.
   (b) Write a program which carries out posterior simulation in the local linear trend model described in Section 8.3.
   (c) Test the programs you have written in parts (a) and (b) using your data set(s).
   (d) Modify your programs to calculate the marginal likelihood for each model and, hence, calculate the Bayes factor comparing the local level model to local linear trend model and test your program using your data set(s). Remember that meaningful marginal likelihoods can only be calculated with informative priors and, hence, choose informative priors of your choice.
   (e) Carry out a prior sensitivity analysis to investigate which aspects of prior elicitation seem to be most important for model comparison involving the local level and local linear trend models.

9
Qualitative and Limited Dependent Variable Models

9.1 INTRODUCTION

The Normal linear regression model is a very powerful one that can be used with a wide variety of data. However, it requires the distribution of y given X to be Normal. There are many applications where this assumption is unreasonable. In this chapter, we will discuss several different types of data where it is not appropriate to use the Normal linear regression model. However, we will show how minor extensions of the Normal linear regression model can be developed which are appropriate for use with such non-standard data types. The types of non-standard data we have in mind are those where the dependent variable is qualitative or limited in some way. Concrete examples will be provided below. However, to provide some intuition consider a transportation economics application where the researcher is interested in investigating why some people choose to travel to work by car and others choose to travel by public transport. The data available to the researcher is based on a survey where commuters are asked whether they traveled by car or public transport and provide personal characteristics (e.g. distance to work, salary level, etc.). If the researcher tried to construct a regression model, the explanatory variables would be these personal characteristics. However, the dependent variable would be *qualitative*. That is, it would be a dummy variable (i.e. 1 if commuter travelled by car, 0 if commuter travelled by public transport). It would not make sense to assume that such a 0-1 dummy variable (conditional on the explanatory variables) followed a Normal distribution.

As a second example, consider a theoretical model which relates the desired investment level of a firm to its characteristics. The corresponding empirical model would have desired investment being the dependent variable and the firm characteristics being explanatory variables. However, in practice, data on desired investment is rarely available. Instead, actual investment is observed. If negative investment is impossible, then actual investment is only equal to desired investment if the latter is positive. Negative values of desired investment become zero values of actual investment. Hence, if the researcher used actual investment as the dependent variable in a regression, she would be using the wrong dependent variable. In this case, the dependent variable would be *censored*. This is an example of a *limited dependent variable*.

As we shall see, both of these examples imply that there is some underlying latent dependent variable for which it is reasonable to assume Normality. In the first example, this latent data is related to the utility associated with each commuting choice. In the second example, it is desired investment. Unfortunately, in neither example do we perfectly observe the latent data. In the first example, we only observe the choice actually made, not the utility associated with the choice. In the second, we observe the latent variable with censoring. However, interpreting these examples as relating to latent data hints at how Bayesian inference can be carried out. Conditional on the latent data, each of the two examples is a Normal linear regression model and the techniques of previous chapters can be used to carry out posterior simulation. If we can derive the posterior of the latent data conditional on the actual data and model parameters, then a Gibbs sampler with data augmentation can be used to carry out Bayesian inference. This is how we proceed in the models considered in this chapter.

In the next section of this chapter, we begin with a more formal statement of the strategy outlined in the preceding paragraph. We then show how this general strategy can be implemented in three models known as *tobit, probit* and *ordered probit*. We then consider the case where the dependent variable is multivariate and focus on a model known as *multinomial probit*. Bayesian analysis of the tobit, probit and ordered probit models involves combining methods for the Normal linear regression model with a model linking latent to observed data. Bayesian analysis of the multinomial probit model is the same except that the Normal linear regression model component is replaced with the seemingly unrelated regressions (SUR) model of Chapter 6 (Section 6.6).

In this chapter, we keep the Normal linear regression as a key component of all our models. This requires us to focus on models in the tobit and probit classes. However, it is worth mentioning that there are many other models that are used for working with qualitative and limited dependent variables which are not as closely related to the Normal linear regression model. Some of these are mentioned in the course of this chapter. For the reader interested in delving deeper into Bayesian analysis of qualitative and limited dependent variable models, a few key references include: Chib (1992) for tobit; Albert and Chib (1993) for many extensions including ordered probit; Geweke, Keane and Runkle (1997), McCulloch and Rossi (1994), McCulloch, Polson and Rossi (2000) and Nobile (2000) for multinomial probit. Multinomial probit models with panel data are also commonly used in the field of marketing. Allenby and Rossi (1999) provides a good introduction to this literature.

9.2 OVERVIEW: UNIVARIATE MODELS FOR QUALITATIVE AND LIMITED DEPENDENT VARIABLES

We will take as our starting point the Normal linear regression model discussed in Chapters 3 and 4. We make one minor change in notation and let $y^* = (y_1^*, \dots, y_N^*)'$ denote the dependent variable. Thus we can write the model as

$$y_i^* = x_i'\beta + \varepsilon_i \tag{9.1}$$

where $x_i = (1, x_{i2}, \dots, x_{ik})'$ or, in matrix notation,

$$y^* = X\beta + \varepsilon \tag{9.2}$$

As in Chapters 3 and 4, we assume

1. ε has a multivariate Normal distribution with mean 0_N and covariance matrix $h^{-1}I_N$;
2. all elements of X are either fixed (i.e. not random variables) or, if they are random variables, they are independent of all elements of ε with a probability density function, $p(X|\lambda)$, where λ is a vector of parameters that does not include β and h.

If y^* were observed, then analysis would proceed as in Chapters 3 or 4. For instance, if a Normal-Gamma natural conjugate prior for β and h were used (or its noninformative limiting case) then analytical posterior results would be obtained from (3.10)–(3.13) in Chapter 3. The marginal likelihood necessary for model comparison is given in (3.34). If an independent Normal-Gamma prior were used for β and h, then the Gibbs sampler which sequentially draws from $p(\beta|y^*, h)$ and $p(h|y^*, \beta)$ could be used to carry out posterior inference. These two densities are specified in (4.6) and (4.8), respectively, of Chapter 4.

In this chapter, we assume that y^* is unobserved, latent data which is related to the actual, observed data, y in some manner. The following sections give some important examples. For the methods described here to work, the relationship between y^* and y should be such that $p(\beta, h|y^*, y) = p(\beta, h|y^*)$ (when working with the natural conjugate prior of Chapter 3) or $p(\beta|y^*, y, h) = p(\beta|y^*, h)$ and $p(h|y^*, y, \beta) = p(h|y^*, \beta)$ (when working with the independent Normal-Gamma prior of Chapter 4). Intuitively, what these conditions are saying is "if you knew y^*, there would be no added information provided by knowing y as well". As we shall see, this holds for all the models in this chapter (and for many others not discussed in this book).

If the relationship between y and y^* satisfies the condition outlined in the preceding paragraph, then a Gibbs sampler with data augmentation can be used to carry out Bayesian inference. In the case of the natural conjugate prior (or its noninformative variant), the posterior simulator sequentially draws from $p(\beta, h|y^*)$ and $p(y^*|y, \beta, h)$. In the case of the independent Normal-Gamma prior, the posterior simulator sequentially draws from $p(\beta|y^*, h)$, $p(h|y^*, \beta)$ and $p(y^*|y, \beta, h)$.

In both cases, the only new derivation that is required is $p(y^*|y, \beta, h)$. That is, $p(\beta, h|y^*)$ is exactly as described in Chapter 3, Section 3.5 and $p(\beta|y^*, h)$ and $p(h|y^*, \beta)$ are exactly as described in Chapter 4, Section 4.2. Accordingly, in the following discussion we will not explicitly write out these latter posterior conditional densities, and focus on $p(y^*|y, \beta, h)$.

9.3 THE TOBIT MODEL

The tobit model is a simple example of a model which can be used when the data is censored. The example given at the beginning of this chapter, with actual investment being a censored observation of desired investment, is a leading case where such censoring occurs. This means that the relationship between y and y^* takes the form

$$\begin{aligned} y_i &= y_i^* && \text{if } y_i^* > 0 \\ y_i &= 0 && \text{if } y_i^* \leq 0 \end{aligned} \tag{9.3}$$

It can immediately be seen that if y^* were known then y would also be known. Hence, $p(\beta, h|y^*) = p(\beta, h|y, y^*)$ and, as described in the preceding section, we can use results from either Chapter 3 or Chapter 4 to take draws from the posterior of the parameters, conditional on y^*. Thus, we need only derive $p(y^*|y, \beta, h)$ and a method for drawing from it to complete the specification of a Gibbs sampler for carrying out posterior inference in the tobit model.

The posterior for the latent data, conditional on the parameters of the model, can be derived in a straightforward fashion. Note first that we have assumed the errors to be independent of one another and, hence, the latent data will exhibit the same property. Thus we can write

$$p(y^*|y, \beta, h) = \prod_{i=1}^{N} p(y_i^*|y_i, \beta, h)$$

and focus on $p(y_i^*|y_i, \beta, h)$. Two cases must be considered: $y_i > 0$ and $y_i = 0$. The first of these is simple: if $y_i > 0$ then we have $y_i^* = y_i$. Formally, if $y_i > 0$, then the conditional posterior for y_i^* is a degenerate density with all probability located at the point $y_i^* = y_i$. The second case can be dealt with by combining (9.1) (i.e. that, unconditionally, y_i^* is Normal) with the fact that $y_i = 0$ implies $y_i^* \leq 0$. That is, y_i^* has a truncated Normal distribution if $y_i = 0$. Formally, we can write $p(y_i^*|y_i, \beta, h)$ as

$$\begin{aligned} y_i^* &= y_i && \text{if } y_i > 0 \\ y_i^*|y_i, \beta, h &\sim N(x_i'\beta, h^{-1})1(y_i^* < 0) && \text{if } y_i = 0 \end{aligned} \tag{9.4}$$

where $1(y_i^* < 0)$ is the indicator function which equals 1 if $y_i^* < 0$, and equals zero otherwise.

Thus, posterior inference in the tobit model can be carried out using a Gibbs sampler which combines results from previous chapters with (9.4). All the tools developed in previous chapters for model comparison and prediction when an MCMC algorithm is used apply here. For instance, the Savage–Dickey density ratio can be used to calculate Bayes factors comparing various models (see Chapter 4, Section 4.2.5). Alternatively, the marginal likelihood for the tobit model can be calculated using the Chib method (see Chapter 7, Section 7.5). Predictive inference can be implemented as described in Chapter 4, Section 4.2.6. Of course, as with any Gibbs sampling algorithm, the MCMC diagnostics of Chapter 4, Section 4.2.4, should be used to confirm convergence and provide insight into the degree of approximation error.

In this section, we have considered the case where the censoring of the dependent variable occurs at zero. It is worth mentioning that allowing censoring to occur at some known value, c, is a trivial extension of the present model. Similarly, allowing for upper and lower known censoring points can be handled. All such extensions merely changing the truncation point(s) in (9.4). The case where censoring is known to occur at some value c, but c is an unknown parameter, is a somewhat more substantive extension of the present model. Nevertheless, it can be handled by adding an extra block to the Gibbs sampler (see Exercise 1). As an example of an application which involves an extension of a tobit model, the interested reader is referred to Li (1999).

9.3.1 Empirical Illustration: The Tobit Model

We use artificial data to illustrate Bayesian inference in the tobit model. We organize the illustration around the question of how important the treatment of censoring is. If censoring occurs, the Normality assumption of the Normal linear regression model is going to be violated. However, if this violation causes only minor problems the researcher might be tempted to stick with the familiar and simpler Normal regression model, rather than move to the more complicated tobit model. We shall see that the importance of the tobit model increases with the degree of censoring.

The artificial data sets are generated by independently drawing x_i from the $U(a, 1)$ distribution and ε_i from the $N(0, 0.25)$ distribution for $i = 1, \dots, 100$. We then set

$$y_i^* = 2x_i + \varepsilon_i$$

and use (9.3) to construct y_i. To investigate the effect of the degree of censoring, we generated four artificial data sets using different values for a. In addition to the tobit model, we also present results for the Normal linear regression model with natural conjugate prior. For both models, we choose noninformative prior hyperparameter values. For the Normal linear regression model, posterior inference can be carried out analytically as described in Chapter 3 (see (3.9)–(3.13) with $\underline{\nu} = 0$ and $\underline{V}^{-1} = 0_{k \times k}$). For the tobit model, we use the Gibbs sampler

Table 9.1 Posterior Results for Slope Coefficient in the Tobit and Normal Linear Regression Models

Data Set		Tobit Model		Normal Linear Regression Model	
a	Prop. of Censored Observations	Mean	Stand. Dev.	Mean	Stand. Dev.
−1	0.52	2.02	0.19	0.95	0.08
−0.5	0.24	1.96	0.15	1.47	0.10
0	0.09	2.09	0.19	1.95	0.17
0.5	0.00	1.97	0.37	1.97	0.37

with data augmentation described above and set $S_0 = 1000$ and $S_1 = 10\,000$. MCMC diagnostics confirm this is an adequate number of burn-in and included replications to ensure convergence and accurate results.

For both models we include an intercept and the artificially generated explanatory variable. Table 9.1 presents the posterior mean and standard deviation of the slope coefficient (i.e. the second element of β) along with the proportion of observations which are censored in each data set. It can be seen that, by increasing a, fewer observations are censored. However, regardless of the degree of censoring, posterior results using the tobit model are quite accurate in that posterior means of the slope coefficient are quite close to 2, the true value used to generate the data sets. However, for the first two data sets, where 52% and 24% of the observations are censored, posterior results from the Normal linear regression model are far different from what they should be. Clearly, there are large benefits to using the tobit model when the degree of censoring is high. However, if the degree of censoring is low, the Normal linear regression model is very similar to the tobit model. The last row to Table 9.1 shows that, if no censoring occurs, the tobit and Normal linear regression models are identical to one another.

9.4 THE PROBIT MODEL

The probit model is commonly used when the dependent variable is a qualitative one indicating an outcome in one of two categories (e.g. an individual either commutes by car or takes public transport). It is usually motivated as arising when an individual is making a choice. Here we use such motivation, but stress that the probit model can be used in other contexts where the dependent variable is a 0-1 dummy.

Assume that an individual has to make a choice between two alternatives. An economist would formalize such a situation by specifying a utility function. Let U_{ji} be the utility that individual i (for $i = 1, \ldots, N$) associates with choice j (for $j = 0, 1$). The individual makes choice 1 if $U_{1i} \geq U_{0i}$ and makes choice

0 otherwise. Thus, the choice depends upon the difference in utilities across the two alternatives and we define this difference as

$$y_i^* = U_{1i} - U_{0i}$$

The probit model assumes that this difference in utilities follows the Normal linear regression model given in (9.1) or (9.2). That is, the individual's utility difference depends on observed characteristics contained in x_i (e.g. distance to work, salary level, etc.) plus an error which is assumed to be Normally distributed. Because of this random error the probit model, and other similar models, are referred to as *random utility models*.

The econometrician does not observe y_i^* directly, but only the choice actually made by individual i. However, just as with the tobit model, y^* can be treated as latent data and a Gibbs sampler with data augmentation can be used to carry out Bayesian inference. For the reasons discussed in Section 9.2, all we need to derive is $p(y^*|y, \beta, h)$.

For the probit model, the relationship between y and y^* takes the form

$$\begin{aligned} y_i &= 1 \quad \text{if } y_i^* \geq 0 \\ y_i &= 0 \quad \text{if } y_i^* < 0 \end{aligned} \tag{9.5}$$

It can immediately be seen that if y^* were known then y would also be known. Hence, $p(\beta, h|y^*) = p(\beta, h|y, y^*)$ and the posterior for the model parameters (conditional on the latent data and depending on the prior selected) takes one of the forms discussed in Chapters 3 and 4.

The form for $p(y^*|y, \beta, h)$ can be derived in a similar manner as in the tobit model. Independence across individuals implies that

$$p(y^*|y, \beta, h) = \prod_{i=1}^{N} p(y_i^*|y_i, \beta, h)$$

and thus, we can focus on $p(y_i^*|y_i, \beta, h)$. The assumptions of the Normal linear regression model imply that $p(y_i^*|\beta, h)$ is Normal. To obtain $p(y_i^*|y_i, \beta, h)$, we combine this Normality result with information contained in y_i. If $y_i = 1$ we obtain a Normal distribution truncated to the left at 0. If $y_i = 0$ we obtain a Normal distribution truncated to the right at 0. To be precise,

$$\begin{aligned} y_i^*|y_i, \beta, h &\sim N(x_i'\beta, h^{-1})1(y_i^* \geq 0) \quad \text{if } y_i = 1 \\ y_i^*|y_i, \beta, h &\sim N(x_i'\beta, h^{-1})1(y_i^* < 0) \quad \text{if } y_i = 0 \end{aligned} \tag{9.6}$$

Posterior inference in the probit model can be carried out using a Gibbs sampler with data augmentation which sequentially draws from (9.6) and $p(\beta, h|y^*)$. As outlined above, results from Chapter 3 or Chapter 4 can be used to draw from the latter density. Model comparison and prediction can be implemented using the standard tools we have used throughout this book.

In addition to parameter estimates, it is often useful to present information about the choice probabilities. These can be derived from the posterior of the parameters by noting that, for any particular values of the parameters,

$$\begin{aligned}\Pr(y_i = 1|\beta, h) &= \Pr(y_i^* \geq 0|\beta, h) \\ &= \Pr(x_i'\beta + \varepsilon_i \geq 0|\beta, h) = \Pr(\sqrt{h}\varepsilon_i \geq -\sqrt{h}x_i'\beta|\beta, h)\end{aligned} \tag{9.7}$$

Since the errors are assumed to be Normally distributed, the last term in (9.7) is simply one minus the cumulative distribution function of the standard Normal (i.e. $\sqrt{h}\varepsilon_i$ is $N(0, 1)$). If we define $\Phi(a)$ as the cumulative distribution function of the standard Normal distribution, then the probability of choosing alternative 1 is $1 - \Phi(-\sqrt{h}x_i'\beta)$.

The terms in (9.7) are functions of the parameters of the model and, hence, their posterior properties can be calculated using Gibbs sampler output in the standard way. That is, using the notation of Chapter 4, Section 4.2.3, the terms in (9.7) are simply $g(\theta)$ for a particular choice of $g()$.

Equation (9.7) illustrates an *identification problem* that was not present in the tobit model. An identification problem is said to occur if multiple values for the model parameters give rise to the same value for the likelihood function. In the probit model, there are an infinite number of values for β and h which yield exactly the same model. This can be seen by noting that $\Pr(x_i'\beta + \varepsilon_i \geq 0|\beta, h) = \Pr(x_i'c\beta + c\varepsilon_i \geq 0|\beta, h)$ for any positive constant c. Since $c\varepsilon_i$ is a $N(0, c^2h^{-1})$, the model is still the same probit model, but with different coefficients and error precision. An equivalent way of showing this would be to write out the likelihood function (see Exercise 2). It can be seen that the value of the likelihood function is the same when the values $(\beta = \beta_0, h = h_0)$ are plugged in as when $\left(\beta = c\beta_0, h = \frac{h_0}{c^2}\right)$ are plugged in (where β_0 and h_0 are any arbitrary values chosen for β and h). In words, the probit model cannot distinguish β and h separately, but only the product $\beta\sqrt{h}$. For the economist, this should not be surprising as the same property exists with utility functions (e.g. if $U(x)$ is a utility function defined over a set of goods, x, then $cU(x)$ will yield identical consumer preferences across goods).

In the probit model, the standard solution to this problem is to set $h = 1$. We adopt this solution in the empirical illustration. An alternative solution is to set one element of β to a fixed number (e.g. set one of the coefficients to 1). However, this alternative solution requires the researcher to know the sign of the relevant coefficient. That is, setting one coefficient to 1 implies that this explanatory variable will have a positive effect on utility (i.e. high values of this variable will increase the probability of making choice 1). In practice, this kind of sign information is rarely available and, hence, the normalization $h = 1$ is usually preferred.

9.4.1 Empirical Illustration: The Probit Model

To illustrate Bayesian inference in the probit model we generate an artificial data set by independently drawing x_i from the $N(0, 1)$ distribution and ε_i from the

$N(0, 1)$ distribution for $i = 1, \ldots, 100$. We then set

$$y_i^* = 0.5x_i + \varepsilon_i$$

and use (9.5) to construct y_i. We estimate the model with an intercept included, use a noninformative prior for β and impose the identification restriction $h = 1$ throughout. The Gibbs sampler with data augmentation described above is used to provide posterior simulator output. We set $S_0 = 1000$ and $S_1 = 10\,000$ and use MCMC diagnostics to confirm that this is an adequate number of burn-in and included replications to ensure convergence and accurate results.

Table 9.2 presents the posterior mean and standard deviation for the intercept and slope coefficients. These indicate that the posterior is located near the true values of 0 and 0.5 used to generate the data. However, due to the identification problem, it is difficult to interpret these coefficients. For instance, β_2 measures the marginal effect of the explanatory variable on the difference in utility between the two alternatives. But this may be hard for a researcher to interpret. Furthermore, this interpretation is subject to the normalization $h = 1$. Identification could have been imposed by setting h to any other number and β would have been scaled proportionately. Put another way, the fact that $E(\beta_2|y) = 0.42$ is positive tells us that increasing the explanatory variable will tend to increase the utility of choice 1 relative to choice 0. Thus, individuals with high values for the explanatory variable will be more likely to make choice 1. However, the precise value of this estimate (i.e. 0.42) does not directly tell us anything. For this reason, with qualitative choice models it is useful to provide direct evidence on how the explanatory variables influence the probability of choosing a particular alternative. The bottom half of Table 9.2 contains a simple example of how this might be done. Three representative individuals have been chosen with high, average and low values of the explanatory variable. Since x_i is $N(0, 1)$, we have chosen values of 2, 0 and -2, respectively. In empirical work involving real data, such individuals can be chosen either through a consideration of the interpretation of the explanatory variables or based on the sample mean and standard deviation. The posterior mean and standard deviation of the probability of making choice 1 are calculated as described in (9.7). Such choice probabilities are simple to

Table 9.2 Posterior Results for Probit Model

	Posterior Mean	Posterior St. Dev.
β_1	-0.10	0.28
β_2	0.42	0.48
	Probability of Making Choice 1	
Individual with $x_i = 2$	0.73	0.28
Individual with $x_i = 0$	0.46	0.11
Individual with $x_i = -2$	0.27	0.20

interpret, such as "The point estimate indicates that there is a 73% chance that an individual with $x_i = 2$ will choose alternative 1". Note that such statements can also be interpreted as predictive choice probabilities for individuals with $x_i = 2, 0$ or -2.

9.5 THE ORDERED PROBIT MODEL

The probit model allows for only two alternatives (e.g. the commuter chooses between car or public transport). However, many empirical applications involve three or more alternatives (e.g. the commuter chooses between car, public transport or bicycle). In the next section, we introduce the multinomial probit model which allows for many alternatives of a general nature. Before we discuss this general model, we introduce the *ordered probit model*, which is a simple extension of the probit model introduced in the previous section. The ordered probit model allows for many alternatives. However, these alternatives must take a particular form. In particular, they must be ordered. For most applications, such an ordering is not reasonable. However, in some cases there is a logical ordering of the alternatives and, if so, then ordered probit may be a sensible model to use. For instance, in marketing surveys, consumers are often asked for their impression of a product and must choose from alternatives Very Bad, Bad, Indifferent, Good, Very Good. In this case, the five choices have a logical ordering from Very Bad through Very Good. In labor economics, ordered probit has been used for the analysis of on-the-job injuries, with injured workers being graded based on the severity of their injuries (e.g. the data sets contain many categories from severe injury through superficial injury).

To describe the ordered probit model, we must generalize the notation of the previous section. The model can be interpreted as a Normal linear regression model where the dependent variable is latent (as in (9.2)). As with the previous models considered in this chapter, the relationship between y^* and y is crucial. With the ordered probit model, y_i can take on values $\{j = 1, \ldots, J\}$, where J is the number of ordered alternatives, and we have

$$y_i = j \text{ if } \gamma_{j-1} < y_i^* \leq \gamma_j \tag{9.8}$$

where $\gamma = (\gamma_0, \gamma_1, \ldots, \gamma_J)'$ is a vector of parameters with $\gamma_0 \leq \cdots \leq \gamma_J$.

In the same manner as in (9.7), we can use the Normality assumption of the regression model for the latent data to work out the probability of choosing a particular alternative. As with the probit model, identifying restrictions are required. Hence, we impose the identifying restriction $h = 1$ for the rest of this section (and discuss further identification issues below). Thus,

$$\Pr(y_i = j|\beta, \gamma) = \Pr(\gamma_{j-1} < y_i^* \leq \gamma_j|\beta, \gamma) = \tag{9.9}$$
$$\Pr(\gamma_{j-1} < x_i'\beta + \varepsilon_i \leq \gamma_j|\beta, \gamma) = \Pr(\gamma_{j-1} - x_i'\beta < \varepsilon_i \leq \gamma_j - x_i'\beta|\beta, \gamma)$$

Since ε_i is $N(0, 1)$, the choice probabilities relate to the cumulative distribution function of the standard Normal distribution. In particular, using notation defined after (9.7) we have

$$\Pr(y_i = j|\beta, \gamma) = \Phi(\gamma_j - x_i'\beta) - \Phi(\gamma_{j-1} - x_i'\beta) \tag{9.10}$$

Thus, ordered probit calculates choice probabilities for any individual by taking a Normal distribution (which integrates to one) and choosing $\gamma_0, \ldots, \gamma_J$ in such a way as to divide up the probability across choices. Using this intuition, it can be seen that more identifying restrictions are required. Consider, for instance, the case where $J = 3$ and probability must be allocated across three alternatives. Visualize a Normal distribution where you are free to choose the mean (i.e. $x_i'\beta$) and four points in the range of the Normal (i.e. γ_0, γ_1, γ_2 and γ_3). There are many different ways of choosing these parameters so as to yield a given allocation of probabilities across alternatives. Suppose, for instance, that x_i contains only an intercept and you want $\Pr(y_i = 1|\beta, \gamma) = 0.025$, $\Pr(y_i = 2|\beta, \gamma) = 0.95$ and $\Pr(y_i = 3|\beta, \gamma) = 0.025$. This can be achieved by setting $\beta = 0$, $\gamma_0 = -\infty$, $\gamma_1 = -1.96$, $\gamma_2 = 1.96$ and $\gamma_3 = \infty$. However, $\beta = 1$, $\gamma_0 = -\infty$, $\gamma_1 = -0.96$, $\gamma_2 = 2.96$ and $\gamma_3 = \infty$ will also yield the same choice probabilities (as will many other combinations of parameter values). Thus, there is an identification problem. The standard way of solving this problem is to set $\gamma_0 = -\infty$, $\gamma_1 = 0$ and $\gamma_J = \infty$.

An alternative way of seeing the necessity of the identifying restriction is to consider the probit model of the previous section. This is equivalent to the ordered probit model with $J = 2$. It can be verified that (9.8) and (9.9) reduce to their probit equivalents in (9.5) and (9.7) if $\gamma_0 = -\infty$, $\gamma_1 = 0$ and $\gamma_2 = \infty$, which are precisely the identifying restrictions introduced in the preceding paragraph.

In the same manner as for the probit model, y_i^* can be interpreted as utility. Because the alternatives are ordered, it is sensible to model the choice probabilities based on this latent utility as being integrals of sequential regions of the Normal distribution. To see this, consider a marketing example where consumers are asked questions relating to the utility obtained from consuming a product and must choose from alternatives Very Bad, Bad, Indifferent, Good or Very Good. Suppose y_i^*, the utility of consumer i, is such that she says the product is 'Bad'. If the utility of this consumer increases very slightly, then the ordering of the categories means she will now say the product is 'Indifferent' (or stay with the 'Bad' choice). With the ordered probit model, there is no way a slight change in utility can induce the consumer to suddenly say the product is 'Very Good'. We stress that restricting utility in this way only makes sense if the categories are ordered. When the consumer is choosing between several unordered alternatives, then the multinomial probit model (to be discussed shortly) is the appropriate choice.

Bayesian inference in the ordered probit model can be done using a Gibbs sampler with data augmentation which sequentially draws from $p(\beta|y^*, \gamma)$, $p(\gamma|y^*, y, \beta)$ and $p(y^*|y, \beta, \gamma)$. As with the probit model, $p(\beta|y_i^*, \gamma)$ will be a Normal density if a Normal or noninformative prior is used for β (e.g., see Chapter 3, Section 3.5, with $h = 1$ imposed). $p(y_i^*|y_i, \beta, \gamma)$ is a truncated Normal density which is a simple extension of (9.6):

$$y_i^*|y_i = j, \beta, \gamma \sim N(x_i'\beta, 1)1(\gamma_{j-1} < y_i^* \leq \gamma_j) \tag{9.11}$$

The new feature of the ordered probit model is $p(\gamma|y_i^*, y_i, \beta)$. We use a flat, improper prior for each of these parameters (i.e. $p(\gamma_j) \propto c$) below, although other priors can be added with minor modifications. It proves easier to draw from the components of γ one at a time. Remembering that identification requires $\gamma_0 = -\infty$, $\gamma_1 = 0$ and $\gamma_J = \infty$, it is convenient to draw from $p(\gamma_j|y^*, y, \beta, \gamma_{(-j)})$ for $j = 2, \ldots, J-1$. The notation $\gamma_{(-j)}$ denotes γ with γ_j removed (i.e. $\gamma_{(-j)} = (\gamma_0, \ldots, \gamma_{j-1}, \gamma_{j+1}, \ldots, \gamma_J)'$. The density $p(\gamma_j|y^*, y, \beta, \gamma_{(-j)})$ can be derived by combining a few simple facts. First, we are conditioning on $\gamma_{(-j)}$ and, thus, we know γ_j must lie in the interval $[\gamma_{j-1}, \gamma_{j+1}]$. Secondly, we are conditioning on both y and y^* and, thus, can figure out which values of the latent data correspond to which values for the actual data. Thirdly, the conditioning arguments provide no other information about γ_j. These facts imply a Uniform distribution:

$$\gamma_j|y^*, y, \beta, \gamma_{(-j)} \sim U(\overline{\gamma}_{j-1}, \overline{\gamma}_{j+1}) \tag{9.12}$$

for $j = 2, \ldots, J-1$, where

$$\overline{\gamma}_{j-1} = \max\{\max\{y_i^* : y_i = j\}, \gamma_{j-1}\}$$

and

$$\overline{\gamma}_{j+1} = \min\{\min\{y_i^* : y_i = j+1\}, \gamma_{j+1}\}$$

The notation $\max\{y_i^* : y_i = j\}$ denotes the maximum value of the latent data over all individuals who have chosen alternative j. $\min\{y_i^* : y_i = j+1\}$ is defined in an analogous fashion.

In summary, posterior inference in the ordered probit model can be carried out using a Gibbs sampler with data augmentation which sequentially draws from (9.11), (9.12) and $p(\beta|y^*)$. This last density is Normal under a noninformative or Normal prior (see Chapter 3, Section 3.5, with $h = 1$ imposed). Model comparison and prediction can be implemented using the standard tools we have used throughout this book. Depending upon which models are being compared, the Savage–Dickey density ratio (see Chapter 4, Section 4.2.5) or the Chib method (see Chapter 7, Section 7.5) may be of particular use with the ordered probit model. Predictive inference can be implemented as described in Chapter 4, Section 4.2.6. The MCMC diagnostics of Chapter 4, Section 4.2.4, should be used to confirm convergence of the Gibbs sampler and provide insight into the degree of approximation error.

The basic issues faced by the empirical researcher when doing Bayesian analysis of probit and ordered probit are very similar. Hence, we do not provide a separate empirical illustration for the ordered probit model.

9.6 THE MULTINOMIAL PROBIT MODEL

There are many cases where individuals choose between several alternatives, but no logical ordering of the alternatives exists. For the reasons discussed in the previous section, it is not appropriate to use ordered probit in these cases. In this section, we introduce the *multinomial probit* model, which is probably the most commonly-used model when several unordered alternatives are involved.

We slightly modify the setup of the earlier sections by assuming that y_i can take on values $\{j = 0, \ldots, J\}$. That is, there are $J + 1$ alternatives indexed by $\{j = 0, \ldots, J\}$ where $J > 1$. To motivate the multinomial probit model, it is useful to extend the random utility framework of Section 9.4. As discussed in that earlier section, the choice an individual makes does not depend upon the absolute utility associated with an alternative, but on the utility relative to other alternatives. Let U_{ji} be the utility of individual i when she chooses alternative j (for $i = 1, \ldots, N$ and $j = 0, \ldots, J$). All information relating to what choices were actually made is provided in utility differences relative to some base alternative. We choose alternative 0 as this base choice and define the latent utility difference variable as

$$y_{ji}^* = U_{ji} - U_{0i}$$

for $j = 1, \ldots, J$. The multinomial probit model assumes this utility difference follows a Normal linear regression model:

$$y_{ji}^* = x_{ji}'\beta_j + \varepsilon_{ji} \tag{9.13}$$

where x_{ji} is a k_j-vector containing explanatory variables that influence the utility associated with choice j (relative to choice 0), β_j is the corresponding vector of regression coefficients and ε_{ji} is the regression error.

Since (9.13) involves J equations, our posterior simulator will combine results for the Seemingly Unrelated Regressions (SUR) model with a method for drawing the latent utility differences. Hence, it proves useful to put (9.13) in the form of a SUR model (see Chapter 6, Section 6.6). Thus, we stack all equations into vectors/matrices as $y_i^* = (y_{1i}^*, \ldots, y_{Ji}^*)'$, $\varepsilon_i = (\varepsilon_{1i}, \ldots, \varepsilon_{Ji})'$,

$$\beta = \begin{pmatrix} \beta_1 \\ \cdot \\ \cdot \\ \beta_J \end{pmatrix}$$

$$X_i = \begin{pmatrix} x_{1i}' & 0 & \cdot & \cdot & 0 \\ 0 & x_{2i}' & 0 & \cdot & \cdot \\ \cdot & \cdot & \cdot & \cdot & \cdot \\ \cdot & \cdot & \cdot & \cdot & 0 \\ 0 & \cdot & \cdot & 0 & x_{Ji}' \end{pmatrix}$$

and define $k = \sum_{j=1}^{J} k_J$ and write

$$y_i^* = X_i\beta + \varepsilon_i \tag{9.14}$$

If we further define

$$y^* = \begin{pmatrix} y_1^* \\ \cdot \\ \cdot \\ y_N^* \end{pmatrix}$$

$$\varepsilon = \begin{pmatrix} \varepsilon_1 \\ \cdot \\ \cdot \\ \varepsilon_N \end{pmatrix}$$

$$X = \begin{pmatrix} X_1 \\ \cdot \\ \cdot \\ X_N \end{pmatrix}$$

and can write the multinomial probit model (in terms of the latent utility differences) as

$$y^* = X\beta + \varepsilon \tag{9.15}$$

Equation (9.15) is in the form of a SUR model, and we use the standard error assumptions for the SUR model. That is, we assume ε_i to be i.i.d. $N(0, H^{-1})$ for $i = 1, \ldots, N$, where H is the $J \times J$ error precision matrix. An alternative way of expressing this is to say ε is $N(0, \Omega)$, where Ω is an $NJ \times NJ$ block-diagonal matrix given by

$$\Omega = \begin{pmatrix} H^{-1} & 0 & \cdot & \cdot & 0 \\ 0 & H^{-1} & \cdot & \cdot & \cdot \\ \cdot & \cdot & \cdot & \cdot & \cdot \\ \cdot & \cdot & \cdot & \cdot & 0 \\ 0 & \cdot & \cdot & 0 & H^{-1} \end{pmatrix} \tag{9.16}$$

The econometrician does not observe y_{ji}^* directly, but rather observes y_i where

$$\begin{aligned} y_i &= 0 \quad \text{if } \max(y_i^*) < 0 \\ y_i &= j \quad \text{if } \max(y_i^*) = y_{ij}^* \geq 0 \end{aligned} \tag{9.17}$$

where $\max(y_i^*)$ is the maximum of the J-vector y_i^*. In words, the individual chooses the alternative which maximizes her utility, and the econometrician observes only this choice.

Remember that the univariate probit model combined the Normal linear regression model with a specification for the latent data, y^*. A Gibbs sampler with data augmentation which sequentially drew from $p(\beta|y^*)$ and $p(y^*|y, \beta)$ was used to carry out Bayesian inference. With the multinomial probit model, we use a similar strategy which combines results from the SUR model with a specification for the latent data, y^*. Thus, a Gibbs sampler can be derived which uses results from Chapter 6, Section 6.6 (to provide $p(\beta|y^*, H)$ and $p(H|y^*, \beta)$) and a truncated multivariate Normal form for $p(y^*|y, \beta, H)$.

Methods for drawing from $p(y^*|y, \beta, H)$ can be developed in a similar manner as for the tobit or probit models. That is, independence across individuals implies that

$$p(y^*|y, \beta, H) = \prod_{i=1}^{N} p(y_i^*|y_i, \beta, H)$$

and, thus, we can focus on $p(y_i^*|y_i, \beta, H)$. Equation (9.14) implies $p(y_i^*|\beta, H)$ is a Normal density. Combining this with the information contained in y_i yields a truncated Normal density. Thus,

$$\begin{aligned} y_i^*|y_i, \beta, H &\sim N(X_i'\beta, H^{-1})1(\max(y_i^*) < 0) && \text{if } y_i = 0 \\ y_i^*|y_i, \beta, H &\sim N(X_i'\beta, H^{-1})1(\max(y_i^*) = y_{ij}^* \geq 0) && \text{if } y_i = j \end{aligned} \quad (9.18)$$

for $j = 1, \ldots, J$.

Econometric analysis of the multinomial probit model (whether Bayesian or frequentist) was held back for many years by computational difficulties relating to the truncated Normal distribution. Bayesian analysis requires taking draws from the truncated multivariate Normal, frequentist analysis required calculating integrals over various regions of the parameter space of the multivariate Normal distribution. Until recently, both of these things were very difficult to do if the number of alternatives was large. However, recent advances in computational algorithms and computer hardware now make it easy to do Bayesian (or frequentist) inference in the multinomial probit model. Early papers which discuss computational issues relating to the multivariate Normal density include Geweke (1991), McCulloch and Rossi (1994) and Geweke, Keane and Runkle (1994). However, the applied Bayesian econometrician can now simply download computer code for drawing from the multivariate Normal subject to linear inequality restrictions from many places. For instance, the website associated with this book has MATLAB code based on Geweke (1991).

Thus, if an independent Normal-Wishart prior is used for β and H, Bayesian inference can be carried out through a posterior simulator which involves sequentially drawing from $p(y^*|y, \beta, H)$ (using (9.18)), $p(\beta|y^*, H)$ (which is Normal, see Chapter 6 (6.50)) and $p(H|y^*, \beta)$ (which is Wishart, see Chapter 6 (6.53)). However, one important new problem arises with the multinomial probit model, and this relates to identification. With the univariate probit model, the restriction $h = 1$ was required to identify this model. This restriction was easy to impose, and actually simplified computation. With the multinomial probit model, imposing the identification restriction is more difficult to do and complicates computation. The reason why the multinomial probit model is unidentified is the same as that presented for the probit model (see Section 9.4). If we define the error covariance matrix as $\Sigma = H^{-1}$ and let σ_{ij} denote the ijth element of Σ, then the standard method of imposing identification is to set $\sigma_{11} = 1$. However, with this restriction imposed $p(H|y^*, \beta)$ no longer has a Wishart distribution, and the results for the SUR model of Chapter 6 can no longer be directly used. Several ways of addressing this problem have been proposed in the literature. For instance, McCulloch and Rossi (1994) simply ignore the identification issue and do not impose $\sigma_{11} = 1$. In this case, $p(H|y^*, \beta)$ is Wishart and computation is straightforward. Instead of presenting empirical results for β, the authors present results for the identified combination of parameters: $\frac{\beta}{\sqrt{\sigma_{11}}}$. However, some researchers feel uncomfortable working with unidentified models. For reasons we will discuss in the empirical illustration, it is hazardous to use noninformative priors, and computational complications can occur (see Nobile (2000) and Hobert and Casella (1996)). Nevertheless, Bayesian analyses of the multinomial probit model, using informative priors, but ignoring the identification restriction are common.

For the researcher who wishes to work only with identified models, McCulloch, Polson and Rossi (2000) offer a useful approach. The underlying regression model in (9.14) assumes that ε_i is $N(0, \Sigma)$. Remember that any joint distribution can be written in terms of a marginal and a conditional distribution. McCulloch, Polson and Rossi (2000) use this insight and partition ε_i as

$$\varepsilon_i = \begin{bmatrix} \varepsilon_{1i} \\ v_i \end{bmatrix}$$

where $v_i = (\varepsilon_{2i}, \ldots, \varepsilon_{Ji})'$. Σ is also partitioned in a conformable manner as

$$\Sigma = \begin{bmatrix} \sigma_{11} & \delta' \\ \delta & \Sigma_v \end{bmatrix} \tag{9.19}$$

The laws of probability imply $p(\varepsilon_i) = p(\varepsilon_{1i})p(v_i|\varepsilon_{1i})$. The properties of the multivariate Normal distribution (see Appendix B, Theorem B.9) can be used to show that $p(\varepsilon_{1i})$ and $p(v_i|\varepsilon_{1i})$ are both Normal. In particular,

$$\varepsilon_{1i} \sim N(0, \sigma_{11}) \tag{9.20}$$

and

$$v_i|\varepsilon_{1i} \sim N\left(\frac{\delta}{\sigma_{11}}\varepsilon_{1i}, \Phi\right) \tag{9.21}$$

where $\Phi = \Sigma_v - \frac{\delta\delta'}{\sigma_{11}}$. Instead of working directly with the $J \times J$ error covariance matrix, Σ, we can work with the parameters σ_{11}, δ and Φ. We can simply set $\sigma_{11} = 1$, put priors over δ and Φ and develop a Gibbs sampler.

It is convenient to assume a Normal prior for δ and a Wishart prior for Φ^{-1}. That is, we take

$$p(\delta, \Phi^{-1}) = p(\delta)p(\Phi^{-1})$$

where

$$p(\delta) = f_N(\delta|\underline{\delta}, \underline{V}_\delta) \tag{9.22}$$

and

$$p(\Phi^{-1}) = f_W(\Phi^{-1}|\underline{\nu}_\Phi, \underline{\Phi}^{-1}) \tag{9.23}$$

With these priors, McCulloch, Polson and Rossi (2000) show that the posterior conditionals necessary for the Gibbs sampler are

$$p(\delta|y^*, \Phi, \beta) = f_N(\delta|\overline{\delta}, \overline{V}_\delta) \tag{9.24}$$

and

$$p(\Phi^{-1}|y^*, \delta, \beta) = f_W(\Phi^{-1}|\overline{\nu}_\Phi, \overline{\Phi}^{-1}) \tag{9.25}$$

The terms in these densities are given by

$$\overline{V}_\delta = \left(\underline{V}_\delta^{-1} + \Phi^{-1}\sum_{i=1}^{N}\varepsilon_{1i}^2\right)^{-1}$$

$$\overline{\delta} = \overline{V}_\delta\left(\underline{V}_\delta^{-1}\underline{\delta} + \Phi^{-1}\sum_{i=1}^{N}v_i\varepsilon_{1i}\right)$$

$$\overline{\Phi}^{-1} = \left[\underline{\Phi} + \sum_{i=1}^{N}(v_i - \varepsilon_{1i}\delta)(v_i - \varepsilon_{1i}\delta)'\right]^{-1}$$

and

$$\overline{\nu}_\Phi = \underline{\nu}_\Phi + N$$

Note that these densities are conditional on y^* and β and, thus, we can use (9.14) and treat $\varepsilon_i = (\varepsilon_{1i}, v_i')'$ as known.

In summary, Bayesian inference in the multinomial probit can be implemented using a Gibbs sampler with data augmentation. If identification issues are ignored, then this Gibbs sampler involves sequentially drawing from (9.15) and from Chapter 6 (6.49) and (6.52). If the identification restriction $\sigma_{11} = 1$ is imposed, then the Gibbs sampler involves sequentially drawing from (9.15), (9.24), (9.25) and from Chapter 6 (6.49). In either case, model comparison and prediction can be implemented using the standard tools we have used throughout this book.

In the empirical example below, these Gibbs sampling algorithms both work very well. However, it should be mentioned that some researchers have found convergence to be quite slow in certain applications. Thus, several Bayesian statisticians have sought to develop more efficient computational algorithms for this and related models. The reader interested in this literature is referred to Liu and Wu (1999), Meng and van Dyk (1999) and van Dyk and Meng (2001).

In this section, we have discussed two different priors for the error covariance matrix in the multinomial probit model. There are several other priors which are commonly used in empirical work (see McCulloch and Rossi, 2000). The multinomial probit model is sometimes criticized for being over-parameterized. That is, if the number of alternatives is large, then Σ contains many parameters. This may lead to inaccurate estimation. This has lead researchers to consider restricted versions of the multinomial model or develop informative priors which impose additional structure on the model. For instance, Allenby and Rossi (1999) restrict Σ to be a diagonal matrix. If such restrictions are sensible, they can greatly reduce the risk of over-parameterization and simplify computation.

9.6.1 Empirical Illustration: The Multinomial Probit Model

In this empirical illustration, we investigate the implications for Bayesian inference of imposing the identifying restriction $\sigma_{11} = 1$. To this end, we estimate the multinomial probit model with and without imposing identification. We use a data set taken from a marketing application. The interested reader is referred to Paap and Franses (2000) and Jain, Vilcassim and Chintagunta (1994) for more details about this application. Here we note only that we use a subset of their data comprising $N = 136$ households in Rome, Georgia. For each household data is available on which of four brands of crackers, Sunshine, Keebler, Nabisco and Private label, was purchased. The data was taken from supermarket optical scanners. Thus, the data set contains four alternatives. 'Private label' is the omitted alternatives (i.e. the latent data, y^*, contains the difference in utility between the other three brands and Private label). For every alternative, we use an intercept and the price of all four brands of crackers in the store at time of purchase as explanatory variables. Thus, $k_1 = k_2 = k_3 = 5$.

For the case where identification is not imposed, we use the independent Normal-Wishart prior with $p(\beta) = f_N(\beta|\underline{\beta}, \underline{V})$ and $p(H) = f_W(H|\underline{\nu}, \underline{H})$ (see Chapter 6 (6.47) and (6.48)). It is not acceptable to use an improper, noninformative, prior.[1] However, we choose the following prior hyperparameter values which are very close to being noninformative: $\underline{\beta} = 0_k$, $\underline{V} = 10\,000 I_k$, $\underline{\nu} = 0.0001$ and $\underline{H} = \frac{1}{\underline{\nu}} I_k$. For the case where identification is imposed, we use the fully noninformative limiting case of the prior $p(\beta) = f_N(\beta|\underline{\beta}, \underline{V})$ combined with (9.22) and (9.23). Thus, $\underline{V}^{-1} = 0_{k \times k}$, $\underline{V}_\delta^{-1} = 0_{(J-1)\times(J-1)}$ and $\underline{\nu}_\Phi = 0$. We should mention, however, that McCulloch, Polson and Rossi (2000) are critical of this 'noninformative' prior. Even though it is noninformative for the parameters δ and Φ, they show it is strongly informative about some aspects of the error

[1] Formally, with an improper prior, the resulting posterior will be improper. To see this, remember that $\frac{\beta}{\sqrt{\sigma_{11}}}$ is identified. β and σ_{11} are not individually identified. Thus, $p\left(y|\beta, \sigma_{11}, \frac{\beta}{\sqrt{\sigma_{11}}} = c\right)$ is simply a constant for any c. The posterior along the line $\frac{\beta}{\sqrt{\sigma_{11}}} = c$ is $p\left(\beta, \sigma_{11}|y, \frac{\beta}{\sqrt{\sigma_{11}}} = c\right) \propto p\left(\beta, \sigma_{11}|\frac{\beta}{\sqrt{\sigma_{11}}} = c\right) p\left(y|\beta, \sigma_{11}, \frac{\beta}{\sqrt{\sigma_{11}}} = c\right)$. If the prior along this line, $p\left(\beta, \sigma_{11}|\frac{\beta}{\sqrt{\sigma_{11}}} = c\right)$, is improper then the posterior will be constant along the line $\frac{\beta}{\sqrt{\sigma_{11}}} = c$. The integral along this line over the interval $(-\infty, \infty)$ is ∞. Thus, the posterior is improper and not a valid p.d.f.

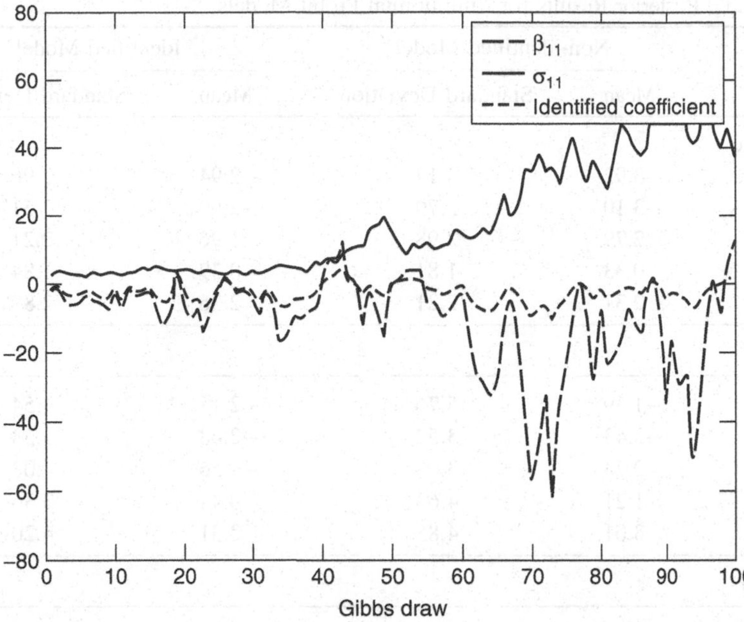

Figure 9.1 Gibbs Draws of Identified and Non-identified Parameters

covariance matrix. In particular, they show it is actually extremely informative about the smallest eigenvalue of Σ.

For the model with non-identified parameters, a Gibbs sampler with data augmentation which sequentially draws from (9.18), (6.49) and (6.52) is used to provide posterior simulator output. For the model with identified parameters, we sequentially draw from (9.18), (6.49), (9.24) and (9.25). For both cases, we use 1000 burn-in replications and 10 000 included replications. MCMC diagnostics indicate the adequacy of these choices.

Before comparing results for the two models, it is instructive to consider the implications of the lack of identification for posterior computation. Figure 9.1 plots 100 of the Gibbs draws of σ_{11}, β_{11} and the identified combination $\frac{\beta_{11}}{\sqrt{\sigma_{11}}}$ (labeled 'Identified coefficient' in Figure 9.1), where β_{11} is the first element of β_1. It can be seen that the draws of σ_{11} and β_{11} wander widely (and their posteriors exhibit enormous variance). However, the draws for the identified coefficient do not wander widely, and always remain close to its posterior mean. Loosely speaking, the likelihood says $\frac{\beta_{11}}{\sqrt{\sigma_{11}}}$ and $\frac{10^{100}\beta_{11}}{10^{100}\sqrt{\sigma_{11}}}$ are equally plausible and, thus, the only thing stopping posterior draws of σ_{11} and β_{11} from wandering off to enormous numbers is the small amount of information contained in the prior. However, the likelihood function provides reasonably precise information about what likely values of $\frac{\beta_{11}}{\sqrt{\sigma_{11}}}$ are. Thus, σ_{11} and β_{11}, although they can wander widely, must wander in a manner which implies $\frac{\beta_{11}}{\sqrt{\sigma_{11}}}$ is fairly well-defined. Figure 9.1

Table 9.3 Posterior Results for Multinomial Probit Models

	Non-identified Model		Identified Model	
	Mean	Standard Deviation	Mean	Standard Deviation
Sunshine				
β_{11}	−3.92	3.17	−2.94	2.96
β_{12}	−3.10	2.79	−2.69	2.51
β_{13}	2.79	2.98	1.96	2.21
β_{14}	0.33	1.81	0.39	1.84
β_{15}	2.31	2.21	2.36	2.80
Keebler				
β_{21}	−1.59	5.76	−2.17	4.54
β_{21}	−2.43	3.52	−2.48	3.33
β_{23}	−2.74	3.98	−1.56	4.03
β_{24}	1.21	4.63	0.44	3.36
β_{25}	3.01	4.85	2.31	4.20
Nabisco				
β_{31}	−2.59	2.43	−2.28	2.59
β_{32}	−0.37	1.02	−0.48	1.27
β_{33}	2.02	2.05	1.73	2.09
β_{34}	−0.33	0.98 502	−0.47	1.36
β_{35}	1.78	1.60	2.21	1.97

also makes it clear why noninformative priors are unacceptable. Without information in the prior, there is nothing stopping draws of β_{11} and σ_{11} going to infinity (and, thus, posterior moments going to infinity).

Because of the identification problem, posterior properties of β and H are hard to interpret in the case where $\sigma_{11} = 1$ is not imposed. Hence, Table 9.3 reports the identified coefficients for both cases (i.e. $\frac{\beta}{\sqrt{\sigma_{11}}}$ for the case where identification is not imposed and β for the case where it is imposed). It can be seen that the identified and non-identified models are leading to very similar posteriors for the identified parameters (at least relative to the very large posterior standard deviations).[2] Remember that β_{1j} for $j = 1, \ldots, 5$ are coefficients which measure how the explanatory variables affect the probability that a consumer will choose the Sunshine brand of crackers. Thus, the point estimate indicates that raising this brand's own price has a negative effect on this probability (i.e. $E(\beta_{12}|y) < 0$). However, raising the price of the other cracker brands (i.e. Keebler, Nabisco or

[2] The fact that the posterior standard deviations are quite large for all parameters is likely due to the very small data set used here. The original data set used in Paap and Franses (2000) and Jain, Vilcassim and Chintagunta (1994) is a panel involving roughly 20 observations per household. Here we are using only one observation per household.

Private label) will increase the probability of choosing the Sunshine brand (i.e. $E(\beta_{13}|y) > 0$, $E(\beta_{14}|y) > 0$ and $E(\beta_{15}|y) > 0$). Similar sensible patterns tend to emerge for the other cracker brands. With some exceptions, raising a brand's own price makes it less likely that brand will be bought, but raising the other brands' prices will make it more likely the brand will be bought.

We could have presented this information in a different way by presenting choice probabilities for several representative cases. For instance, we could have calculated the predictive choice distribution when all prices are set at their sample means, then calculated the predictive choice distribution with the price of one brand set at a higher level. The comparison between these two cases would have shed light on how price effects the probability of choosing a particular brand.

9.7 EXTENSIONS OF THE PROBIT MODELS

There are many extensions of the probit, ordered probit and multinomial probit models which brevity precludes us discussing in any detail. Perhaps the most important of these involve the use of panel data. Marketing is a field where such data is often available. With supermarket scanners, it is possible to observe the choices made during many supermarket visits by many people. Furthermore, individuals vary in their utility functions. This suggests the use of the random coefficients model in Chapter 7 (Section 7.4) for the latent utility differences used with the probit model. For instance, in the case where two alternatives are possible (i.e. a particular product is either purchased or not purchased) and panel data is available, the equation for the Normal linear regression model with latent dependent variable (see (9.1)) can be written as

$$y^*_{it} = x'_{it}\beta_i + \varepsilon_{it} \tag{9.26}$$

The methods developed in Chapter 7, involving a hierarchical prior for β_i to model consumer heterogeneity, can be used to derive $p(\beta_i|y^*)$ for $i = 1, \ldots, N$. The expression for $p(y^*_{it}|y_{it}, \beta_i)$ is a simple extension of (9.6). Thus, a Gibbs sampler with data augmentation for the random coefficients panel probit can be set up in a straightforward fashion (see Exercise 4). Bayesian inference in the random coefficients panel multinomial probit model can be carried out in a similar fashion by combining derivations for the multinomial probit (Section 9.6), random coefficients (Chapter 7, Section 7.4) and SUR (Chapter 6, Section 6.6) models. The reader interested in learning more about Bayesian inference in marketing applications using extensions of multinomial probit models is referred to Allenby and Rossi (1999) and Rossi, McCulloch and Allenby (1996). The panel probit model can also be extended to allow for treatment of time series considerations (e.g. to allow for autocorrelated errors). Geweke, Keane and Runkle (1997) discusses a particular so-called *multinomial multiperiod probit model*.

The fact that Bayesian analysis of panel probit and multinomial probit models involves combining Gibbs sampler blocks from various simpler models illustrates yet again an important message of this book. Bayesian inference is usually implemented using posterior simulators and these often work in terms of blocks of parameters conditional on other blocks. The modular nature of such posterior simulators means combining pieces of different models is particularly easy. Hence, any of the probit (or tobit) models can be combined with any of the models of the previous chapters. So, for instance, a probit (or tobit) model with nonlinear regression function, or with autocorrelated errors, or with heteroskedasticity, etc., can be developed by combining results from the present chapter with results from another chapter of this book. In the next chapter, we will discuss even more extensions of the probit model: mixtures-of-Normals probit and semiparametric probit. All such extensions exploit the modular nature of the Gibbs sampler.

9.8 OTHER EXTENSIONS

In this chapter, we have focused on models where suitably-defined latent data can be assumed to follow a Normal linear regression model. However, there is a myriad of other popular models which cannot be put into this form. Most of these can be categorized as involving a linear regression model with errors following a non-Normal distribution. For instance, if the dependent variable is a count (e.g. the number of patents filed by a firm, the number of deaths following a particular health intervention), then assuming Normal errors is often unreasonable and a regression model involving the Poisson distribution is commonly used. If the dependent variable is a duration (e.g. the number of weeks a spell of unemployment lasts), then it is common to use regression-based models involving the exponential or Weibull distributions. Some of these extensions will be briefly discussed in the last chapter of this book (see Chapter 12, Section 12.4.3).

With regards to qualitative choice data, the class of *logit models* is a popular competitor to the probit class. In fact, there are logit counterparts to every probit model considered in this chapter (e.g. in addition to the logit model, there is the *rank ordered logit model* and the *multinomial logit model*). These models have the same random utility model motivation as do probit models and the only difference is that they assume a *logistic distribution* for the model errors instead of the Normal assumption of probit models. We will not define the logistic distribution here. However, a crucial property of this distribution is that an analytical expression exists for its cumulative distribution function. This property does not hold for the Normal distribution. Thus, until the development of posterior simulators such as that described in Geweke (1991) (see Section 9.6), computational difficulties precluded estimation of the multinomial probit model if the number of alternatives was at all large. The multinomial logit model, with an analytical expression for the c.d.f. which forms a key part of the likelihood function, did not suffer from this problem.

Bayesian inference in variants of the multinomial logit model is well-developed. Bayesian inference in the logit model is described in Zellner and Rossi (1984). The multinomial case, with emphasis on informative prior elicitation and computation, is discussed in Koop and Poirier (1993). Bayesian inference in the rank-ordered logit model is discussed in Koop and Poirier (1994). Poirier (1994) discusses logit models with noninformative priors.

The logit class of models offers an attractive alternative to the probit class and, of course, the usual Bayesian tools (e.g. posterior odds and posterior predictive p-values) can be used to choose between them. However, when there are more than two alternatives, the multinomial logit model has a property which is unattractive in many applications. The choice probabilities implied by the multinomial logit model must satisfy an *Independence of Irrelevant Alternatives* (or IIA) property. This means that the ratio of probabilities of any two choices will be the same, regardless of what the other alternatives are. Suppose, for instance, that the commuter can choose between taking the bus ($y = 1$) or the car ($y = 0$) and the alternatives are equally likely (i.e. $\frac{p(y=0)}{p(y=1)} = 1$). Now suppose that a bicycle lane is constructed so that the commuter can now bicycle to work. The IIA property says that the addition of this alternative does not alter the fact that $\frac{p(y=0)}{p(y=1)} = 1$. In this example, it is possible (although unlikely) that the IIA property is reasonable. Originally we have assumed $p(y = 0) = p(y = 1) = 0.50$. Suppose there is a 20% chance that the commuter will cycle to work following the introduction of the bicycle lane. This is consistent with $p(y = 0) = p(y = 1) = 0.40$, which still implies $\frac{p(y=0)}{p(y=1)} = 1$. However, to illustrate where the IIA property is unreasonable, econometricians tweak the previous example to create the so-called Red Bus-Blue Bus problem. This assumes that the commuter originally has a choice between taking a Red Bus ($y = 1$) or a car ($y = 0$) to work. If a new bus company with blue buses starts operating on the commuter's route ($y = 2$), then it is not reasonable to assume IIA holds. Suppose, for instance, that initially $p(y = 0) = p(y = 1) = 0.50$ and, thus, $\frac{p(y=0)}{p(y=1)} = 1$. Since the Blue Bus is virtually identical to the Red Bus, the introduction of this alternative would likely leave the commuter just as likely to take the car to work and, thus, $p(y = 0) = 0.50$ and $p(y = 1) = p(y = 2) = 0.25$. Hence, the introduction of the new alternative implies $\frac{p(y=0)}{p(y=1)} = 2$. Such changes are a violation of IIA, and are not allowed for by the multinomial logit model.

Several variants of logit models have been developed to surmount this restrictive property of the multinomial logit model. One popular variant is the *nested logit model*, which assumes a nested structure for the decision-making process. For instance, in the Red Bus-Blue Bus example, the econometrician would use a logit model for a commuter who makes a choice between car and public transportation. If the latter choice is made, then a second logit model is used for the commuter's choice between the Red Bus and the Blue Bus. Hence, a logit model is nested within another logit model. Poirier (1996) discusses Bayesian inference in nested logit models.

9.9 SUMMARY

In this chapter, we have introduced several models which can be used when the dependent variable is qualitative or censored. We have focused on cases which can be written in terms of a Normal linear model with a latent dependent variable. Conditional on this latent data, results from Chapters 3 and 4 for the Normal linear regression model apply. Bayesian inference in a wide class of models can be carried out by developing a Gibbs sampler with data augmentation which combines results for the Normal linear regression model (conditional on latent data) with a distribution for the latent data (conditional on the observed data and parameters).

The first model of this form discussed in this chapter was the tobit model. This model involves a dependent variable which is censored. The form of the censoring determines the distribution for the latent data conditional on the observed data and model parameters.

The rest of the models discussed in this chapter fall in the class of probit models. These can be used when the dependent variable is qualitative in nature. Qualitative dependent variables often occur in applications where an individual chooses from a set of alternatives. Thus, they can be motivated in terms of the individual making a utility maximizing choice. The difference in utility between each alternative and a base alternative determines which choice the individual makes. If this unobserved utility difference is the dependent variable in a Normal linear regression model, then a probit model results. This utility difference becomes the latent data in a Gibbs sampler with data augmentation.

This chapter considered several probit models. It began with the standard probit model (which involves two alternatives), before proceeding to the ordered probit model (which involves many alternatives which are ordered) and the multinomial probit model (which involves many unordered alternatives). Model extensions relevant for panel data were discussed. The chapter concluded with a brief discussion of other related models (e.g. in the logit class) which cannot be written in terms of a Normal linear regression model with latent dependent variables.

9.10 EXERCISES

Except for Exercise 2, the exercises in this chapter are closer to being small projects than standard textbook questions. Except where otherwise noted, feel free to use whatever priors you want. It is always useful to test any programs you write using artificial data. Feel free to experiment with different artificial data sets. Remember that some data sets and MATLAB programs are available on the website associated with this book.

1. Consider the tobit model with an unknown censoring point. All model details and assumptions are as in the body of the chapter (see (9.1) and (9.2)), except

that (9.3) is replaced by

$$y_i = y_i^* \quad \text{if } y_i^* > c$$
$$y_i = 0 \quad \text{if } y_i^* \leq c$$

and c is an unknown parameter known to lie in the interval (0, 1).

(a) Assuming a $U(0, 1)$ prior for c, show how the Gibbs sampler with data augmentation described in Section 9.3 can be modified to allow for posterior inference about c.

(b) Write a program which uses your result from part (a) to carry out Bayesian inference in the tobit model with unknown censoring point. Use your program on an artificial data set of the sort described in Section 9.3.1.

Note: $p(c|y^*, \beta, h)$ might not have a convenient form, but c is a scalar and will be defined over a restricted interval so even very crude methods might work well. For instance, an independence chain Metropolis–Hastings algorithm which uses a $U(0, 1)$ candidate generating density would be very simple to program. Investigate whether such an algorithm is efficient. Can you come up with a more efficient algorithm?

2. (a) Derive the likelihood function for the probit model. Hint: Remember that the likelihood function is $\prod_{i=1}^{N} p(y_i|\beta, h)$ and use (9.7).

(b) Verify that the identification problem discussed in Section 9.4 occurs with the probit model.

3. (a) Write a program which carries out posterior simulation for the simple probit model of Section 9.3 (or take the program from the website associated with this book and make sure you understand it) using an informative Normal prior for β. Use artificial data to confirm your program is working.

(b) Extend your program of part (a) to calculate Bayes factors for testing whether individual coefficients are equal to zero using both the Savage–Dickey density ratio (see Chapter 4, Section 4.2.5) and the Chib method (see Chapter 7, Section 7.5). How many Gibbs sampler replications are necessary to ensure that these two methods give the same estimate of the Bayes factor (to two decimal places)?

4. (a) Derive the conditional posterior distributions necessary to set up a Gibbs sampler with data augmentation for the random coefficients panel probit model introduced (9.26).

(b) Program up the Gibbs sampler described in part (a) and test your program using artificial data.

(c) How would your answer to part (a) and program in part (b) change for the random coefficients panel multinomial probit model?

(d) Using artificial data or the full cracker data set (available on the website associated with this book or at the *Journal of Applied Econometrics* data archive, http://qed.econ.queensu.ca/jae/, listed under Paap and Franses, 2000) carry out Bayesian inference in the random coefficients panel probit model.

10 Flexible Models: Nonparametric and Semiparametric Methods

10.1 INTRODUCTION

All the models considered in previous chapters involved making assumptions about functional forms and distributions. For instance, the Normal linear regression model involved the assumptions that the errors were Normally distributed, and the relationship between the dependent and explanatory variables was a linear one. Such assumptions are necessary to provide the likelihood function, which is a crucial component of Bayesian analysis. However, economic theory rarely tells us precisely what functional forms and distributional assumptions we should use. For instance, in a production example economic theory often tells us that a firm's output in increasing in its inputs and eventually diminishing returns to each input will exist. Economic theory will not say "a constant elasticity of substitution production function should be used". In practice, a careful use of the model comparison and fit techniques described in previous chapters (e.g. posterior predictive p-values and posterior odds ratios) can often be used to check whether the assumptions of a particular likelihood function are reasonable. However, in light of worries that likelihood assumptions may be inappropriate and have an effect on empirical results, there is a large and growing non-Bayesian literature on *nonparametric* and *semiparametric methods*.[1] To motivate this terminology, note that likelihood functions depend on parameters and, hence, making particular distributional or functional form assumptions yields a *parametric likelihood function*. The idea underlying the nonparametric literature is to try and get rid of the such parametric assumptions either completely (in the case of nonparametric methods) or partially (in the case of semiparametric methods).[2]

[1] Horowitz (1998) and Pagan and Ullah (1999) provide good introductions to this literature.

[2] This noble goal of "letting the data speak" is often hard to achieve in practice since it is necessary to place some structure on a problem in order to get meaningful empirical results. Nonparametric methods do involve making assumptions, so it is unfair to argue that likelihood-based inference 'makes assumptions' while nonparametric inference 'lets the data speak'. The issue at heart of the

Bayesian inference is always based on a parametric likelihood function and, hence, in a literal sense we should not refer to Bayesian 'nonparametric' or 'semiparametric' methods. This is the reason why the main title to this chapter is 'Flexible Models'. Nevertheless, there are many Bayesian models which are similar in spirit to non-Bayesian nonparametric methods and, thus, there is a large and growing literature which uses the name Bayesian nonparametrics. This field is too large to attempt to survey in a single chapter and, hence, we focus on two sorts of Bayesian nonparametric approaches which are particularly simple and can be done using the methods of previous chapters. The interested reader is referred to Dey, Muller and Sinha (1998) for a broader overview of Bayesian nonparametrics.

To motivate the Bayesian nonparametric approaches discussed here, it is useful to consider the assumptions underlying the Normal linear regression model. The researcher may wish to relax the assumption of a linear relationship (i.e. relax a functional form assumption) or relax the assumption of Normal errors (i.e. relax a distributional assumption). The two approaches described here relate to these two aspects. The section called *Bayesian non- and semiparametric regression* relaxes functional form assumptions, and the section on modeling with mixtures of Normals relaxes distributional assumptions. As we shall see, we can do Bayesian semiparametric regression using only techniques from Chapter 3 on the Normal linear regression model with natural conjugate prior. Modeling with mixtures of Normals can be done using a Gibbs sampler which is an extension of the one introduced in Chapter 6 (Section 6.4) for the regression model with Student-t errors.

10.2 BAYESIAN NON- AND SEMIPARAMETRIC REGRESSION

10.2.1 Overview

In Chapter 5, we discussed the nonlinear regression model

$$y_i = f(X_i, \gamma) + \varepsilon_i$$

where X_i is the ith row of X, $f(\cdot)$ is a known function which depends upon X_i and a vector of parameters, γ. In this section, we begin with a very similar starting point in that we write the nonparametric regression model as

$$y_i = f(X_i) + \varepsilon_i \tag{10.1}$$

but $f(\cdot)$ is an *unknown function*. Throughout this section, we make the standard assumptions that

distinction between nonparametric and likelihood-based methods is what kind of assumptions are made. For instance, a nonlinear regression model makes the assumption "the relationship between y and x takes a specific nonlinear form", whereas a nonparametric regression model makes assumptions relating to the smoothness of the regression line. The question of which sort of assumptions are more sensible can only be answered in the context of a particular empirical application.

1. ε is $N(0_N, h^{-1}I_N)$.
2. All elements of X are either fixed (i.e. not random variables) or, if they are random variables, they are independent of all elements of ε with a probability density function $p(X|\lambda)$, where λ is a vector of parameters that does not include any of the other parameters in the model.

Before discussing nonparametric regression, it is worth mentioning that nonlinear regression methods using extremely flexible choices for $f(X_i, \gamma)$ allow the researcher to achieve a goal similar to the nonparametric econometrician without the need for any new methods. For instance, by using one of the common series expansions (e.g. a Taylor series, Fourier or Muntz–Szatz expansion) one can obtain a parametric form for $f(X_i, \gamma)$ which is sufficiently flexible to approximate any unknown function. The choice of a truncation point in the series expansion allows the researcher to control the accuracy of the approximation.[3]

Nonparametric regression methods hinge on the idea that $f()$ is a smooth function. That is, if X_i and X_j are close to one another, then $f(X_i)$ and $f(X_j)$ should also be close to one another. Nonparametric regression methods, thus, estimate the nonparametric regression line by taking local averages of nearby observations. Many nonparametric regression estimators of $f(X_i)$ have the form

$$\widehat{f}(X_i) = \sum_{j \in N_i} w_j y_j \tag{10.2}$$

where w_j is the weight associated with the jth observation and N_i denotes the neighborhood around X_i. Different approaches vary in how the weights and neighborhood are defined. Unfortunately, if there are many explanatory variables, then nonparametric methods suffer from the so-called *curse of dimensionality*. That is, nonparametric methods average over 'nearby' observations to approximate the regression relationship. For a fixed sample size, as the dimension of X_i increases 'nearby' observations become further and further apart, and nonparametric methods become more and more unreliable. Thus, it is rare to see the nonparametric regression model in (10.1) directly used in applications involving many explanatory variables. Instead, various models which avoid the curse of dimensionality are used. In this section, we discuss two such models, beginning with the *partial linear model*.

10.2.2 The Partial Linear Model

The partial linear model divides the explanatory variables into some which are treated parametrically (z) and some which are treated nonparametrically (x). If x is of low dimension, then the curse of dimensionality can be overcome. The choice of which variables receive a nonparametric treatment is an application-specific one. Usually, x contains the most important variable(s) in the analysis for which it is crucial to correctly measure their marginal effect(s). Here we

[3]Koop, Osiewalski and Steel (1994) is a paper which implements such an approach.

assume x is a scalar, and briefly discuss below how extensions to nonscalar x can be handled.

Formally, the partial linear model is given by

$$y_i = z_i\beta + f(x_i) + \varepsilon_i \tag{10.3}$$

where y_i is the dependent variable, z_i is a vector of k explanatory variables, x_i is a scalar explanatory variable and $f(\cdot)$ is an unknown function. Note that z_i does not contain an intercept, since $f(x_1)$ plays the role of an intercept. We refer to $f()$ as the *nonparametric regression line*.

The basic idea underlying Bayesian estimation of this model is that $f(x_i)$ for $i = 1, \ldots, N$ can be treated as unknown parameters. If this is done, (10.3) is a Normal linear regression model (albeit one with more explanatory variables than observations). A Bayesian analysis of this model using a natural conjugate prior can be done exactly as described in Chapter 3. Thus, it is simple and straightforward to carry out Bayesian inference in the partial linear model.

We begin by ordering observations so that $x_1 \leq x_2 \leq \cdots \leq x_N$. Since the data points are independent of one another, their precise ordering is irrelevant and choosing to order observations in ascending order makes the definition of what a 'nearby' observation is clear. Stack all variables into matrices in the usual way as $y = (y_1, \ldots, y_N)'$, $Z = (z_1,' \ldots, z'_N)'$ and $\varepsilon = (\varepsilon_1, \ldots, \varepsilon_N)'$. If we let $\gamma = (f(x_1), \ldots, f(x_N))'$,

$$W = [Z : I_N]$$

and $\delta = (\beta', \gamma')'$, then we can write (10.3) as

$$y = W\delta + \varepsilon \tag{10.4}$$

Note first that γ is an N-vector containing each point on the nonparametric regression line. At this stage, we have not placed any restrictions on the elements of γ. Hence, we are being nonparametric in the sense that $f(x_i)$ can be anything and $f()$ is a completely unrestricted unknown function. Secondly, (10.4) is simply a regression model with explanatory variables in the $N \times (N + k)$ matrix W. However, (10.4) is an unusual regression model, since there are more unknown elements in δ than there are observations, i.e. $N + k \geq N$. An implication of this is that a perfect fit is available such that the sum of squared errors is zero. For instance, if we had an estimate of δ of the form

$$\widehat{\delta} = \begin{pmatrix} 0_k \\ y \end{pmatrix}$$

then the resulting errors would all be zero. Note that $\widehat{\delta}$ implies the points on the nonparametric regression line are estimated as $\widehat{f(x_i)} = y_i$. Hence, this estimate implies no smoothing at all of the nonparametric regression line. In terms of (10.2), the implied weights are $w_i = 1$ and $w_j = 0$ for $j \neq i$. Such an estimator is unsatisfactory. Prior information can be used to surmount this pathology.

In the nonparametric regression literature, estimators are based on the idea that $f()$ is a smooth function. That is, x_i and x_{i-1} are close to one another, then $f(x_i)$ should also be close to $f(x_{i-1})$. In a Bayesian analysis, such information can be incorporated in a prior. There are many ways of doing this, but here we implement one simple approach discussed in Koop and Poirier (2002). We assume a natural conjugate Normal-Gamma prior for β, γ and h. By adopting such a choice, we are able to obtain simple analytical results which do not require posterior simulation methods. To focus on the nonparametric part of the partial linear model, we assume the standard noninformative prior for h and β:

$$p(\beta, h) \propto h \tag{10.5}$$

For the coefficients in the nonparametric part of the model, we use the partially informative prior (see Chapter 3, Exercise 4) on the first differences of γ:

$$R\delta \sim N(0_{N-1}, h^{-1}V(\eta)) \tag{10.6}$$

where $V(\eta)$ is a positive definite matrix which depends upon a hyperparameter η (which will be explained later), and $R = [0_{(N-1)\times k} : D]$, where D is the $(N-1) \times N$ first-differencing matrix:

$$D = \begin{bmatrix} -1 & 1 & 0 & 0 & \cdots & \cdots & 0 \\ 0 & -1 & 1 & 0 & \cdots & \cdots & 0 \\ \cdots & \cdots & \cdots & \cdots & \cdots & \cdots & \cdots \\ \\ 0 & \cdots & \cdots & 0 & 0 & -1 & 1 \end{bmatrix} \tag{10.7}$$

Note that this structure implies that we only have prior information on $f(x_i) - f(x_{i-1})$. The fact that we expect nearby points on the nonparametric regression line to be similar is embedded in (10.6) through the assumption that $E[f(x_i) - f(x_{i-1})] = 0$. $V(\eta)$ can be used to control the expected magnitude of $f(x_i) - f(x_{i-1})$ and, thus, the degree of smoothness in the nonparametric regression line.

In this discussion of prior information, it is worth mentioning that the researcher sometimes wants to impose inequality restrictions on the unknown function describing the nonparametric regression line. For instance, the researcher may know that $f()$ is a monotonically increasing function. This is simple to do using the techniques described in Chapter 4 (Section 4.3).

Before presenting the posterior for this model, a brief digression on two points is called for. First, the perceptive reader may have noticed that the structure of the partial linear model is almost identical to the local level model of Chapter 8. In fact, if we omit the parametric term (i.e. drop Z) and change the i subscripts in this chapter to t subscripts, then this nonparametric regression model is identical to the state space model. This is not surprising once one recognizes that both models have ordered data and the structure in the state equation of (8.5) is identical to that of the prior given in (10.6). The fact that state space methods can be used to carry out nonparametric regression has been noted in several places (e.g. Durbin and Koopman, 2001). Everything written in Chapter 8 (Section 8.2) is thus relevant here. For instance, empirical Bayes methods can be used as described in Section 8.2.3 if the researcher does not wish to elicit prior hyperparameters such as η. Secondly, the reader with a mathematical training may be bothered by the fact that we have referred to (10.6) as controlling 'the degree of smoothness' in the nonparametric regression line through prior information about first differences. Usually, the degree of smoothness of a function is measured by its second derivative, which would suggest we use prior information about second differences (i.e. $[f(x_{i+1}) - f(x_i)] - [f(x_i) - f(x_{i-1})]$). Prior information about second differences can be incorporated in a trivial fashion by redefining D in (10.7) to be a second-differencing matrix.

It is straightforward to prove (see Chapter 3, Exercise 4), that the posterior for the Normal linear regression model with partially noninformative Normal-Gamma prior is

$$\delta, h|y \sim NG(\widetilde{\delta}, \widetilde{V}, \widetilde{s}^{-2}, \widetilde{\nu}) \tag{10.8}$$

where

$$\widetilde{V} = (R'V(\eta)^{-1}R + W'W)^{-1} \tag{10.9}$$

$$\widetilde{\delta} = \widetilde{V}(W'y) \tag{10.10}$$

$$\widetilde{\nu} = N \tag{10.11}$$

and

$$\widetilde{\nu s^2} = (y - W\widetilde{\delta})'(y - W\widetilde{\delta}) + (R\widetilde{\delta})'V(\eta)^{-1}(R\widetilde{\delta}) \tag{10.12}$$

Furthermore, the posterior is a valid p.d.f., despite the fact that the number of explanatory variables in the regression model is greater than the number of observations. Intuitively, prior information about the degree of smoothness in the nonparametric regression function suffices to correct the perfect fit pathology noted above.

In an empirical study, interest usually centers on the nonparametric part of the model. Using (10.8) and the properties of the multivariate Normal distribution (see Appendix B, Theorem B.9), it follows that

$$E(\gamma|y) = [M_Z + D'V(\eta)^{-1}D]^{-1}M_Z y \tag{10.13}$$

where $M_Z = I_N - Z(Z'Z)^{-1}Z'$. Equation (10.13) can be used as an estimate of $f()$, and we refer to it as the 'fitted nonparametric regression line'. To aid in interpretation, note that M_Z is a matrix which arises commonly in frequentist studies of the linear the regression model. $M_Z y$ are the OLS residuals from the regression of y on Z. Hence, (10.13) can be interpreted as removing the effect of y on Z (i.e. since $M_Z y$ are residuals) and then smoothing the result using the matrix $[M_Z + D'V(\eta)^{-1}D]^{-1}$. Note also that in the purely nonparametric case (i.e. Z does not enter the model), if the prior in (10.6) becomes noninformative (i.e. $V(\eta)^{-1} \rightarrow 0_{N-1,N-1}$), then $E(\gamma|y) = y$ and the nonparametric part of the model merely fits the observed data points (i.e. there is no smoothing).

So far, we have said nothing about $V(\eta)$, and many different choices are possible. A simple choice, reflecting only smoothness considerations (i.e. $f(x_i) - f(x_{i-1})$ is small), would be to take $V(\eta) = \eta I_{N-1}$.[4] This prior depends only upon the scalar hyperparameter η, which can be selected by the researcher to control the degree of smoothness. To provide more intuition on how the Bayesian posterior involves an averaging of nearby observations, it is instructive to look at $E(\gamma_i|y, \gamma^{(i)})$, where $\gamma^{(i)} = (\gamma_1, \ldots, \gamma_{i-1}, \gamma_{i+1}, \ldots, \gamma_N)$. For the pure nonparametric regression case (i.e. where Z does not enter), it can be shown that:

$$E(\gamma_i|y, \gamma^{(i)}) = \frac{1}{2+\eta}(\gamma_{i-1} + \gamma_{i+1}) + \frac{\eta}{2+\eta} y_i$$

for $i = 2, \ldots, N-1$. $E(\gamma_i|y, \gamma^{(i)})$ is a weighted average of y_i and the closest points on the nonparametric regression curve above and below i (i.e. γ_{i-1} and γ_{i+1}). Since η controls the degree of smoothness we wish to impose on $f(\cdot)$, it makes sense that as $\eta \rightarrow \infty$ we obtain $E(\gamma_i|y, \gamma^{(i)}) = y_i$ (i.e. no smoothing whatsoever). As $\eta \rightarrow 0$ we obtain $E(\gamma_i|y, \gamma^{(i)}) = \frac{1}{2}(\gamma_{i-1} + \gamma_{i+1})$. Furthermore, it can be shown that $var(\gamma_i|y, \gamma^{(i)}) = \frac{\sigma^2\eta}{2+\eta}$ which goes to zero as $\eta \rightarrow 0$. Thus, the limiting case of $\eta \rightarrow 0$ yields $\gamma_i = \frac{1}{2}(\gamma_{i-1} + \gamma_{i+1})$, and the nonparametric regression component is merely a straight line.

In summary, Bayesian inference in the partial linear model can be carried out using the familiar Normal linear regression model with natural conjugate prior if we treat the unknown points on the nonparametric regression line as parameters. Despite the fact that the number of explanatory variables in the partial linear model is greater than the number of observations, the posterior is proper. Model comparison and prediction can be done in exactly the same manner as in Chapter 3.

In many cases, the researcher may be willing to choose a particular value for η. Or, as in the following application, empirical Bayes methods as described in Chapter 8 (Section 8.2.3) can be used to estimate η. However, it is worthwhile to briefly mention another method for selecting a value for η in a data-based fashion. This new method, which is commonly used by nonparametric statisticians, is referred to as *cross-validation*. The basic idea of cross-validation is that some of the data is withheld. The model is estimated using the remaining data and used to predict the withheld data. Models are compared on the basis of how well they predict the withheld data.[5] In the present context, we could define a

[4] In small data sets, the distance between x_i and x_{i-1} may be large, and it might be desirable to incorporate this into the prior. A simple way of doing this would be to use a prior involving $V(\eta)$ being a diagonal matrix with (i,i)th elements equal to $v_i = \eta(x_i - x_{i-1})$.

[5] It is worth mentioning that cross-validation can be used as a model comparison/evaluation tool for any model, not just nonparametric ones.

cross-validation function as

$$CV(\eta) = \frac{1}{N}\sum_{i=1}^{N}(y_i - E(y_i|y^{(i)}))^2$$

where $y^{(i)} = (y_1, \ldots, y_{i-1}, y_{i+1}, \ldots, y_N)'$. That is, we delete one observation at a time and calculate the fitted nonparametric regression line using the remaining data. We then use $(y_i - E(y_i|y^{(i)}, \eta))^2$ as a metric of how well the resulting nonparametric regression line fits the left out data point. η is chosen so as to minimize the cross-validation function.

Empirical Illustration: The Partial Linear Model

To illustrate Bayesian inference in the partial linear model, we use an artificial data set using a very nonlinear data generating mechanism. For $i = 1, \ldots, 100$ we generate

$$y_i = x_i \cos(4\pi x_i) + \varepsilon_i \tag{10.14}$$

where ε_i is i.i.d. $N(0, 0.09)$ and x_i is i.i.d. $U(0, 1)$. The data is then re-ordered so that $x_1 \leq x_2 \leq \cdots \leq x_{100}$.

For simplicity, we assume a purely nonparametric model (i.e. do not include Z). The partially informative prior, given in (10.5) and (10.6), requires us to select a value for η. Once a value for η is selected, posterior inference about the nonparametric regression line can be done based on (10.8)–(10.13). Here we use the empirical Bayes methods described in Chapter 8 (Section 8.2.3) to estimate η. As stressed in Section 8.2.3, (very weak) prior information about η, γ_1 or h is required to do empirical Bayes methods in this model. Here we use prior information about η, and assume

$$\eta \sim G(\underline{\mu}_\eta, \underline{\nu}_\eta)$$

and choose nearly noninformative values of $\underline{\nu}_\eta = 0.0001$ and $\underline{\mu}_\eta = 1.0$.

Remember that empirical Bayes estimation involves finding the maximum of the marginal likelihood times $p(\eta)$ (see Chapter 8 (8.21)). With the partially informative prior, the integrating constant is not defined. However, insofar as we are interested in comparing models with different values for η, such integrating constants are irrelevant, since they cancel out in the Bayes factors. These considerations suggest that we should choose the value of η which maximizes

$$p(\eta|y) \propto p(y|\eta)p(\eta) \propto (|\widetilde{V}||R'V(\eta)^{-1}R|)^{\frac{1}{2}}(\widetilde{\nu s^2})^{-\frac{\widetilde{\nu}}{2}} f_G(\eta|\underline{\mu}_\eta, \underline{\nu}_\eta)$$

We do the one-dimensional maximization of $p(\eta|y)$ through a grid search. The reader who finds this brief discussion of implementing empirical Bayes methods confusing is urged to re-read Chapter 8 (Section 8.2.3) for a more thorough explanation.

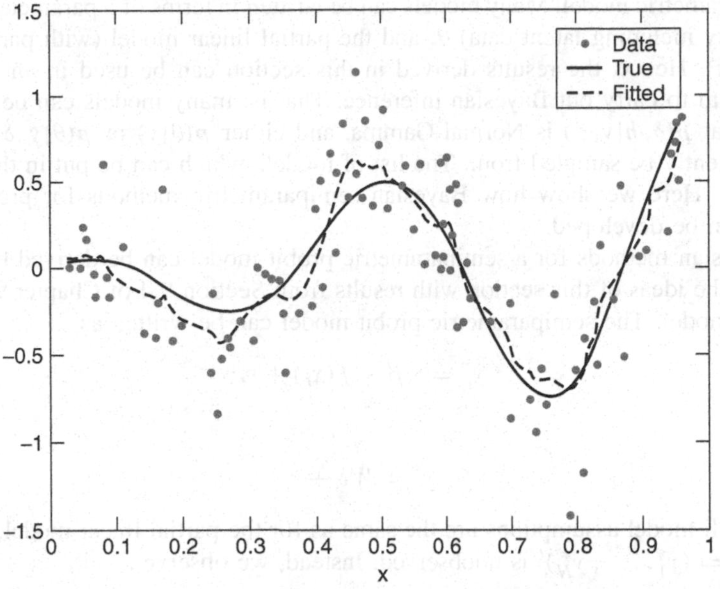

Figure 10.1 True and Fitted Nonparametric Regression Lines

The value of η chosen by the empirical Bayes procedure is 0.1648. Figure 10.1 plots the fitted nonparametric regression line using this choice for η along with the actual data and the true regression line given in (10.14) used to generate the data. It can be seen that the fitted nonparametric regression line tracks the (very nonlinear) shape of the true regression line quite well. If an empirical application requires a smoother curve, then a prior on the second differences, $[f(x_{i+1}) - f(x_i)] - [f(x_i) - f(x_{i-1})]$, can be used.

Since Chapter 3 already contains an empirical illustration using the Normal linear regression model with natural conjugate prior, no further empirical results will be presented here. Of course, all the tools presented there can be used to carry out further posterior inference (e.g. HPDIs can be presented at each point on the nonparametric regression line), model comparison (e.g. Bayes factors comparing this model to a parametric model can be calculated) or prediction. Furthermore, the perceptive reader may have noticed that this figure looks very similar to Figure 8.2, and it is worth stressing yet again that state space and nonparametric regression methods are very similar.

Extensions: Semiparametric Probit and Tobit

In this book we have emphasized the modular nature of Bayesian modeling, especially in terms of posterior simulation. In the present context, the partial linear regression model can serve as one component of a more complicated non- or semiparametric model. Many models can be written in terms of a parameter vector (possibly including latent data) θ, and the partial linear model (with parameters δ and h). Hence, the results derived in this section can be used in an MCMC algorithm to carry out Bayesian inference. That is, many models can be written such that $p(\delta, h|y, \theta)$ is Normal-Gamma, and either $p(\theta|y)$ or $p(\theta|y, \delta, h)$ can conveniently be sampled from. The list of models which can be put in this form is huge. Here we show how Bayesian semiparametric methods for probit and tobit can be developed.

Bayesian methods for a semiparametric probit model can be derived by combining the ideas of this section with results from Section 9.4 of Chapter 9 on the probit model. The semiparametric probit model can be written as

$$y_i^* = z_i\beta + f(x_i) + \varepsilon_i \tag{10.15}$$

or

$$y^* = W\delta + \varepsilon \tag{10.16}$$

where all model assumptions are the same as for the partial linear model, except that $y^* = (y_1^*, \ldots, y_N^*)'$ is unobserved. Instead, we observe

$$\begin{aligned} y_i &= 1 \text{ if } y_i^* \geq 0 \\ y_i &= 0 \text{ if } y_i^* < 0 \end{aligned} \tag{10.17}$$

Bayesian inference for this model proceeds by noting that $p(\delta, h|y^*)$ is precisely that given in (10.8)–(10.13) except that y is replaced by y^* in these formulae. In fact, if we make the usual identifying assumption that $h = 1$, the conditional posterior distribution for δ is simply Normal. Furthermore,

$$p(y^*|y, \delta, h) = \prod_{i=1}^{N} p(y_i^*|y_i, \delta, h)$$

and $p(y_i^*|y_i, \delta, h)$ is truncated Normal (see Chapter 9, Section 9.4). Hence, a simple Gibbs sampler with data augmentation which involves only the Normal and truncated Normal distributions can be used to carry out Bayesian inference.

To be precise, the MCMC algorithm involves sequentially drawing from

$$\delta|y^* \sim N(\widetilde{\delta}, \widetilde{V}) \tag{10.18}$$

and, for $i = 1, \ldots, N$,

$$\begin{aligned} y_i^*|y_i, \delta, \beta &\sim N(z_i\beta + \gamma_i, 1)1(y_i^* \geq 0) \quad \text{if } y_i = 1 \\ y_i^*|y_i, \delta, \beta &\sim N(z_i\beta + \gamma_i, 1)1(y_i^* < 0) \quad \text{otherwise} \end{aligned} \tag{10.19}$$

where $1(A)$ is the indicator function which equals 1 if condition A is true and otherwise equals 0.

Bayesian methods for a semiparametric tobit model can be derived along similar lines to semiparametric probit by combining the techniques for the partial linear model with those for parametric tobit models (see Chapter 9, Section 9.3). Comparable to (10.16) and (10.17), the semiparametric tobit model can be written as

$$y_i^* = z_i\beta + f(x_i) + \varepsilon_i \tag{10.20}$$

or

$$y^* = W\delta + \varepsilon \tag{10.21}$$

where $y^* = (y_1^*, \ldots, y_N^*)'$ is unobserved. In the tobit model, we observe

$$\begin{aligned} y_i &= y_i^* \quad \text{if } y_i^* > 0 \\ y_i &= 0 \quad\;\; \text{if } y_i^* \leq 0 \end{aligned} \tag{10.22}$$

Bayesian inference for this model proceeds by noting that our results for the partial linear model provide us with $p(\delta, h|y^*)$. Furthermore,

$$p(y^*|y, \delta, h) = \prod_{i=1}^{N} p(y_i^*|y_i, \delta, h)$$

and $p(y_i^*|y_i, \delta, h)$ is either simply y_i or truncated Normal. Hence, a simple Gibbs sampler with data augmentation can be used to carry out Bayesian inference. Formally, the MCMC algorithm involves sequentially drawing from

$$\delta, h|y^* \sim NG(\widetilde{\delta}, \widetilde{V}, \widetilde{s}^{-2}, \widetilde{\nu}) \tag{10.23}$$

and, for $i = 1, \ldots, N$,

$$\begin{aligned} y_i^* &= y_i \quad \text{if } y_i > 0 \\ y_i^*|y_i, \delta, \beta, h &\sim N(z_i\beta + \gamma_i, h^{-1})1(y_i^* < 0) \quad \text{if } y_i = 0 \end{aligned} \tag{10.24}$$

Hence, Bayesian semiparametric probit or tobit (as well as many other models) can be carried out in a straightforward fashion using MCMC methods that combine the results for the partial linear model with some other model component.

It is also worth mentioning briefly that there is a myriad of other ways to do Bayesian non- or semiparametric regression. One particular class of model which does much the same thing as nonparametric regression is the class of *spline models*. We do not discuss them here, but refer the interested reader to Green and Silverman (1994), Silverman (1985), Smith and Kohn (1996) or Wahba (1983). There are many other methods for flexible modeling on a regression function which are not discussed in this book. The interested reader is referred to Dey, Muller and Sinha (1998) for a discussion of some of these models and methods.

10.2.3 An Additive Version of the Partial Linear Model

Thus far we have assumed x_i to be a scalar in the partial linear model. In this scalar case, the prior used to impose smoothness on the nonparametric regression line involved simply reordering the observations so that $x_1 \leq \cdots \leq x_N$. As discussed at the beginning of this chapter, when x_i is a vector the curse of dimensionality may preclude sensible nonparametric inference. However, if x_i is of low dimension, then it may be possible to implement Bayesian inference by using a nearest neighbor algorithm to measure the distance between observations. The data can then be reordered according to the distance between observations and the posterior given in (10.8) used to carry out Bayesian inference. For instance, a common definition of the distance between observations i and j is

$$dist_{i,j} = \sum_{l=1}^{p}(x_{il} - x_{jl})^2$$

where $x_i = (x_{i1}, \ldots, x_{ip})'$ is a p-vector. The procedure for ordering the data involves selecting a first observation (e.g. the observation with the minimum value for the first element of x). The second observation is the one which is closest to the first observation. The third observation is the one closest to the second (after deleting the first observation), etc. Once the data have been ordered, the Bayesian procedure described above can be used. However, if p is large (e.g. $p > 3$), then this procedure may work very poorly (and may be sensitive to the choice of first observation and the definition of distance between observations). Accordingly, many variants of the partial linear model have been proposed which place restrictions on $f()$ to break the curse of dimensionality. Here we describe one common model, and develop Bayesian methods for carrying out econometric inference.

The additive version of the partial linear model is given by

$$y_i = z_i\beta + f_1(x_{i1}) + f_2(x_{i2}) + \cdots + f_p(x_{ip}) + \varepsilon_i \tag{10.25}$$

where $f_j(\cdot)$ for $j = 1, \ldots, p$ are unknown functions. In other words, we are restricting the nonparametric regression line to be additive in p explanatory variables:

$$f(x_i) = f_1(x_{i1}) + f_2(x_{i2}) + \cdots + f_p(x_{ip})$$

In many applications, such an additivity assumption may be sensible, and it is definitely much more flexible than the linearity assumption of standard regression methods.

Extending the notation described between (10.3) and (10.4), we can write this model as

$$y = Z\beta + \gamma_1 + \gamma_2 + \cdots + \gamma_p + \varepsilon \tag{10.26}$$

where $\gamma_j = (\gamma_{1j}, \ldots, \gamma_{Nj})' = [f_j(x_{1j}), \ldots, f_j(x_{Nj})]'$. In other words, the N points on the nonparametric regression line corresponding to the jth explanatory variable are stacked in γ_j for $j = 1, \ldots, p$. The data are ordered according to the first explanatory variable so that $x_{11} \leq x_{21} \leq \cdots \leq x_{N1}$. We refer to this ordering below as the 'correct' ordering.

In the case where x was a scalar, we used the simple intuition that, if we ordered the data points so that $x_1 \leq x_2 \leq \cdots \leq x_N$, then it was sensible to put a prior on $f(x_i) - f(x_{i-1})$. Here we have p explanatory variables which can be used to order the observations, so there is not one simple ordering which can be adopted. However, remember that the ordering information was only important as a way of expressing prior information about the degree of smoothness of the nonparametric regression line. If we express prior information for each of $\gamma_1, \ldots, \gamma_p$ with observations ordered according to its own explanatory variable, then transform back to the correct ordering, we can carry out Bayesian inference in a manner virtually identical to that in Section 10.2.2. To emphasize the intuition, let me repeat the econometric strategy in slightly different words. With independent data, it does not matter how the data is ordered, provided all variables are ordered in the same way. Here we have our observations ordered as $x_{11} \leq x_{21} \leq \cdots \leq x_{N1}$. However, prior information on the degree of smoothness for $f_j()$ should be elicited with observations ordered so that $x_{1j} \leq x_{2j} \leq \cdots \leq x_{Nj}$. But this means that, for $j = 2, \ldots, p$, the prior will be elicited with the observations ordered incorrectly (i.e. the correct ordering does not have $x_{1j} \leq x_{2j} \leq \cdots \leq x_{Nj}$, but rather has $x_{11} \leq x_{21} \leq \cdots \leq x_{N1}$). How do we solve this problem? After eliciting each prior, we simply re-order the data back to the correct ordering. Once we have done this, we are back in the familiar world of the Normal linear regression model with natural conjugate prior.

To write out this strategy formally, some new notation is required. Remember that our previous notation (e.g. $\gamma_1, \ldots, \gamma_p$) used an ordering of observations such that $x_{11} \leq x_{21} \leq \cdots \leq x_{N1}$. Define $\gamma_j^{(j)}$ as being equal to γ_j with observations ordered according to the jth explanatory variable (i.e. all data is ordered so that $x_{1j} \leq x_{2j} \leq \cdots \leq x_{Nj}$ for $j = 2, \ldots, p$). For individual elements of $\gamma_j^{(j)}$ we use the notation

$$\gamma_j^{(j)} = \begin{pmatrix} \gamma_{1j}^{(j)} \\ \gamma_{2j}^{(j)} \\ \cdot \\ \cdot \\ \gamma_{Nj}^{(j)} \end{pmatrix} = \begin{pmatrix} \gamma_{1j}^{(j)} \\ \gamma_j^{(j*)} \end{pmatrix}$$

That is, we have isolated out the first point on the jth component of the nonparametric regression line ($\gamma_{1j}^{(j)}$) from all the remaining points which we stack in an $(N-1)$-vector $\gamma_j^{(j*)}$. We define a similar notation when the observations are ordered according to the first explanatory variable with $\gamma_j^{(*)}$ equaling γ_j with one element deleted. This element is the one corresponding to the smallest value of the jth explanatory variable.

Before formally deriving the requisite posterior, it is important to note that there is an identification problem with the additive model, in that constants may be added and subtracted appropriately without changing the likelihood. For instance, the models $y_i = f_1(x_{i1}) + f_2(x_{i2}) + \varepsilon_i$ and $y_i = g_1(x_{i1}) + g_2(x_{i2}) + \varepsilon_i$ are equivalent if $g_1(x_{i1}) = f_1(x_{i1}) + c$ and $g_2(x_{i2}) = f_2(x_{i2}) - c$, where c is any constant. Insofar as interest centers on the marginal effect of each variable on y (i.e. on the shapes of $f_j(x_{ij})$ for $j = 1, \ldots, p$) or the overall fit of the nonparametric regression model, the lack of identification is irrelevant. Here we impose identification in a particular way, but many other choices can be made, and the interpretation of the empirical results will not change in any substantive way. We impose identification by setting $\gamma_{1j}^{(j)} = 0$ for $j = 2, \ldots, p$ (i.e. all except the first additive functions are restricted to have intercepts equalling zero).

For γ_1, β and h we use the same partially informative prior as before. In particular, the noninformative prior for β and h is given in (10.5) and, for γ_1 (i.e. the nonparametric regression line corresponding to the first explanatory variable) we use the prior on the degree of smoothness

$$D\gamma_1 \sim N(0_{N-1}, h^{-1}V(\eta_1)) \tag{10.27}$$

where D is the first-differencing matrix defined in (10.7). For $\gamma_j^{(j)}$ for $j = 2, \ldots, p$ the smoothness prior can be written as

$$D\gamma_j^{(j)} \sim N(0_{N-1}, h^{-1}V(\eta_j)) \tag{10.28}$$

Alternatively, since we impose the identifying assumption $\gamma_{j1}^{(j)} = 0$, we can write (10.28) as

$$D^*\gamma_j^{(j*)} \sim N(0_{N-1}, h^{-1}V(\eta_j)) \tag{10.29}$$

where D^* is an $(N-1)\times(N-1)$ matrix equal to D with the first column removed. Note that, as desired (10.28) and (10.29) imply that if $x_{i-1,j}$ and x_{ij} are close to one another, then $f_j(x_{i-1,j})$ and $f_j(x_{i,j})$ should also be close to one another. As discussed previously, other priors can be used (e.g. D can be replaced with the second-differencing matrix) with minimal changes in the following posteriors.

The prior in (10.28) is for $j = 2, \ldots, p$, and is expressed using the observations ordered in an incorrect manner (i.e. they are ordered as $x_{1j} \leq x_{2j} \leq \cdots \leq x_{Nj}$), so we have to re-order them before proceeding further. Hence, we define D_j, which is equivalent to D except that the rows and columns are re-ordered so that observations are ordered correctly (i.e. as $x_{11} \leq x_{21} \leq \cdots \leq x_{N1}$). We also introduce the notation D_j^* which is comparable to D^*. That is, D_j^* is equal to D_j with the column corresponding to the first point on the nonparametric regression line removed.

A concrete example of how this works might help. Suppose we have $N = 5$ and two explanatory variables which have values:

$$X = \begin{bmatrix} 1 & 3 \\ 2 & 4 \\ 3 & 1 \\ 4 & 2 \\ 5 & 5 \end{bmatrix}$$

The data has been ordered in the correct manner so that the first explanatory variable is in ascending order, $x_{11} \leq \cdots \leq x_{51}$. However, when observations are ordered in this way, the second explanatory variable is not in ascending order. The prior given in (10.28), written for the observations ordered according to $x_{12} \leq \cdots \leq x_{52}$, must be rearranged to account for this. This involves creating a rearranged version of D:

$$D_2 = \begin{bmatrix} 0 & 0 & -1 & 1 & 0 \\ 1 & 0 & 0 & -1 & 0 \\ -1 & 1 & 0 & 0 & 0 \\ 0 & -1 & 0 & 0 & 1 \end{bmatrix}$$

It can be verified that $D_2\gamma_2$ defines the distance between neighboring values for the second explanatory variable and, thus, it is sensible to put smoothness prior on it. The identification restriction implies $\gamma_{32} = 0$ and, hence,

$$D_2^* = \begin{bmatrix} 0 & 0 & 1 & 0 \\ 1 & 0 & -1 & 0 \\ -1 & 1 & 0 & 0 \\ 0 & -1 & 0 & 1 \end{bmatrix}$$

In summary, with the additive model we use the same smoothness prior on each of p unknown functions. Since the observations are ordered so that $x_{11} \leq \cdots \leq x_{N1}$, the smoothness prior for γ_1 can be written using the first difference matrix D. However, for $\gamma_2, \ldots, \gamma_p$ the same smoothness prior must be written in terms of a suitably rearranged version of D. We label these rearranged first difference matrices D_j for $j = 2, \ldots, p$. Imposing the identification restriction involves removing the appropriate column of D_j, and we label the resulting matrix D_j^*.

One more piece of notation relating to the imposition of the identification restriction is required. Let I_j^* equal the $N \times N$ identity matrix with one column deleted. The column deleted is for the observation which has the lowest value for the jth explanatory variable.

With this notation established, we can proceed in a similar manner as for the partial linear model. The model can be written as a Normal linear regression model:

$$y = W\delta + \varepsilon \tag{10.30}$$

where

$$W = [Z : I_N : I_2^* : \ldots : I_p^*]$$

and $\delta = (\beta', \gamma_1', \gamma_2^{(*)'}, \ldots, \gamma_p^{(*)'})'$ contains $K = k + N + (p-1) \times (N-1)$ regression coefficients. The prior for this model can be written in compact notation as

$$R\delta \sim N(0_{p(N-1)}, h^{-1}\underline{V}) \tag{10.31}$$

where

$$R = \begin{bmatrix} 0_{(N-1)\times k} & D & 0 & \cdot & \cdot & 0 \\ 0_{(N-1)\times k} & 0 & D_2^* & \cdot & \cdot & \cdot \\ \cdot & \cdot & \cdot & \cdot & 0 & \cdot \\ \cdot & \cdot & \cdot & 0 & \cdot & 0 \\ 0_{(N-1)\times k} & \cdot & \cdot & 0 & \cdot & D_p^* \end{bmatrix}$$

and

$$\underline{V} = \begin{bmatrix} V(\eta_1) & \cdot & \cdot & 0 \\ 0 & \cdot & \cdot & \cdot \\ \cdot & \cdot & \cdot & 0 \\ 0 & \cdot & 0 & V(\eta_p) \end{bmatrix}$$

At this point, it is useful to stress that, although the notation has become complicated due to questions of identification and the ordering of observations, this is still simply a Normal linear regression model with natural conjugate prior. Thus, all of the familiar results and techniques for this model are relevant, and we have

$$\delta, h|y \sim NG(\widetilde{\delta}, \widetilde{V}, \widetilde{s}^{-2}, \widetilde{\nu}) \tag{10.32}$$

where

$$\widetilde{V} = (R'\underline{V}^{-1}R + W'W)^{-1} \tag{10.33}$$

$$\widetilde{\delta} = \widetilde{V}(W'y) \tag{10.34}$$

$$\widetilde{\nu} = N \tag{10.35}$$

and

$$\widetilde{\nu}\widetilde{s}^2 = (y - W\widetilde{\delta})'(y - W\widetilde{\delta}) + (R\widetilde{\delta})'\underline{V}^{-1}(R\widetilde{\delta}) \tag{10.36}$$

Bayesian inference in this additive model is complicated by the fact that it is potentially difficult to elicit prior hyperparameters in a data-based fashion. Note that the prior allows for a different degree of smoothing in each unknown function (i.e. we have η_j for $j = 1, \ldots, p$). In some cases, the researcher may have prior information that allows her to choose values for each η_j. However, in many cases it will be sensible to smooth each unknown function by the same amount (i.e. setting $\eta_1 = \cdots = \eta_p \equiv \eta$ will be reasonable), and only one prior hyperparameter needs to be chosen. Empirical Bayesian inference can be carried out exactly as for the partial linear model. Model comparison and prediction can be carried out using the familiar methods for the Normal linear regression model.

Illustration: An Additive Model

To illustrate Bayesian inference in the partial linear model with additive nonparametric regression line, we generate artificial data from

$$y_i = f_1(x_{i1}) + f_2(x_{i2}) + \varepsilon_i$$

for $i = 1, \ldots, 100$, where ε_i is i.i.d. $N(0, 0.09)$ and x_{i1} and x_{12} are i.i.d. $U(0, 1)$. We take

$$f_1(x_{i1}) = x_i \cos(4\pi x_{i1})$$

and

$$f_2(x_{i2}) = \sin(2\pi x_{i2})$$

The partially informative prior in (10.31) requires elicitation of prior hyperparameters η_1 and η_2. We set $\eta \equiv \eta_1 = \eta_2$, and use the same empirical Bayesian methods as for the partial linear model to select a value for η. This value for η is then used to make posterior inference about the two components of the nonparametric regression line using (10.32)–(10.36).

As in the previous section, we use (very weak) prior information about η. In particular, we assume

$$\eta \sim G(\underline{\mu}_\eta, \underline{\nu}_\eta)$$

and choose nearly noninformative values of $\underline{\nu}_\eta = 0.0001$ and $\underline{\mu}_\eta = 1.0$. We choose the value of η which maximizes

$$p(\eta|y) \propto p(y|\eta)p(\eta) \propto (|\widetilde{V}||R'\underline{V}^{-1}R|)^{\frac{1}{2}}(\widetilde{\nu s^2})^{-\frac{\widetilde{\nu}}{2}} f_G(\eta|\underline{\mu}_\eta, \underline{\nu}_\eta)$$

The value of η chosen by the empirical Bayes procedure is 0.4210. Figures 10.2a and b plot the fitted and true nonparametric regression lines for each of the two additive functions in our nonparametric regression model (i.e. $E(\gamma_j|y)$ and $f_j(x_{ij})$ for $j = 1, 2$). These figures indicate that we are successfully estimating $f_j(\cdot)$. Remember that the identifying restriction means we can only estimate the functions up to an additive constant. This is reflected in the slight shifting of thc two components of the fitted nonparametric regression lines in Figures 10.2a and b. As noted in our illustration of the partial linear model, if the researcher requires a smoother curve, then a prior on the second differences, $[f(x_{i+1}) - f(x_i)] - [f(x_i) - f(x_{i-1})]$, can be used. Furthermore, in a serious empirical application other posterior features (e.g. HPDIs), model comparison tools (e.g. Bayes factors comparing this model to a parametric model) or predictive distributions could be presented.

Extensions

With the partial linear model we noted that many extensions were possible that would allow for Bayesian inference in, for example, semiparametric probit or

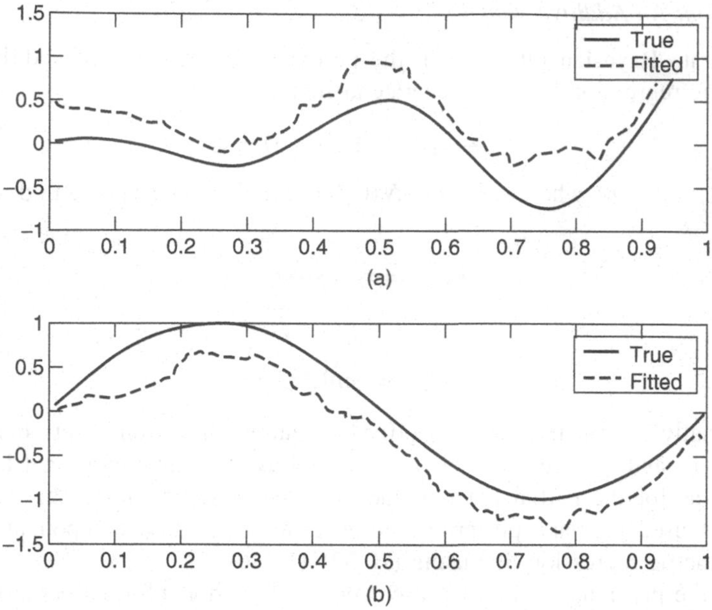

Figure 10.2 True and Fitted Lines for (a) First and (b) Second Additive Term

tobit models. With the additive variant of the partial linear model, the exact same extensions can be done in the same manner.

10.3 MIXTURES OF NORMALS MODELS

10.3.1 Overview

The partial linear model and its additive variant allowed for *the regression line* to have an unknown functional form. There are also many techniques for allowing *whole distributions* to have unknown forms. Here, we describe one such set of techniques. The basic idea underlying the model in this section is that a very flexible distribution can be obtained by mixing together several distributions. The resulting flexible distribution can be used to approximate the unknown distribution of interest. In this section, we discuss mixtures of Normal distributions, since these are commonly used and simple to work with. However, it should be mentioned that any set of distributions can be mixed, resulting in a more flexible distribution than would be obtained by simply choosing a single distribution.

The models considered in this section are not 'nonparametric', in the sense that they cannot become any unknown distribution. This is because they are so-called *finite mixtures of Normals*. For instance, a distribution which mixes five different Normal distributions, although very flexible, cannot accommodate any possible distribution. Thus, finite mixtures of Normals should be considered only as an extremely flexible modeling strategy. However, we note that *infinite mixtures* are, to all intents and purposes, nonparametric. Infinite mixtures of Normals will not be discussed here. Robert (1996), which is Chapter 24 of *Markov Chain Monte Carlo in Practice,* provides an introduction to this area of Bayesian statistics. A particular infinite mixture model involving *Dirichlet process priors* is very popular. Escobar and West (1995) and West, Muller and Escobar (1994) provide thorough discussions of this model.

We have already seen one particular example of a mixture of Normals model. Chapter 6 (Section 6.4) considered the case of the linear regression model with independent Student-t errors, and showed how it could be obtained using a particular mixture of Normals. Since the Student-t distribution is more flexible than the Normal (i.e. the Normal is a special case of the Student-t which arises when the degrees of freedom parameter goes to infinity), Section 6.4 provides a simple example of how mixing Normals can lead to a more flexible distribution. Here we consider more general mixtures of Normals in the context of the linear regression model. However, the basic concepts can be used anywhere the researcher wishes to make a flexible distributional assumption. For instance, the panel data models in Chapter 7 assumed hierarchical priors having particular distributions for the individual effects (e.g. Normal in (7.7) and exponential in (7.46) for the stochastic frontier model). Mixtures of Normals can be used to make these hierarchical priors more flexible. Posterior simulation can be done by combining the relevant components of the Gibbs sampler outlined below with the appropriate Gibbs sampler from Chapter 7. Geweke and Keane (1999) offers another nice use of mixtures of Normals in that it develops a mixtures of Normals probit model.

10.3.2 The Likelihood Function

The linear regression model can be written as

$$y = X\beta + \varepsilon \tag{10.37}$$

where the notation is the same as in previous chapters (e.g. see Chapter 3, Section 3.2). The likelihood function for the Normal linear regression model was based on the assumptions that

1. ε_i is i.i.d. $N(0, h^{-1})$ for $i = 1, \ldots, N$.
2. All elements of X are either fixed (i.e. not random variables) or, if they are random variables, they are independent of all elements of ε with a probability density function, $p(X|\lambda)$ where λ is a vector of parameters that does not include β and h.

Here we replace the first assumption with one where ε_i is a mixture of m different distributions. That is,

$$\varepsilon_i = \sum_{j=1}^{m} e_{ij} \left(\alpha_j + h_j^{\frac{1}{2}} \eta_{ij} \right) \tag{10.38}$$

where η_{ij} is i.i.d. $N(0, 1)$ for $i = 1, \ldots, N$, $j = 1, \ldots, m$ and e_{ij}, α_j and h_j are all parameters. e_{ij} indicates the component in the mixture that the ith error is drawn from. That is $e_{ij} = 0$ or 1 for $j = 1, \ldots, m$ and $\sum_{j=1}^{m} e_{ij} = 1$. Since η_{ij} is Normal, it follows that $(\alpha_j + h_j^{-\frac{1}{2}} \eta_{ij})$ is a Normal random variable with mean α_j and precision h_j. Thus, (10.38) specifies that the regression error is a weighted average of m different distributions. Each of these component distributions is $N(\alpha_j, h_j^{-1})$. This motivates the terminology *mixture of Normals*. The special case where $\alpha_j = 0$ for all j is referred to as a *scale mixture of Normals*. The special case where $h_1 = \cdots = h_m$ is referred to as a *mean (or location) mixture of Normals*. The mixture of Normals used in Chapter 6 (Section 6.4) was a scale mixture of Normals involving a particular hierarchical prior. To simplify notation, we stack these new parameters into vectors in the usual way: $\alpha = (\alpha_1, \ldots, \alpha_m)'$, $h = (h_1, \ldots, h_m)'$, $e_i = (e_{i1}, \ldots, e_{im})'$ and $e = (e_1', \ldots, e_N')'$.

In practice, it is unknown which component the ith error is drawn from and, thus, we let p_j for $j = 1, \ldots, m$ be the probability of the error being drawn from the jth component in the mixture. That is, $p_j = P(e_{ij} = 1)$. Formally, this means that e_i are i.i.d. draws from the Multinomial distribution (see Appendix B, Definition B.23):

$$e_i \sim M(1, p) \tag{10.39}$$

where $p = (p_1, \ldots, p_m)'$. Remember that, since p is a vector of probabilities, we must have $0 \leq p_j \leq 1$ and $\sum_{j=1}^{m} p_j = 1$.

As with many models, there is some arbitrariness as to what gets labeled 'prior' information and what gets labeled 'likelihood' information. Equation (10.39) could be interpreted as a hierarchical prior for e_i. However, following standard practice, here we refer to β, h, α and p as the parameters of the model and $p(y|\beta, h, \alpha, p)$ as the likelihood function. The component indicators, e_i for $i = 1, \ldots, N$, will be treated as latent data (and will prove useful in the Gibbs sampling algorithm outlined below). Since p_j is the probability of the error being drawn from the jth component in the Normal mixture, it can be seen that the likelihood function is

$$p(y|\beta, h, \alpha, p) = \frac{1}{(2\pi)^{\frac{N}{2}}} \prod_{i=1}^{N} \left\{ \sum_{j=1}^{m} p_j \sqrt{h_j} \exp\left[-\frac{h_j}{2} (y_i - \alpha_j - \beta' x_i)^2 \right] \right\} \tag{10.40}$$

where x_i is a k-vector containing the explanatory variables for individual i.

10.3.3 The Prior

As with any Bayesian model, any prior can be used. Here we describe a commonly-used prior which allows for convenient computation and is flexible enough to accommodate a wide range of prior beliefs. However, before describing precise forms for prior densities, there are two underlying issues which must be discussed.

First, the mixtures of Normals model is an example of a model where the likelihood function is unbounded.[6] This means that the standard frequentist theory underlying maximum likelihood estimation breaks down. For the Bayesian, the pathology implies that the researcher should not use a noninformative prior. Bayesian inference with an informative prior, however, can be done in the usual way.[7]

Secondly, there is an identification problem in this model, in that multiple sets of parameter values are consistent with the same likelihood function. For instance, consider a mixture with two components (i.e. $m = 2$). The probabilities associated with each component are $p_1 = 0.25$ and $p_2 = 0.75$. The first distribution in the mixture has $\alpha_1 = 2.0$ and $h_1 = 2.0$, while the second has $\alpha_2 = 1.0$ and $h_2 = 1.0$. This distribution is identical to one where the labeling of the two components is reversed. That is, it is exactly the same as one with parameter values $p_1 = 0.75$, $p_2 = 0.25$, $\alpha_1 = 1.0$, $h_1 = 1.0$, $\alpha_2 = 2.0$ and $h_2 = 2.0$. Because of this, it is necessary for the prior to impose a *labelling restriction*, such as

$$\alpha_{j-1} < \alpha_j \tag{10.41}$$

$$h_{j-1} < h_j \tag{10.42}$$

or

$$p_{j-1} < p_j \tag{10.43}$$

for $j = 2, \ldots, m$. Only one such restriction need be imposed. Here (10.41) will be chosen, although imposing (10.42) or (10.43) will only cause minor modification in the following material.

We begin with a prior for β and h, which is a simple extension of the familiar independent Normal-Gamma prior (see Chapter 4, Section 4.2). In particular,

$$\beta \sim N(\underline{\beta}, \underline{V}) \tag{10.44}$$

and we assume independent Gamma priors for h_j for $j = 2, \ldots, m$,

$$h_j \sim G(\underline{s}_j^{-2}, \underline{\nu}_j) \tag{10.45}$$

[6]To see this, set β to $\widehat{\beta}$, the OLS estimate, h_j^{-1} to the OLS estimate of the error variance and $\alpha_j = 0$ for $j = 2, \ldots, m$. For some $c > 0$ set $p_1 = c$ and $p_j = \frac{1-c}{m-1}$ for $j = 2, \ldots, m$. If $\alpha_1 = (y_1 - \widehat{\beta}' x_1)$, then the likelihood function goes to infinity as $h_1 \to \infty$.

[7]A proof of this statement is provided in Geweke and Keane (1999) for the prior used in this section.

The Dirichlet distribution (see Appendix B, Definition B.28) is a flexible and computationally convenient choice for parameters such as p which lie between zero and one and sum to one (remember that $0 \leq p_j \leq 1$ and $\sum_{j=1}^{m} p_j = 1$). Thus, we take

$$p \sim D(\underline{p}) \tag{10.46}$$

where $\underline{p}$ is an m-vector of prior hyperparameters. Appendix B, Theorem B.17 lists some properties which show how $\underline{p}$ can be interpreted.

Here we impose the labeling restriction through α. Hence, we assume the prior for this parameter vector to be Normal with the restrictions in (10.41) imposed:

$$p(\alpha) \propto f_N(\alpha|\underline{\alpha}, \underline{V}_\alpha)1(\alpha_1 < \alpha_2 < \cdots < \alpha_m) \tag{10.47}$$

Remember that $1(A)$ is the indicator function equalling 1 if condition A holds and otherwise equalling zero.

10.3.4 Bayesian Computation

As with many other models in this book, Bayesian inference can be carried out using a Gibbs sampler with data augmentation. Intuitively, if we knew which component in the mixture each error was drawn from, then the model would reduce to the Normal linear regression model with independent Normal-Gamma prior (see Chapter 4, Section 4.2). Thus, treating e as latent data will greatly simplify things. This intuition motivates a Gibbs sampler which sequentially draws from the full posterior conditional distributions $p(\beta|y, e, h, p, \alpha)$, $p(h|y, e, \beta, p, \alpha)$, $p(p|y, e, \beta, h, \alpha)$, $p(\alpha|y, e, \beta, h, p)$ and $p(e|y, \beta, h, p, \alpha)$. Below we derive the precise form for each of these distributions. These derivations are relatively straightforward, involving multiplying the appropriate prior times $p(y|e, \beta, h, \alpha, p)$ and re-arranging the result. Using methods comparable to those used to derive (10.40), it can be shown that

$$p(y|e, \beta, h, \alpha, p) = \frac{1}{(2\pi)^{\frac{N}{2}}} \prod_{i=1}^{N} \left\{ \sum_{j=1}^{m} e_{ij}\sqrt{h_j} \exp\left[-\frac{h_j}{2}(y_i - \alpha_j - \beta' x_i)^2\right] \right\} \tag{10.48}$$

Conditional on e, $p(\beta|y, e, h, p, \alpha)$ and $p(h_j|y, e, \beta, p, \alpha)$ for $j = 1, \ldots, m$ simplify, and results from Chapter 4, Section 4.2 can be applied directly. In particular, $p(\beta|y, e, h, p, \alpha)$ does not depend upon p and

$$\beta|y, e, h, \alpha \sim N(\overline{\beta}, \overline{V}) \tag{10.49}$$

where

$$\overline{V} = \left(\underline{V}^{-1} + \sum_{i=1}^{N}\sum_{j=1}^{m} e_{ij} h_j x_i x_i'\right)^{-1}$$

and

$$\overline{\beta} = \overline{V}\left(\underline{V}^{-1}\underline{\beta} + \sum_{i=1}^{n}\sum_{j=1}^{m} e_{ij}h_j x_i[y_i - \alpha_j]\right)$$

Furthermore, for $j = 1, \ldots, m$, the posterior conditionals for the h_js are independent of one another and simplify to

$$h_j|y, e, \beta, \alpha \sim G(\overline{s}_j^{-2}, \overline{\nu}_j) \quad (10.50)$$

where

$$\overline{\nu}_j = \sum_{i=1}^{N} e_{ij} + \underline{\nu}_j$$

and

$$\overline{s}_j^2 = \frac{\sum_{i=1}^{N} e_{ij}(y_i - \alpha_j - x_i'\beta)'(y_i - \alpha_j - x_i'\beta) + \underline{\nu}_j \underline{s}_j^2}{\overline{\nu}_j}$$

To aid in interpretation, remember that e_{ij} is an indicator variable equalling 1 if the ith error comes from the jth component in the mixture. Hence, $\sum_{i=1}^{N} e_{ij}$ simply counts the number of observations in the jth component, the term $\sum_{i=1}^{N}\sum_{j=1}^{m} e_{ij}h_j x_i x_i'$ is comparable to the term $hX'X$ in Chapter 4 (4.4), but for the ith observation it picks out the appropriate h_j. Other terms have similar intuition.

Noting that α_j enters in the role of an intercept from a Normal linear regression model in (10.48) and (10.47) describes a Normal prior (subject to the labelling restrictions), it can be seen that the conditional posterior of α is Normal (subject to the labeling restrictions). In particular,

$$p(\alpha|y, e, \beta, h) \propto f_N(\alpha|\overline{\alpha}, \overline{V}_\alpha)1(\alpha_1 < \alpha_2 < \cdots < \alpha_m) \quad (10.51)$$

where

$$\overline{V}_\alpha = \left(\underline{V}_\alpha^{-1} + \sum_{i=1}^{N}\left\{\sum_{j=1}^{m} e_{ij}h_j\right\} e_i e_i'\right)^{1}$$

and

$$\overline{\alpha} = \overline{V}_\alpha\left[\underline{V}_\alpha^{-1}\underline{\alpha} + \sum_{i=1}^{N}\left\{\sum_{j=1}^{m} e_{ij}h_j\right\} e_i(y_i - \beta' x_i)\right]$$

These formulae may look somewhat complicated, but they are calculated using methods which are minor modifications of those used for the Normal linear regression model. The term $\{\sum_{j=1}^{m} e_{ij}h_j\}$ picks out the relevant error precision for observation i.

Multiplying (10.46) by (10.48) yields the kernel of the conditional posterior, $p(p|y, e, \beta, h, \alpha)$. Straightforward manipulations show that this only depends upon e, and has a Dirichlet distribution

$$p \sim D(\overline{\rho}) \tag{10.52}$$

where

$$\overline{\rho} = \underline{\rho} + \sum_{i=1}^{N} e_i$$

Remember that e_i shows which component in the mixture the ith error is drawn from. It is an m-vector containing all zeros except for a 1 in the appropriate location. Thus $\sum_{i=1}^{N} e_i$ is an m-vector containing the number of observations drawn from each Normal distribution in the mixture.

The last block in the Gibbs sampler is $p(e|y, \beta, h, p, \alpha)$. The rules of conditional probability imply $p(e|y, \beta, h, p, \alpha) \propto p(y|e, \beta, h, p, \alpha)p(e|\beta, h, p, \alpha)$. The prior independence assumptions imply $p(e|\beta, h, p, \alpha) = p(e|p)$ and, thus, $p(e|y, \beta, h, p, \alpha)$ can be obtained by multiplying (10.48) by (10.39) and re-arranging. If this is done, we find that $p(e|y, \beta, h, p, \alpha) = \prod_{i=1}^{N} p(e_i|y, \beta, h, p, \alpha)$, and each of the $p(e_i|y, \beta, h, p, \alpha)$ is a Multinomial density (see Appendix B, Definition B.23). To be precise,

$$e_i|y, \beta, h, p, \alpha \sim$$

$$M\left(1, \left[\frac{p_1 f_N(y_i|\alpha_1 + \beta' x_i, h_1^{-1})}{\sum_{j=1}^{m} p_j f_N(y_i|\alpha_j + \beta' x_i, h_j^{-1})}, \ldots, \frac{p_m f_N(y_i|\alpha_m + \beta' x_i, h_m^{-1})}{\sum_{j=1}^{m} p_j f_N(y_i|\alpha_j + \beta' x_i, h_j^{-1})}\right]'\right) \tag{10.53}$$

Posterior inference in the linear regression model with mixture of Normals errors can be carried out using a Gibbs sampler which sequentially draws from (10.49), (10.50), (10.51), (10.52) and (10.53).

10.3.5 Model Comparison: Information Criteria

All the model comparison methods described in previous chapters can be used with mixtures of Normals. With this class of model, an important issue is the selection of m, the number of components in the mixture. This can be done by calculating the marginal likelihood for a range of values for m and choosing the value which yields the largest marginal likelihood. Either the Gelfand–Dey method (see Chapter 5, Section 5.7) or the Chib method (see Chapter 7, Section 7.5) can be used to calculate the marginal likelihood. A minor complication arises, since both these methods require the evaluation of prior densities, and the labeling restriction means that (10.47) only gives us the prior kernel for α. However, the necessary integrating constant can be calculated using prior simulation. A crude prior simulator would simply take draws from $f_N(\alpha|\underline{\alpha}, \underline{V}_\alpha)$ and calculate the proportion of draws which satisfy $\alpha_1 < \alpha_2 < \cdots < \alpha_m$. One over this proportion is the required integrating constant. More efficient prior simulators can be developed using algorithms for drawing from the truncated Normal.

However, calculating marginal likelihoods can be computationally demanding and care has to be taken with prior elicitation (e.g. marginal likelihoods are usually not defined when using noninformative priors). Accordingly, interest exists in shortcut methods for summarizing the data evidence in favor of a model. Motivated by this consideration, various *information criteria* have been developed. In this section, a few of these will be introduced. Their advantage is that they are easy to calculate, and typically do not depend on prior information. Their disadvantage is that it is hard to provide a rigorous justification for their use. That is, the logic of Bayesian inference says that a model should be evaluated based on the probability that it generated the data. Hence, for the pure Bayesian, the posterior model probability should be the tool for model comparison. Information criteria do not have such a formal justification (at least from a Bayesian perspective). However, as noted below, they can often be interpreted as approximations to quantities which have a formal Bayesian justification.

Information criteria can be used with any model. Accordingly, let us temporarily adopt the general notation of Chapter 1, where θ is a p-vector of parameters and $p(y|\theta)$, $p(\theta)$ and $p(\theta|y)$ are the likelihood, prior and posterior, respectively. Information criteria typically have the form

$$IC(\theta) = 2\ln[p(y|\theta)] - g(p) \tag{10.54}$$

where $g(p)$ is an increasing function of p. The traditional use of information criteria involves evaluating $IC(\theta)$ at a particular point (e.g. the maximum likelihood value for θ) for every model under consideration, and choosing the model with the highest information criteria. Most information criteria differ in the functional form used for $g(p)$. This is a function which rewards parsimony. That it, it penalizes models with excessive parameters.

In Bayesian circles, the most common information criterion is the *Bayesian Information Criterion* (or BIC)

$$BIC(\theta) = 2\ln[p(y|\theta)] - p\ln(N) \tag{10.55}$$

As shown in Schwarz (1978), twice the log of the Bayes factor comparing two models is approximately equal to the difference in BICs for the two models. Two other popular information criteria are the *Akaike Information Criterion* (or AIC), given by

$$AIC(\theta) = 2\ln[p(y|\theta)] - 2p \tag{10.56}$$

and the *Hannan–Quinn Criterion* (or HQ)

$$HQ(\theta) = 2\ln[p(y|\theta)] - pc_{HQ}\ln[\ln(N)] \tag{10.57}$$

In (10.57) c_{HQ} is a constant. HQ is a consistent model selection criterion[8] if $c_{HQ} > 2$.

These are the most popular of the many information criteria which exist. There are many places where the interested reader can find out more. The discussion and citations in Poirier (1995, p. 394) provide a good starting point. Kass and Raftery (1995) is a fine survey paper on Bayes factors which, among many other things, draws out the relationship between Bayes factors and information criteria. Carlin and Louis (2000) include much relevant discussion, including a newly developed information criterion called the *Deviance Information Criterion*, which is designed to work well when models involve latent data and hierarchical priors. In this book, we note only that a quick and dirty method of model selection is to choose the model with the highest value for an information criterion. In the following empirical illustration, we investigate how effective this strategy is in selecting the number of components in a Normal mixture.

10.3.6 Empirical Illustration: A Mixture of Normals Model

We illustrate the mixtures of Normals model using two artificial data sets. To focus on the mixtures of Normals aspect of the model, we do not include any explanatory variables (i.e. β does not enter the model). The two data sets are, thus, all generated from

$$y_i = \varepsilon_i$$

where ε_i takes the mixtures of Normals form of (10.38). All three data sets have $N = 200$. The data sets are given by:

1. Data Set 1 has $m = 2$. The first Normal has $\alpha_1 = -1$, $h_1 = 16$ and $p_1 = 0.75$. The second Normal has $\alpha_2 = 1$, $h_2 = 4$ and $p_2 = 0.25$.
2. Data Set 1 has $m = 3$. The first Normal has $\alpha_1 = -1$, $h_1 = 4$ and $p_1 = 0.25$. The second Normal has $\alpha_2 = 0$, $h_2 = 16$ and $p_2 = 0.5$. The third Normal has $\alpha_3 = 1$, $h_3 = 16$ and $p_3 = 0.25$.

Histograms of these data sets are given in Figures 10.3a and b. These figures are included to show just how flexible mixtures of Normals can be. By mixing just two or three Normals together, we can get distributions which are very non-Normal. Mixtures of Normals can be used to model skewed, fat-tailed or multi-modal distributions.

We use a prior which is proper but very near to being noninformative. In particular, using the prior in (10.45), (10.46) and (10.47) we set $\underline{\alpha} = 0_m$, $\underline{V}_\alpha = (10\,000^2)I_m$, $\underline{s}_j^{-2} = 1$, $\underline{\nu}_j = 0.01$ and $\underline{p} = \iota_m$, where ι_m is an m-vector of ones. Bayesian inference is carried out using the Gibbs sampler involving (10.49), (10.50), (10.51), (10.52) and (10.53). For each data set, Bayesian inference is

[8]A consistent model selection criterion is one which chooses the correct model with probability one as sample size goes to infinity.

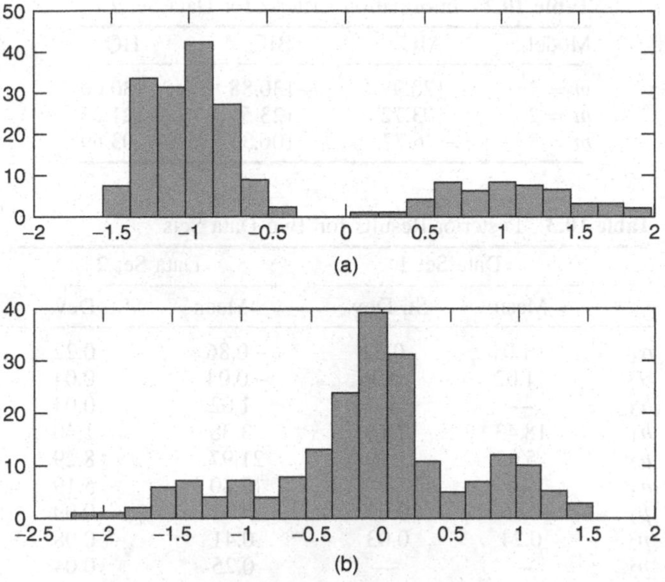

Figure 10.3 Histogram of (a) Data Set 1, (b) Data Set 2

done using $m = 1, 2$ and 3. The information criteria are evaluated at the posterior mean of the parameters in the model. The Gibbs sampler was run for 11 000 replications, with 1000 burn-in replications discarded and 10 000 replications retained. MCMC diagnostics indicate that this is an adequate number of replications to ensure convergence of the Gibbs sampler.

Tables 10.1 and 10.2 contain information criteria for Data Sets 1 and 2, respectively. The information criteria are consistent with one another and conclusive. For Data Set 1 (which was generated with $m = 2$), all of the information criteria select $m = 2$ as the preferred model. For Data Set 2, the information criteria all select the correct value of $m = 3$. Thus, at least for these data sets, information criteria do seem to be useful for selecting the number of components in a Normal mixture.

Table 10.3 presents posterior means and standard deviations of all parameters for the selected model for each data set. A comparison of posterior means with the values used to generate the data sets indicate that we are obtaining very reliable

Table 10.1 Information Criteria for Data Set 1

Model	AIC	BIC	HQ
$m = 1$	−174.08	−183.98	−183.09
$m = 2$	92.41	77.62	74.40
$m = 3$	−52.24	−81.92	−79.25

Table 10.2 Information Criteria for Data Set 2

Model	AIC	BIC	HQ
$m = 1$	-120.99	-130.88	-130.00
$m = 2$	-103.72	-123.51	-121.23
$m = 3$	-76.77	-106.35	-103.69

Table 10.3 Posterior Results for Two Data Sets

	Data Set 1		Data Set 2	
	Mean	St. Dev.	Mean	St. Dev.
α_1	-1.01	0.02	-0.86	0.22
α_2	1.02	0.06	-0.04	0.04
α_3	—	—	1.02	0.04
h_1	18.43	2.14	3.38	1.40
h_2	5.65	1.19	21.97	8.29
h_3	—	—	17.80	5.19
p_1	0.76	0.03	0.33	0.09
p_2	0.24	0.03	0.41	0.08
p_3–	—	—	0.25	0.04

estimates of all parameters. An examination of posterior standard deviations indicates that the parameters are reasonably precisely estimated, despite having only a moderately large sample size.

10.4 EXTENSIONS AND ALTERNATIVE APPROACHES

As we have stressed throughout, virtually any model in this book can be used as a component of a larger model. In many cases, posterior simulation for the larger model can be done using a Gibbs sampler, where one or more blocks of the Gibbs sampler can be lifted directly from the simpler model discussed in this book. We have shown how such a strategy can be used to develop posterior simulators for semiparametric probit and tobit models. A myriad of other such extensions are also possible. Similar extensions exist for the mixtures of Normals linear regression model considered above. There are many obvious extensions of models from previous chapters (e.g. mixtures of Normals nonlinear regression or any of the panel data models can be extended to have mixtures of Normals errors). Mixtures of Normals can also be used to make hierarchical priors more flexible. The possibilities are virtually limitless.

Bayesian nonparametrics is, at present, a very active research area and there are many approaches we have not discussed (e.g. Dirichlet process priors, wavelets, splines, etc.). Dey, Muller and Sinha (1998), *Practical Nonparametric and Semiparametric Bayesian Statistics*, provides an introduction to many such approaches in this rapidly developing field.

10.5 SUMMARY

In this chapter, Bayesian inference in several flexible models has been discussed. These models are designed to achieve similar goals as the non- or semiparametric models so popular in the non-Bayesian literature. There are numerous Bayesian nonparametric methods, but in this chapter we focus on simple models which are straightforward extensions of models discussed in previous chapters. The chapter is divided into sections containing models involving nonparametric regression (i.e. the regression line has an unknown functional form) and models involving a flexible error distribution.

The first model considered was the partial linear model. This is a regression model where some explanatory variables enter in a linear fashion and another one enters in a nonparametric fashion. We showed how this model can be put in the form of a Normal linear regression model with natural conjugate prior and, thus, analytical results from Chapter 3 apply directly. We showed how the partial linear model could be used as a component of a more complicated model (e.g. a semiparametric probit or tobit model), and how a Gibbs sampler could be constructed in a straightforward manner. We next considered the partial linear model where $p > 1$ explanatory variables were treated in a nonparametric fashion. Although such a model can be analyzed by ordering the data using a distance function, such an approach may not work well if p is more than 2 or 3. Hence, an additive version of the partial linear model was discussed. This model can also be put in the form of a Normal linear regression model with natural conjugate prior and, thus, a posterior simulator is not required. For modeling flexible error distributions, mixtures of Normals are powerful tools. We showed how a Gibbs sampler with data augmentation can be used to carry out posterior inference in the linear regression model with mixture of Normals errors.

No new tools for Bayesian computation were developed in this chapter. Bayesian quantities such as posterior and predictive means, Bayes factors, etc., can all be calculated using methods described in previous chapters. The only new tool introduced is a model selection technique involving information criteria. Information criteria do not have a rigorous Bayesian interpretation (other than as an approximation). However, they are typically very easy to calculate (and do not depend upon prior information) and, thus, are popular in practice. In this chapter, we showed how information criteria can be used to select the number of components in a Normal mixture.

10.6 EXERCISES

The exercises in this chapter are closer to being small projects than standard textbook questions. Remember that some data sets and MATLAB programs are available on the website associated with this book. The house price data set is available on this website, or in the *Journal of Applied Econometrics* Data Archive listed under Anglin and Gencay (1996) (http://qed.econ.queensu.ca/jae/1996-v11.6/anglin-gencay/).

1. The empirical illustration in Chapter 3 (Section 3.9) used the Normal linear regression model with natural conjugate prior with the house price data set. Please refer to this illustration for data definitions.
   (a) Use the house price data set and the partial linear model to investigate whether the effect of lot size on house price is nonlinear. Experiment with different priors (including empirical Bayes methods).
   (b) Calculate the Bayes factor comparing the partial linear model to the Normal linear regression model for this data set.
   (c) Carry out a prior sensitivity analysis to investigate how robust your answer in part (b) is to prior choice.
2. The additive version of the partial linear model may be too restrictive in some applications. This motivates interest in specifications which are more general, but less likely to suffer from the curse of dimensionality than the partial linear model. One such specification includes interaction terms between explanatory variables. In the case where $p = 2$, we would have the model

$$y_i = f_1(x_{i1}) + f_2(x_{i2}) + f_3(x_{i1}x_{i2}) + \varepsilon_i$$

   (a) Describe how the methods of Section 10.2.3 can be extended to allow for Bayesian analysis of this model.
   (b) Using artificial data sets of your choice, investigate the empirical performance of the methods developed in part (a).
3. Chapter 9 included empirical illustrations of the tobit and probit models (see Sections 9.3.1 and 9.4.1, respectively).
   (a) Re-do these empirical illustrations using semiparametric tobit and probit as described in the present chapter. Note: the empirical illustrations describe the artificial data sets used (see also the website associated with this book).
   (b) Describe how Bayesian inference in a semiparametric ordered probit model could be developed.
   (c) Write a program which carries out Bayesian inference in the semiparametric ordered probit model and investigate its performance using artificial data.
4. The empirical illustration of the mixtures of Normals model (Section 10.3.6) used information criteria to select the number of elements in the mixture.
   (a) Write a program (or modify the one on the website associated with this book) which uses marginal likelihoods to choose the number of elements in the mixture.
   (b) Using an informative prior, investigate the performance of your program using the artificial data sets described in Section 10.3.6.
   (c) Repeat part (b) using different data sets, and compare results obtained using marginal likelihoods with those obtained using information criteria.

11 Bayesian Model Averaging

11.1 INTRODUCTION

In each of the previous chapters, we have focused on a particular class of models. Techniques of wide validity (e.g. of posterior computation) were then developed in the context of a particular model. In this chapter, we reverse the process and motivate and discuss a general technique before applying it to a model. This technique is called *Bayesian Model Averaging* (*BMA*), and it comes with a new tool of posterior computation called *Markov Chain Monte Carlo Model Composition* (MC^3). We devote an entire chapter to BMA, since it can be of enormous importance for empirical work, is difficult to implement in a non-Bayesian manner,[1] and is an area where many advances have been made in recent years. In short, this is an empirically important area where Bayesian econometrics is of particular benefit.

The concept of Bayesian model averaging was briefly introduced in Chapter 2, and can be described very simply using the rules of probability. Using our standard notation, we let M_r for $r = 1, \ldots, R$ denote the R different models under consideration. Each model depends upon a vector of parameters θ_r and is characterized by prior $p(\theta_r|M_r)$, likelihood $p(y|\theta_r, M_r)$ and posterior $p(\theta_r|y, M_r)$. Using these familiar building blocks (see Chapter 1 (1.5)), posterior model probabilities, $p(M_r|y)$, for $r = 1,2, \ldots, R$, can be obtained. Let ϕ be a vector of parameters which has a common interpretation in all models. That is, ϕ is a function of θ_r for each of $r = 1, \ldots, R$. Typically, ϕ will be the focus of interest in the empirical study. For instance, in an application involving stochastic frontier models, the researcher may consider many different models (e.g. having different inefficiency distributions or having different functional forms for the production function). However, each model is used to make inferences about the same thing: the efficiencies of each firm. In this case, ϕ would be the vector containing all

[1]Frequentist econometrics does not treat models (or their truth) as random variables and, thus, the concept of model averaging cannot be given a rigorous statistical foundation. There are, however, various *ad hoc* frequentist methods of model averaging (e.g. see Sala-i-Martin, 1997).

these efficiencies. As another example, the researcher might be interested in the effect education has on economic growth. The researcher will typically have numerous regression models involving education and various combinations of other explanatory variables (e.g. investment, openness to trade, etc.). However, the coefficient on education would be the focus of interest in every regression model and, thus, ϕ would be this coefficient.

The logic of Bayesian econometrics says that all that is known about ϕ is summarized in its posterior: $p(\phi|y)$. Furthermore, the rules of probability say

$$p(\phi|y) = \sum_{r=1}^{R} p(\phi|y, M_r)p(M_r|y) \tag{11.1}$$

Alternatively, if $g(\phi)$ is a function of ϕ, the rules of conditional expectation imply that

$$E[g(\phi)|y] = \sum_{r=1}^{R} E[g(\phi)|y, M_r]p(M_r|y) \tag{11.2}$$

In words, the logic of Bayesian inference says that one should obtain results for every model under consideration and average them. The weights in the averaging are the posterior model probabilities. Thus, once the researcher has worked out the posterior and the marginal likelihood (and, if necessary, methods of computation for both) for each model, Bayesian model averaging is conceptually straightforward. However, implementing Bayesian model averaging can be difficult, since R, the number of models under consideration, is often huge. If $E[g(\phi)|y, M_r]$ and $p(M_r|y)$ must be calculated by posterior simulation, then it may simply be impossible to work with more than a few models. Accordingly, the Bayesian model averaging literature has focused on cases where these quantities can be approximated or calculated analytically. As we shall see, even in these cases, there are many applications where R is simply too big for every possible model to be incorporated in averages such as (11.2). This has led to various algorithms which do not require dealing with every possible model. The most popular of these algorithms is MC^3. Intuitively, whereas posterior simulators such as MCMC draw from the posterior distribution of the parameters, MC^3 takes draws from the posterior distribution of models.

These issues are most clearly understood in the context of a specific model and, accordingly, in the following section we discuss Bayesian model averaging in the Normal linear regression model.

11.2 BAYESIAN MODEL AVERAGING IN THE NORMAL LINEAR REGRESSION MODEL

11.2.1 Overview

There are many empirical applications involving the linear regression model where a large number of explanatory variables exist which might possibly influence the dependent variable. Examples abound in the economic growth literature, where there is a myriad of competing theories of economic growth (and, hence, explanatory variables). In a time series example, uncertainty over how long the lag length should be often implies many potential explanatory variables. Thus, the researcher is often faced with the situation where numerous potential explanatory variables exists. She may expect that many of these explanatory variables are irrelevant, but does not know which ones. What should she do in this case? She may be tempted to simply include all potential variables in a regression. However, this approach is unsatisfactory, since including irrelevant variables tends to decrease the accuracy of the estimation, making it difficult to uncover effects that really are there.[2] Thus, the traditional approach is to do a sequence of tests with the aim of selecting a single best model which has all the irrelevant variables omitted. The problems associated with the presentation of results from a single model selected on the basis of a sequence of tests has long been recognized in the statistical literature. Statistical discussions of these problems are provided in many places. For instance, Poirier (1995, pp. 519–523) provides a theoretical discussion of the problems with so-called pre-test estimators. Draper (1995) and Hodges (1987) are also important references in this field.

Here, we provide a brief, intuitive, description of key issues addressed in these papers. First, each time a test is carried out, there is a possibility that a mistake will be made (i.e. that the researcher will reject the 'better' model for a 'not so good' one). The probability of making a mistake quickly increases as sequences of tests are carried out. Secondly, even if a sequential testing procedure does lead to selection of the 'best' model, standard decision theory implies that it is rarely desirable to simply present results for this model and ignore all evidence from the 'not quite so good' models. By doing so, *model uncertainty* is ignored. That is, the researcher is not absolutely sure that any estimate of ϕ is precisely correct for two reasons. First, she does not know exactly what the parameters of the model are (i.e. *parameter uncertainty* exists), and secondly, she does not know exactly which model is the correct one (i.e. model uncertainty exists). Traditional posterior inference, as used in previous chapters, deals with parameter uncertainty. However, model uncertainty can be very important as well. By ignoring model uncertainty, sequential testing procedures can lead to seriously misleading inferences. Fortunately, the logic of Bayesian econometrics tells us exactly how to deal with model uncertainty. Bayesian model averaging, using (11.1) and/or (11.2), is the correct way of handling multiple models.

In the case where there is a myriad of potential explanatory variables, alternative models can be defined through the set of explanatory variables they include. However, if K is the number of potential explanatory variables, this means that

[2]In frequentist statistical procedures, including irrelevant explanatory variables will tend to increase standard errors on all coefficients and, thus, make it difficult to find significant effects.

there are 2^K possible models (i.e. models are defined by the inclusion or exclusion of each explanatory variable). If K is at all large, then the number of possible models is astronomical. For instance, it is not at all uncommon to have 30 potential explanatory variables and, if so, there are $2^{30} > 10^9$ models. Even if the computer could analyze each model in 0.001 of a second, it would take almost two years to analyze all the models! Thus, directly doing Bayesian model averaging by explicitly calculating every term in (11.1) or (11.2) is usually impossible. However, as we shall see, so-called MC^3 algorithms have been developed to surmount this problem.

A second problem relates to prior information. Bayesian model averaging is often used in empirical applications where the researcher has many potential explanatory variables, but is unsure about which ones are important and which ones are not. In such cases, it is rare for the researcher to have substantial prior information. Even if she does, deriving priors for 2^K models may be an impossible task. Accordingly, it would be nice if noninformative priors could be used. However, as we have seen (see Chapter 3, Section 3.6.2), posterior model probabilities cannot be meaningfully calculated with improper noninformative priors. Thus, many researchers have attempted to develop proper priors which can be automatically used by the researcher, without requiring subjective input or fine tuning for each individual model. In this chapter, we describe one class of so-called *benchmark priors* developed in Fernandez, Ley and Steel (2001a,b). Raftery, Madigan and Hoeting (1997) adopt a similar approach. However, it is worth mentioning that many other approaches exist. The reader interested in learning more about Bayesian model averaging might want to look at Hoeting, Madigan, Raftery and Volinsky (1999), or the Bayesian Model Averaging website (http://www.research.att.com/~volinsky/bma.html).

11.2.2 The Likelihood Function

The likelihood function for each model is based on the Normal linear regression model, which we have considered in previous chapters (e.g. Chapter 3, (3.7)). Since we have written out this likelihood so many times before, we will not do it again here. As a reminder of notation, remember that we have data for $i = 1, \ldots, N$ individuals, and the observations of the dependent variable are placed in an N-vector $y = (y_1, \ldots, y_T)'$. We have $r = 1, \ldots, R$ models, denoted by M_r. These are all Normal linear regression models which differ in their explanatory variables. However, we write the explanatory variables in a slightly different manner than in previous chapters. In particular, it is standard to assume that all models contain an intercept. Hence, all the potential explanatory variables are stacked in a $N \times K$ matrix X but, unlike in previous chapters, the first column of this matrix is not an intercept. In accordance with this notation, we will write our linear regression models as

$$y = \alpha \iota_N + X_r \beta_r + \varepsilon \tag{11.3}$$

where ι_N is a $N \times 1$ vector of ones, X_r is a $N \times k_r$ matrix containing some (or all) columns of X. The N-vector of errors, ε, is assumed to be $N(0_N, h^{-1}I_T)$.[3] Since there are 2^K possible subsets of X, there are 2^K possible choices for X_r and, thus, $R = 2^K$.

11.2.3 The Prior

The choice of prior can be quite important when doing Bayesian model averaging. A proper prior is required to yield meaningful posterior odds. However, we want a prior which does not require detailed input from the researcher. Given the computational demands of Bayesian model averaging, we use a Normal-Gamma natural conjugate prior. As discussed in detail in Chapter 3, analytical results exist for posterior moments and posterior model probabilities. The choice of hyperparameters is crucial and several different recommendations exist in the literature. Here we describe the strategy recommended in Fernandez, Ley and Steel (2001b).

It is important first to remember the rule of thumb mentioned throughout this book: *When comparing models using posterior odds ratios, it is acceptable to use noninformative priors over parameters which are common to all models. However, informative, proper priors should be used over all other parameters.* This consideration suggests that it is acceptable to use the standard noninformative prior for h,

$$p(h) \propto \frac{1}{h} \tag{11.4}$$

and for the intercept,

$$p(\alpha) \propto 1 \tag{11.5}$$

To make absolutely certain that the noninformative prior for the intercept has the same implications for every model, Fernandez, Ley and Steel (2001b) recommend standardizing all the explanatory variables by subtracting off their means. This will have no effect on the slope coefficients, β_r, but ensures that the intercept can be interpreted in the same manner in every model as measuring the mean of y.[4] We follow this recommendation below.

With these prior choices made, we need only worry about the prior for β_r. Remember that the natural conjugate Normal-Gamma prior implies (see Chapter 3, Section 3.4)

$$\beta_r|h \sim N(\underline{\beta}_r, h^{-1}\underline{V}_r) \tag{11.6}$$

[3]The notation is a bit sloppy here. Formally, we should put r subscripts on each intercept and error precision. However, since these parameters are common to all models and have the same interpretation in all models, we simply write them as α and h.

[4]To be precise, if explanatory variables are measured as deviations from means then, by construction, they will have mean zero. Since the error also has mean zero, this implies the mean of the dependent variable is the intercept.

It is common practice to center priors over the hypothesis that explanatory variables have no effect on the dependent variable. In the case where we have many potential explanatory variables, but suspect that many of them may be irrelevant, the case for centering the prior in this way is strengthened. Thus, we set

$$\underline{\beta}_r = 0_{k_r}$$

It remains to choose $\underline{V}_r$. To this end, we use something called a *g-prior*, which involves setting

$$\underline{V}_r = [g_r X_r' X_r]^{-1} \tag{11.7}$$

The g-prior was first introduced in Zellner (1986), and the reader is referred to this paper for a detailed motivation. Here we note that this is a commonly-used benchmark prior, which requires only that the researcher elicit the scalar prior hyperparameter g_r. The prior is slightly unusual since it depends upon X_r, the matrix of explanatory variables in the model. However, since we are conditioning on X_r in the likelihood function (e.g. see Chapter 3, Section 3.3) and the posterior, we are not violating any of the rules of conditional probability by conditioning on X_r in the prior as well. In contrast, conditioning on y in the prior would violate the rules of conditional probability. The g-prior says that the prior covariance of β_r is proportional to the comparable data-based quantity. To see this, remember that with a noninformative prior the key term relating to the posterior covariance matrix would be $\overline{V} = (X_r' X_r)^{-1}$ (see Chapter 3 (3.20)), which is proportional to (11.17).[5] There are other justifications for the g-prior (see Zellner, 1986), but having a prior with similar properties as the data information is sensible. Eliciting prior covariance matrices such as $\underline{V}_r$ is often very hard to do (especially the off-diagonal elements). The g-prior simplifies this task by reducing the choice to a single hyperparameter.

In summary, we are using noninformative priors for α and h and, for the slope coefficients, we are setting

$$\beta_r | h \sim N(0_{k_r}, h^{-1}[g_r X_r' X_r]^{-1})$$

We have not said anything yet about choosing g_r, but will do so once the posterior has been presented.

11.2.4 The Posterior and Marginal Likelihood

The posterior for the Normal linear regression model with natural conjugate prior has been presented in Chapter 3, and we will not repeat the derivations of that chapter. The crucial parameter vectors are β_r, and it is straightforward to adapt previous results to show that (see Chapter 3 (3.14)–(3.16)) the posterior for β_r

[5]The reader with a knowledge of frequentist econometrics may gain motivation by noting that the variance of the OLS estimator is $\sigma^2(X'X)^{-1}$.

follows a multivariate t distribution with mean

$$E(\beta_r|y, M_r) \equiv \overline{\beta}_r = \overline{V}_r X_r' y \tag{11.8}$$

covariance matrix

$$var(\beta_r|y, M_r) = \frac{\overline{\nu}\overline{s}_r^2}{\overline{\nu} - 2}\overline{V}_r \tag{11.9}$$

and $\overline{\nu} = N$ degrees of freedom. Furthermore,

$$\overline{V}_r = [(1 + g_r)X_r' X_r]^{-1} \tag{11.10}$$

and

$$\overline{s}_r^2 = \frac{\frac{1}{g_r+1} y' P_{X_r} y + \frac{g_r}{g_r+1}(y - \overline{y}\iota_N)'(y - \overline{y}\iota_N)}{\overline{\nu}} \tag{11.11}$$

where

$$P_{X_r} = I_N - X_r(X_r' X_r)^{-1} X_r'$$

These formulae are slightly different from those in Chapter 3 because we have integrated out the intercept, α.

Using the g-prior, the marginal likelihood for model r is

$$p(y|M_r) \propto \left(\frac{g_r}{g_r + 1}\right)^{\frac{k_r}{2}} \left[\frac{1}{g_r + 1} y' P_{X_r} y + \frac{g_r}{g_r + 1}(y - \overline{y}\iota_T)'(y - \overline{y}\iota_T)\right]^{-\frac{N-1}{2}} \tag{11.12}$$

The posterior model probabilities can be calculated in the standard way using (see Chapter 1 (1.5))

$$p(M_r|y) = cp(y|M_r)p(M_r) \tag{11.13}$$

where c is a constant which is the same for all models. This constant will cancel out in all relevant formulae below and can be ignored. Alternatively, the fact that $\sum_{r=1}^{R} p(M_r|y) = 1$ can be used to evaluate c.

We allocate equal prior model probability to each model, and set

$$p(M_r) = \frac{1}{R}$$

Thus, we can ignore $p(M_r)$ and just use the marginal likelihood (normalized) in our Bayesian model averaging exercise. To be precise.

$$p(M_r|y) = \frac{p(y|M_r)}{\sum_{j=1}^{R} p(y|M_j)} \tag{11.14}$$

However, it is worth mentioning that some authors recommend different choices for $p(M_r)$. For instance, many researchers prefer parsimony, and feel that simpler models should be preferred to more complex ones, all else being equal.

As discussed in Chapters 2 and 3, posterior odds ratios do include a reward for parsimony. However, some recommend including an additional reward for parsimony through $p(M_r)$. This can be done in a straightforward fashion by allowing $p(M_r)$ to depend upon k_r.

These formulae allow us to provide some additional motivation for the g-prior. The value $g_r = 0$ corresponds to a perfectly noninformative prior. The value $g_r = 1$ implies that prior and data information are weighted equally in the posterior covariance matrix (see (11.9) and (11.10)). Using this intuition, most researchers would say that $g_r = 1$ is too big. Thus, one strategy a researcher could follow is to try a range of values for g_r between 0 and 1. However, the most common strategy is to choose g_r based on some measure such as an information criterion (see Chapter 10, Section 10.3.5). It would be too much of a digression to discuss in detail the results which relate Bayes factors calculated using g-priors to various information criteria. The interested reader is referred to Fernandez, Ley and Steel (2001b). There are many results of the form "if g_r takes a certain form, then the log of the resulting Bayes factor approximates a certain information criterion". For instance, if $g_r = \frac{1}{[\ln(N)]^3}$ then the log Bayes factor mimics the Hannan–Quinn criterion with $c_{HQ} = 3$ (see Chapter 10 (10.57)) for large values of N. Fernandez, Ley and Steel (2001b), after extensive experimentation with artificial data, recommend choosing

$$g_r = \begin{cases} \frac{1}{K^2} & \text{if } N \leq K^2 \\ \frac{1}{N} & \text{if } N > K^2 \end{cases} \tag{11.15}$$

We use (11.15) in the following empirical example.

11.2.5 Bayesian Computation: Markov Chain Monte Carlo Model Composition

In theory, the results in the previous section should be all we need to carry out model averaging. However, in practice, the number of models under consideration (usually $R = 2^K$) makes it impossible to evaluate (11.8)–(11.14) for every possible model. Accordingly, there have been many algorithms developed which carry out Bayesian model averaging without evaluating every possible model. In this section, we describe one commonly-used algorithm, initially developed in Madigan and York (1995).

To provide some intuition for how many Bayesian model averaging algorithms work, it is useful to consider what posterior simulation algorithms such as MCMC do. These algorithms take draws from the parameter space. These draws are made to mimic draws from the posterior by taking many draws from regions of the parameter space where posterior probability is high and few draws from regions where the posterior probability is low. Thus, MCMC algorithms do not draw from every region of the parameter space, but focus on regions of high posterior probability. In Bayesian econometrics, models are random variables (albeit discrete ones), just like parameters. Hence, posterior simulators which draw from the model space instead of the parameter space can be derived. These algorithms do not need to evaluate every model, but rather focus on the models of high posterior model probability. The name Markov Chain Monte Carlo Model Composition, or MC^3, is motivated by the fact that the algorithm is drawing from model space.

The most common MC^3 algorithm for sampling from model space is based on a Metropolis–Hastings algorithm (see Chapter 5, Section 5.5). It simulates a chain of models, which we denote by $M^{(s)}$ for $s = 1, \ldots, S$. $M^{(s)}$ is the model drawn at replication s (i.e. $M^{(s)}$ is one of $M_1, \ldots, M_R$). In the case under consideration (i.e. the Normal linear regression model where models are defined by which explanatory variables are included), this MC^3 algorithms draws candidate models from a particular distribution over model space and then accepts them with a certain probability. If a candidate model is not accepted, then the chain remains at the current model (i.e. $M^{(s)} = M^{(s-1)}$). In other words, it is exactly like a traditional Metropolis–Hastings algorithm, except that models are being drawn instead of parameters.

In Chapter 5 (Section 5.5.2) the Random Walk Chain Metropolis–Hastings algorithm was introduced. This generated candidate draws from regions of the parameter space in the neighborhood of the current draw. The MC^3 algorithm developed by Madigan and York (1995) does exactly the same thing in model space. That is, a candidate model, M^*, is proposed which is drawn randomly (with equal probability) from the set of models, including (i) the current model, $M^{(s-1)}$, (ii) all models which delete one explanatory variable from $M^{(s-1)}$, and (iii) all models which add one explanatory variable to $M^{(s-1)}$. If candidate models are generated in this way, the acceptance probability has the form

$$\alpha(M^{(s-1)}, M^*) = \min\left[\frac{p(y|M^*)p(M^*)}{p(y|M^{(s-1)})p(M^{(s-1)})}, 1\right] \tag{11.16}$$

$p(y|M^{(s-1)})$ and $p(y|M^*)$ can be calculated using (11.12). In the common case where equal prior weight is allocated to each model, $p(M^*) = p(M^{(s-1)})$ and these terms cancel out of (11.16). In this case, the Bayes factor comparing M^* to $M^{(s-1)}$ is the only quantity which must be calculated in (11.16).

Posterior results based on the sequence of models generated from the MC^3 algorithm can be calculated by averaging over draws in the standard MCMC manner (see Chapter 4 (4.11)). So, for instance, (11.2) can be approximated by $\widehat{g}_{S_1}$, where

$$\widehat{g}_{S_1} = \frac{1}{S_1}\sum_{s=S_0+1}^{S} E[g(\phi)|y, M^{(s)}] \tag{11.17}$$

As with our previous MCMC algorithms, $\widehat{g}_{S_1}$ converges to $E[g(\phi)|y]$ as S_1 goes to infinity (where $S_1 = S - S_0$). As with other MCMC algorithms, a starting value for the chain, $M^{(0)}$, must be chosen and, hence, S_0 burn-in replications should be discarded to eliminate the effects of this choice. In most cases involving the Normal linear regression model with conjugate prior $E[g(\phi)|y, M^{(s)}]$ can be evaluated analytically and, hence, (11.17) can be calculated in a straightforward manner. Using similar reasoning, the frequencies with which models are drawn can be used to calculate Bayes factors. For instance, if the MC^3 algorithm draws the model M_r A times and the model M_s B times, then the ratio $\frac{A}{B}$ will converge to the Bayes factor comparing M_r to M_s.

As with any posterior simulator, it is important to verify convergence of the algorithm and to estimate the accuracy of approximations such as (11.17). Fernandez, Ley and Steel (2001b) suggest a simple way of doing this: based on a reduced set of models (e.g. every model visited by the MC^3 algorithm) calculate $p(M_r|y)$ analytically (using (11.12)) and using the MC^3 algorithm. If the algorithm has converged, then the two ways of calculating posterior model probabilities should yield the same result. The relationship between the analytical and MC^3 results give an idea of approximation error and simple diagnostics can be constructed to check for convergence. For instance, Fernandez, Ley and Steel (2001b) suggest calculating the correlation between the analytical and MC^3 posterior model probabilities and taking enough replications to ensure this correlation is above 0.99.

11.2.6 Empirical Illustration: Model Uncertainty in Growth Regressions

In the economic growth literature, researchers are interested in finding out which variables influence economic growth. A common tool is the cross-country growth regression involving data from many countries. A measure of output growth is used as the dependent variable, and there are many potential explanatory variables. In this empirical illustration, we consider Bayesian model averaging in such a framework. The empirical example closely follows Fernandez, Ley and Steel (2001a),[6] and we use their data set which covers $N = 72$ countries and contains $K = 41$ potential explanatory variables. The dependent variable is average per capita GDP growth for the period 1960–1992. For the sake of brevity, we do not list all of the explanatory variables here (see the original paper for a detailed description of all the data). However, Table 11.1 provides a list of short form names for all explanatory variables, which should be enough to provide a rough idea of what each explanatory variable is measuring. The point to stress here is that the list of potential explanatory variables is very long, reflecting economic factors (e.g. investment, exports), political factors (e.g. political rights, rule of law), cultural and religious factors (e.g. fraction of population belonging to different religious groups), demographic factors (e.g. population growth), geographic factors, education, etc.

[6] I would like to thank the authors of this paper for their help with this empirical illustration. The data was taken from the *Journal of Applied Econometrics* data archive (www.econ.queensu.ca/jae). In addition, the authors' website, http://mcmcmc.freeyellow.com/, contains the papers, data and Fortran code.

Table 11.1 Bayesian Model Averaging Results

Explanatory Variable	BMA Post. Prob.	Posterior Mean	Posterior St. Dev.
Primary School Enrollment	0.207	0.004	0.010
Life expectancy	0.935	0.001	3.4×10^{-4}
GDP level in 1960	0.998	−0.016	0.003
Fraction GDP in Mining	0.460	0.019	0.023
Degree of Capitalism	0.452	0.001	0.001
No. Years Open Economy	0.515	0.007	0.008
% of Pop. Speaking English	0.068	-4.3×10^{-4}	0.002
% of Pop. Speaking Foreign Lang.	0.067	2.9×10^{-4}	0.001
Exchange Rate Distortions	0.081	-4.0×10^{-6}	1.7×10^{-5}
Equipment Investment	0.927	0.161	0.068
Non-equipment Investment	0.427	0.024	0.032
St. Dev. of Black Market Premium	0.049	-6.3×10^{-7}	3.9×10^{-6}
Outward Orientation	0.039	-7.1×10^{-5}	5.9×10^{-4}
Black Market Premium	0.181	−0.001	0.003
Area	0.031	-5.0×10^{-9}	1.1×10^{-7}
Latin America	0.207	−0.002	0.004
Sub-Saharan Africa	0.736	−0.011	0.008
Higher Education Enrollment	0.043	−0.001	0.010
Public Education Share	0.032	0.001	0.025
Revolutions and Coups	0.030	-3.7×10^{-6}	0.001
War	0.076	-2.8×10^{-4}	0.001
Political Rights	0.094	-1.5×10^{-4}	0.001
Civil Liberties	0.127	-2.9×10^{-4}	0.001
Latitude	0.041	9.1×10^{-7}	3.1×10^{-5}
Age	0.083	-3.9×10^{-6}	1.6×10^{-5}
British Colony	0.037	-6.6×10^{-5}	0.001
Fraction Buddhist	0.201	0.003	0.006
Fraction Catholic	0.126	-2.9×10^{-4}	0.003
Fraction Confucian	0.989	0.056	0.014
Ethnolinguistic Fractionalization	0.056	3.2×10^{-4}	0.002
French Colony	0.050	2.0×10^{-4}	0.001
Fraction Hindu	0.120	−0.003	0.011
Fraction Jewish	0.035	-2.3×10^{-4}	0.003
Fraction Muslim	0.651	0.009	0.008
Primary Exports	0.098	-9.6×10^{-4}	0.004
Fraction Protestant	0.451	−0.006	0.007
Rule of Law	0.489	0.007	0.008
Spanish Colony	0.057	2.2×10^{-4}	1.5×10^{-3}
Population Growth	0.036	0.005	0.046
Ratio Workers to Population	0.046	-3.0×10^{-4}	0.002
Size of Labor Force	0.072	6.7×10^{-9}	3.7×10^{-8}

The prior is given in (11.4)–(11.7), with the choice of g_r given in (11.17). The posterior and marginal likelihood formulae necessary to implement Bayesian model averaging are given in (11.8)–(11.14). We have $R = 2^{41}$ models, which is far too many to evaluate. Hence, we use the MC^3 algorithm described in the previous section. The following results are based on taking 1 100 000 draws and discarding the first 100 000 as burn-in replications (i.e. $S_0 = 100\,000$ and $S_1 = 1\,000\,000$).

Elements in the column of Table 11.1 labeled 'BMA Post. Prob.' can be interpreted as the probability that the corresponding explanatory variable should be included. It is calculated as the proportion of models drawn by the MC^3 algorithm which contain the corresponding explanatory variable. Informally, this is a useful diagnostic for deciding whether an individual explanatory variable does have an important role in explaining economic growth. It can be seen that several variables (i.e. Life expectancy, GDP level in 1960, Equipment Investment and Fraction Confucian) do have an important role in explaining economic growth. Regardless of which other explanatory variables are included, these variables almost always exhibit strong explanatory power. However, for the rest of the explanatory variables, there is some uncertainty as to whether they do have important roles to play in explaining economic growth. And, for many of the explanatory variables, there is strong evidence that they should not be included.

The other two columns of Table 11.1 contain posterior means and standard deviations for each regression coefficient, averaged across models (as in (11.2), implemented using (11.17)). Remember that models where a particular explanatory variable is excluded are interpreted as implying a zero value for its coefficient. Hence, the average in (11.17) involves some terms where $E[g(\phi)|y, M^{(s)}]$ is calculated, and others where the value of zero is used. With the exception of the few variables with high BMA posterior probability, most of the posterior means are small relative to their standard deviations. Thus, Bayesian model averaging is indicating a high degree of uncertainty about which factors explain economic growth. Undoubtedly, this is an honest representation of the fact that the data are not informative enough to send a clear story.

The MC^3 algorithm allows for the calculation of posterior model probabilities by simply counting the proportion of draws taken from each model. For the top ten models, the column of Table 11.2 labeled '$p(M_r|y)$ MC^3 estimate' contains posterior model probabilities calculated in this way. The column labeled '$p(M_r|y)$ Analytical' contains the exact values calculated using (11.12) and (11.14). It can be seen that the posterior model probability is widely scattered across models with no single model dominating. In fact, the top ten models account for only a little more than 4% of the total posterior model probability. The numbers in tables such as this (perhaps with many more models) allow for an assessment of the convergence of the MC^3 algorithm. The numbers in the last two columns of Table 11.2 are slightly different from one another. These differences are small enough for present purposes and we can be confident that the numbers in Table 11.1 are approximately correct. However, for a research paper

Table 11.2 Posterior Model Probabilities for Top 10 Models

	$p(M_r\|y)$ Analytical	$p(M_r\|y)$ MC^3 estimate
1	0.0089	0.0088
2	0.0078	0.0080
3	0.0052	0.0052
4	0.0035	0.0035
5	0.0032	0.0035
6	0.0029	0.0029
7	0.0028	0.0028
8	0.0028	0.0025
9	0.0028	0.0025
10	0.0024	0.0023

a higher degree of accuracy is typically called for and the number of replications should be increased. Here, we have used 1 000 000 replications, which sounds like a large number. However, the number of possible models is 2^{41} and, with 1 000 000 replications, we have visited only a tiny fraction of them.[7]

The posterior model probabilities in the last column of Table 11.2 are calculated using the models actually visited by the MC^3 algorithm. This raises a minor issue. Posterior model probabilities (unlike Bayes factors) require the calculation of $\sum_{j=1}^{R} p(y|M_j)$ (see (11.14)). However, in Table 11.2 we have taken the sum over the models visited by the MC^3 algorithm (not all R models). The omitted models will be those with low marginal likelihoods, so the effect of omitting them will usually be very small, especially if enough replications are taken to ensure convergence. Nevertheless, it does mean that the posterior model probabilities in Table 11.2 are slight over-estimates. George and McCulloch (1997) suggest a simple way of correcting for this. They recommend choosing a predetermined subset of the models. Index this set of models as $j = 1, \ldots, R^*$. $\sum_{j=1}^{R^*} p(y|M_j)$ can be calculated analytically. The MC^3 visit frequencies can be used to provide an estimate of this same quantity. The ratio of the analytical result to the estimate can be used to correct posterior model probabilities such as those given in Table 11.2.

To help see why Bayesian model averaging is so important, Table 11.3 presents posterior results for the single most preferred model (i.e. the one with highest marginal likelihood). The column labeled 'BMA Post. Prob.' repeats information from Table 11.1. The other two columns contain the posterior mean and standard deviation for each regression coefficient. Table 11.2 showed that the posterior model probability associated with this preferred model is only 0.0089. Hence, if we were to base our results on Table 11.3, we would be putting all our faith in a model which we are more than 99% sure is not the correct one!

[7] Fernandez, Ley and Steel (2001a) take many more draws to ensure convergence of the MC^3 algorithm. This accounts for the slight differences between the results they present and those presented here.

Table 11.3 Posterior Results for the Preferred Model

Explanatory Variable	BMA Post. Prob.	Posterior Mean	Posterior St. Dev.
Life expectancy	0.935	0.001	1.9×10^{-4}
GDP level in 1960	0.998	−0.017	0.002
Degree of Capitalism	0.452	0.003	7.9×10^{-4}
Equipment Investment	0.927	0.159	0.039
Non-equipment Investment	0.427	0.064	0.019
Sub-Saharan Africa	0.736	−0.013	0.003
Fraction Confucian	0.989	0.058	0.011
Fraction Muslim	0.651	0.010	0.003
Fraction Protestant	0.451	−0.011	0.004
Rule of Law	0.489	0.017	0.004

A comparison of Tables 11.1 and 11.3 indicates that, by choosing a single model, the researcher can be seriously misled. That is, a key output of cross-country growth regressions is an estimate of the marginal effect of each explanatory variable on economic growth. For some variables (e.g. Life expectancy and GDP level in 1960) the posterior means are roughly the same in the two tables. However, for some explanatory variables, the posterior means are quite different. For instance, the marginal effect of the Non-equipment Investment variable is estimated at 0.024 in Table 11.1, but is almost three times as big in Table 11.3. Furthermore, the posterior standard deviations in Table 11.3 are all much smaller than those in Table 11.1. Intuitively, the strategy of selecting a single model ignores uncertainty about whether that model is the correct one. By ignoring model uncertainty, Table 11.3 provides a misleadingly over-confident view of the accuracy of results. An examination of results for the Rule of Law variable illustrate this point nicely. The econometrician who chose a single model would report that the estimate of the marginal effect of the Rule of Law on economic growth was quite substantial (0.017) and very different from zero (i.e. the point estimate is more than four standard deviations from zero). However, the econometrician who used Bayesian model averaging would report that the marginal effect was more modest (0.007) and not very different from zero (i.e. the point estimate is less than one standard deviation from zero). Thus, the econometrician who chose a single model could come to a very misleading conclusion about the effect of an extremely crucial variable on economic growth.

11.3 EXTENSIONS

So far we have shown how Bayesian model averaging can be implemented in one particular model (the Normal linear regression model) using one particular computational technique (an MC^3 algorithm). However, it is worth mentioning that the literature on Bayesian model averaging is replete with discussion of other models and other methods. Hoeting, Madigan, Raftery and Volinsky (1999) provide a good overview of much of this discussion (and their list of citations provides even more). The Bayesian Model Averaging website (http://www.research.att.com/~volinsky/bma.html) provides links to many papers and software.

Bayesian model averaging in models other than the Normal linear regression model is usually implemented using approximations. That is, Bayesian model averaging involves marginal likelihoods (or Bayes factors) and an MC^3 algorithm. In the Normal linear regression model with natural conjugate prior, analytical formulae exist for the marginal likelihoods. For many other models, such analytical formulae do not exist, but accurate approximations to marginal likelihoods (or Bayes factors) do exist. These approximations can then be used along with the MC^3 algorithm described above to carry out Bayesian model averaging.

The MC^3 approach described in this chapter is simple and popular. However, various authors have proposed alternative approaches which can, in some cases, be less computationally demanding and have better convergence properties. Clyde, Desimone and Parmigiani (1996) present an interesting approach involving importance sampling. Other algorithms with names like *reversible jump Markov Chain Monte Carlo* are popular. Although such techniques have not been commonly used in the econometrics literature, there are several accessible papers in the Bayesian statistics literature (e.g. Clyde, 1999) for those interested in learning more about computation in this rapidly developing field. A few key references for the reader interested in learning more about related approaches include Carlin and Chib (1995), Carlin and Louis (2000, pp. 211–225) and Phillips and Smith (1996).

In this chapter, we have focused on model *averaging*. However, similar techniques are often used for model *selection*. We have argued above that, by selecting a single model when alternative plausible models exist, misleading inferences can result since model uncertainty is ignored. However, there are some cases where selecting a single model may be desirable. For instance, Bayesian model averaging typically requires a parameter (or function of parameters) to exist which has the same interpretation in every model (what we have called ϕ at the beginning of this chapter). If such a parameter does not exist, then it does not make much sense to average across models. Of course, the researcher can simply use the techniques described in this chapter and choose the model with highest posterior probability. However, many other approaches exist. A few key recent references for the interested reader include Brown, Fearn and Vannucci (1999), Chipman, George and McCulloch (1998), Clyde (1999) and George and McCulloch (1993).

Of course, the list of references in this section is not meant to be exhaustive. Nevertheless, the reader who looks at the papers mentioned here and examines their bibliographies will get a good overview of the literature up to year 2000. However, this field is a rapidly changing one, with influential new papers coming out every year. Hence, the reader interested in delving deeply in the field may wish to search through recent journal issues and check the Bayesian model averaging website for updated information.

11.4 SUMMARY

This chapter discussed Bayesian model averaging. The basic idea of Bayesian model averaging is straightforward: when many plausible models exist, the researcher should not base inference on a single model, but rather average over all models. The laws of probability imply that the weights in the model averaging should be posterior model probabilities. However, implementing Bayesian model averaging can be computationally demanding, since it is quite common for the number of models to be astronomical. A leading example is the linear regression model with many potential explanatory variables. If models are defined based on whether a particular explanatory variable is included or excluded, then 2^K models exist (where K is the number of potential explanatory variables). In the light of this problem, the Bayesian model averaging literature has focused on classes of models where analytical results exist and has developed various MC^3 algorithms which draw from the model space (instead of the parameter space). In this chapter, we have focused on one particular model class and MC^3 algorithm (although other models and approaches are briefly mentioned in a section on Extensions). In particular, we have described how to do Bayesian model averaging using the Normal linear regression model with natural conjugate prior using an MC^3 algorithm first developed by Madigan and York (1995). The MC^3 algorithm shares some similarities with a Random Walk Chain Metropolis–Hastings algorithm, in that it involves drawing models in the neighborhood of the current draw and accepting them with a certain probability. For a given model, the natural conjugate prior means that analytical posterior and marginal likelihood results existed. Hence, it is computationally feasible to draw the huge numbers of models necessary to achieve convergence of the MC^3 algorithm.

Since Bayesian model averaging often involves working with huge numbers of models, elicitation of an informative prior for each model can be a difficult or impossible task. Accordingly, we describe a particular benchmark prior, based on the g-prior, which can be used automatically without requiring the researcher to subjectively select values for prior hyperparameters.

This chapter also included an empirical illustration of Bayesian model averaging based on Fernandez, Ley and Steel (2001a). Using cross-country growth regressions, the illustration showed how to implement Bayesian model averaging. Furthermore, it showed the importance of Bayesian model averaging in that some key results using Bayesian model averaging are very different from those obtained based on a single preferred model.

11.5 EXERCISES

The exercises in this chapter are closer to being small projects than standard textbook questions. Remember that some data sets and MATLAB programs are available on the website associated with this book.

1. The empirical illustration in this chapter used only one choice for g_r and did not discuss the issue of prior sensitivity analysis. Do the empirical illustration over again for other choices for this prior hyperparameter. You may choose whatever values you want. Choices used by other researchers include $g_r = \frac{k_r}{N}$, $g_r = \frac{K^{\frac{1}{k_r}}}{N}$, $g_r = \sqrt{\frac{1}{N}}$, $g_r = \sqrt{\frac{k_r}{N}}$, $g_r = \frac{1}{[\log(N)]^3}$ and $g_r = \frac{\log(k_r+1)}{\log(N)}$. Are empirical results sensitive to prior choice for this data set? The data (along with MATLAB code for doing the empirical illustration) can be found on the website associated with this book. The data is also available from the *Journal of Applied Econometrics* data archive (www.econ.queensu.ca/jae) under Fernandez, Ley and Steel (2001a).
2. Another popular approach which addresses issues similar to Bayesian model averaging is described in George and McCulloch (1993). This approach involves using the Normal linear regression model with independent Normal-Gamma prior (see Chapter 4), with one minor alteration. This alteration is in the prior for the regression coefficients and is useful for the case where a large number of explanatory variables exist, but the researcher does not know which ones are likely to be important. To capture this, the prior for each regression coefficient is a mixture of two Normals, both with mean zero. One of the terms in the mixture has very small variance (i.e. it says the coefficient is virtually zero), and the other has a large variance (i.e. it allows the coefficient to be large). To be precise, for each coefficient β_j for $j = 1, \ldots, K$, the prior is

$$\beta_j|\gamma_j \sim (1-\gamma_j)N(0, \tau_j^2) + \gamma_j N(0, c_j^2\tau_j^2)$$

   where c_j and τ_j are known prior hyperparameters with τ_i being small and c_j large. In addition, $\gamma_j = 0$ or 1 with $P(\gamma_j = 1) = p_j$ and $0 \leq p_j \leq 1$.
   (a) Using a particular prior for each p_j (e.g. a Uniform prior or the prior given in Chapter 10 (10.46)), derive a Gibbs sampler with data augmentation this model. Hint: This prior involves a two-component Normal mixture and the derivations of Chapter 10 (Section 10.3) are relevant here. If you are having trouble, you may wish to start with a simpler variant of this model where $\tau_j \equiv \tau, c_j \equiv c$ and $\gamma_j \equiv \gamma$ (i.e. the same prior is used for every coefficient).
   (b) Using the cross-country growth data set (see Exercise 1 for details) and your answer to part (a), carry out Bayesian inference using this approach. With regards to the issue of prior elicitation, for the error precision you may wish to make a noninformative choice. The choice of the prior hyperparameters τ_j and c_j is a bit more tricky, since defining what is 'small' and 'large' depends upon the interpretation of the marginal effect (which depends upon the units the explanatory variables are measured in). To surmount this problem, you may wish to standardize each explanatory variable by subtracting off its mean and dividing by its standard deviation. This ensures that each coefficient measures the effect on the dependent variable

of a one standard deviation change in the explanatory variable. With variables standardized in this way, in most applications it makes sense to set $\tau_j \equiv \tau$ and $c_j \equiv c$. Start by setting $\tau = 0.0001$ and $c = 1000$, but experiment with various values for both. The ambitious reader may wish to look at George and McCulloch (1993, Section 2.2) for more sophisticated ways of choosing these prior hyperparameters.

12 Other Models, Methods and Issues

12.1 INTRODUCTION

In a book such as this, it is not possible to cover all of the models and methods used by Bayesian econometricians. Indeed, one of the messages of this book is that we should not seek to create a complete set of off-the-shelf models and associated methods for the Bayesian econometrician to choose from when confronting a particular empirical application. The logic of Bayesian inference can be used with any model, and the tools introduced in the present book (e.g. of posterior simulation) have wide applicability. Thus, the reader of this book should not come away thinking "I know the models I need to do Bayesian econometric research", but rather, "I know the necessary tools to do Bayesian econometric research in any model I should wish to design". When confronting a new data set, the researcher should first think of what appropriate models might be (i.e. likelihoods and priors). The appropriate models might be of a standard design used by previous researchers, but they might not be. Once a set of models is chosen, the researcher should use the rules of probability to work out the posteriors, marginal likelihoods and predictives. Finally, methods for working with all of these should be developed. These methods will usually involve posterior simulation, and the present book has introduced a variety of methods (e.g. the Gibbs sampler or Metropolis–Hastings algorithms). These methods of posterior simulation have been introduced in the context of particular models, but we stress that they are widely applicable.

The argument of the previous paragraph suggests that the present book need not be complete and cover every possible model and method. Hence, the book could end here. However, it is useful to have one final chapter which briefly describes some other common models, methods and issues. The purpose of this chapter is not to be thorough, but rather offer a brief discussion of the basic issues involved, and then direct the reader to the relevant literature. Even with this chapter, we will be far from mentioning every popular model and tool in the Bayesian literature. The choices made in this chapter reflect to a large extent the personal opinions of the author in terms of what models are or will be popular, what new methods of posterior simulation might become popular in the future, etc.

This chapter is organized in sections involving models, methods (primarily of posterior simulation) and other issues (primarily relating to prior elicitation and identification).

12.2 OTHER METHODS

In this book, we have gone through several methods of posterior simulation. For most of the models we have used Gibbs sampling with data augmentation. Such methods work very well in most applications. However, especially in models involving latent data, convergence issues can be a problem. That is, the necessary number of burn-in and included replications may have to be so large that the computational burden can be quite high (or even prohibitive). This problem tends to arise when the number of parameters in the model is large and the correlation between the parameters is quite high. Since latent data can be interpreted as a vector of parameters, the number of parameters in many of the models discussed in this book is quite high. The Bayesian statistics literature is replete with algorithms which attempt to create more efficient posterior simulation algorithms for particular classes of models. For instance, in Chapter 9 we cited Liu and Wu (1999), Meng and van Dyk (1999) and van Dyk and Meng (2001) as papers which sought to develop more efficient algorithms for posterior simulation in the multinomial probit model. The methods discussed in these papers also have wider relevance for other MCMC algorithms with data augmentation. In our discussion of state space models (see Chapter 8, Section 8.3.1), we introduced a particular algorithm due to DeJong and Shephard (1995). This algorithm was designed to allow for more efficient posterior simulation than other, simpler, algorithms used previously. As general advice, when working with a model with latent data, it is wise to be very careful about convergence issues (using MCMC diagnostics) and, if necessary, do further research in the Bayesian statistical literature to try and find more efficient algorithms which can be adapted to the model under study.

Several of the posterior simulation algorithms used in this book were either Metropolis–Hastings or Metropolis-within-Gibbs algorithms. Such algorithms take draws from a candidate distribution and then accept them with a certain probability. If the candidate is not accepted, then the chain remains at the current draw (see Chapter 5, Section 5.5). If the candidate distribution is poorly chosen, then the acceptance probability can be low and the chain of draws can stay fixed at a single value for a long time. In such cases, the chain can take a long time to converge and the computational cost can be large. The easy advice to give in this case is simply to choose a better candidate distribution for generating draws. However, this can be hard to do, especially if the number of parameters is large.

The problems described in the previous two paragraphs (i.e. Gibbs sampler convergence and choice of candidate generating densities) are not too daunting in many models, particularly the regression-based models which are the focus of the present book. However, should they occur, there are a few approaches which may be worth investigating. These will be briefly described in this section.

The posterior simulators used in this book all involve taking draws from common densities for which computer code exists in many places. These include the Normal, Gamma, Wishart, etc. However, there are several methods which are more general, and can be used to draw from a wide variety of non-standard densities. If one of these methods can be used to directly draw from a posterior (or conditional posterior) distribution, then it may no longer be necessary to use a Metropolis–Hastings (or Metropolis-within-Gibbs) algorithm (which may be plagued by low acceptance probabilities).

Three popular methods are called *acceptance sampling*, the *Griddy–Gibbs sampler* and *adaptive rejection sampling*. Acceptance sampling is based on the following theorem.

Theorem 12.1: Acceptance Sampling

Let $p^*(\theta|y)$ be the kernel of a posterior density (i.e. $p(\theta|y) = cp^*(\theta|y)$) and $q(\theta)$ be the kernel of a density it is convenient to draw from. $q(\theta)$ is referred to as the *source density* kernel, and it must be defined over the same support as $p(\theta|y)$. Assume the source and posterior density kernels satisfy

$$0 \leq \frac{p^*(\theta|y)}{q(\theta)} \leq a \tag{12.1}$$

where a is a finite constant. Then an algorithm which takes random draws from $q(\theta)$ and accepts them with probability

$$\frac{p^*(\theta|y)}{q(\theta)a} \tag{12.2}$$

will yield random draws from $p(\theta|y)$.

An examination of this theorem indicates that the success of acceptance sampling depends upon finding a source density which satisfies the bound in (12.1). It is not always possible to do so. Furthermore, the source density must be such that the acceptance probability in (12.2) does not tend to be too small. It is not hard to find examples where the acceptance probability is so minuscule that acceptance sampling is simply not computationally feasible. Nevertheless, acceptance sampling can be useful in many cases, and is a commonly-used tool. In fact, the Normal and Gamma random number generators provided by common computer packages typically use acceptance sampling.

The Griddy–Gibbs sampler is discussed in Ritter and Tanner (1992). It is an approximate method whose basic idea can be described quite simply. Suppose you have a non-standard posterior (or conditional posterior in a Gibbs sampler) that you wish to draw from. This can be approximated by taking a grid of points and evaluating the posterior at each point. These can be used to compute a multinomial approximation to the posterior. Since the multinomial is easy to draw from, approximate draws from the non-standard posterior can be obtained in a straightforward manner. In other words, the Griddy–Gibbs sampler involves taking a discrete approximation to the non-standard posterior (or conditional posterior) of interest.

Adaptive rejection sampling is a popular method for drawing from log-concave densities (see Gilks and Wild (1992) for details). Gilks (1996) and Gilks and Roberts (1996) are two good papers for the reader interested in looking more deeply into other methods for drawing from non-standard posteriors or, in general, improving performance of MCMC algorithms.

Posterior computation can be slow in, for example, Metropolis-within-Gibbs algorithms if some of the blocks have non-standard form. Damien, Wakefield and Walker (1999) describe a method which introduces latent data in a particular way which turns such a non-standard problem into a Gibbs sampler with data augmentation where all of the blocks are easy to draw from. We shall not describe their approach here, but note only that it seems to be a very useful approach for a wide variety of models involving hierarchical priors which are slight generalizations of those discussed in this book. For instance, the algorithm seems to work well for random effects panel data models with nonlinear regression functions or non-standard dependent variables (e.g. where the dependent variable is a duration or a count). There have also been some recent advances in importance sampling (see Richard and Zhang, 2000) which are more efficient than implementations used previously. *Efficient importance sampling* is well-suited to models with latent data, and involves using a sequence of least squares regressions to find an importance function that accurately approximates the underlying posterior density.

Various approximations also exist that allow the researcher to avoid posterior simulation altogether. Most simply, in many cases, posterior distributions converge to Normality as $N \rightarrow \infty$. A typical theorem in this *Bayesian asymptotic* literature would take the following form.

Theorem 12.2: A Bayesian Central Limit Theorem

Under suitable regularity conditions, as $N \rightarrow \infty$, the posterior distribution can be approximated by

$$\theta | y \sim N(\widetilde{\theta}, [I(\widetilde{\theta})]^{-1}) \tag{12.3}$$

where $\widetilde{\theta}$ is the posterior mode[1] and $I(\widetilde{\theta})$ is the observed information matrix evaluated at the point $\widetilde{\theta}$. The observed information matrix is defined as the negative of the Hessian of the log of the posterior (i.e. the matrix of second derivatives of the log posterior with respect to the elements of θ).

[1] $\widetilde{\theta}$ can also be the maximum likelihood estimate and the same theorem holds. Intuitively, in regularly behaved cases, as sample size increases, data information becomes more and more predominant over prior information. Thus, the prior is irrelevant asymptotically.

This theorem can be proven under a variety of regularity conditions (see Poirier (1995, p. 306) for details). Loosely speaking, these regularity conditions come down to excluding pathologies in the prior (e.g. the prior must not attach zero weight to regions of the parameter space near the maximum likelihood estimate) and excluding certain cases where the dimensionality of the parameter space increases with sample size. Hence, Theorem 12.2 can be used to carry out approximate Bayesian inference in most models. From a computational perspective, all that is required is a program to maximize the posterior (or the likelihood function) and evaluate the Hessian. Most relevant computer software has programs which maximize arbitrary functions and calculate Hessians (although it is usually better for the user to program up the Hessian analytically).

Normal approximations are not that commonly used by Bayesian econometricians since the resulting approximation error can be quite large. It is much more common to use *Laplace approximations*, which are much more accurate. A particular variant, introduced in Tierney and Kadane (1986) is particularly accurate. Formally, the Tierney–Kadane approach is an asymptotic one where the approximation error is of order $\frac{1}{N^2}$ (as opposed to $\frac{1}{N}$ or even $\frac{1}{\sqrt{N}}$ in other asymptotic approximations).

The Laplace approximation says that if $f(\theta)$ is a smooth positive function of a k-vector θ and $h(\theta)$ is a smooth function with a unique minimum $\widetilde{\theta}$, then the integral

$$I = \int f(\theta)e^{-Nh(\theta)}d\theta \tag{12.4}$$

can be approximated by

$$\widetilde{I} = f(\widetilde{\theta})\left(\frac{2\pi}{N}\right)^{\frac{k}{2}}|\widetilde{\Sigma}|^{\frac{1}{2}}\exp\left[-Nh(\widetilde{\theta})\right] \tag{12.5}$$

where $\widetilde{\Sigma}$ is the negative inverse of the Hessian of $h(\theta)$ evaluated at $\widetilde{\theta}$. The approximation error is of order $\frac{1}{N}$.

Various Bayesian features of interest (e.g. posterior means of parameters or functions of parameters and marginal likelihoods) can be put in the form of (12.4) by making suitable choices for $f(\theta)$ and $h(\theta)$. However, Tierney and Kadane (1986) note that the accuracy of the approximation can be improved even more in many Bayesian cases. Consider, for instance, the question of approximating $E[g(\theta)|y]$, where $g()$ is a positive function of the parameters of the model. Using our general notation for prior, likelihood and model parameters, this can be written as

$$E[g(\theta)|y] = \frac{\int \exp[\log\{g(\theta)\} + \log\{p(y|\theta)p(\theta)\}]d\theta}{\int \exp[\log\{p(y|\theta)p(\theta)\}]d\theta} \tag{12.6}$$

The numerator and denominator can both be approximated using (12.5). In terms of the notation of (12.4), the numerator implies $f(\theta) = 1$ and $h(\theta) =$

$-\frac{\log\{g(\theta)\}+\log\{p(y|\theta)p(\theta)\}}{N}$. The denominator implies $f(\theta) = 1$ and $h(\theta) = -\frac{\log\{p(y|\theta)p(\theta)\}}{N}$. Tierney and Kadane (1986) show that the fact that $f(\theta)$ is the same in numerator and denominator implies that part of the approximation error cancels out when the ratio in (12.6) is taken and, as a result, the approximation error is of order $\frac{1}{N^2}$. As with the Normal approximation, the Laplace approximation requires the maximization of functions and the calculation of Hessians. In practice, many researchers (e.g. Koop and Poirier, 1993) have found the Tierney–Kadane approximation to be highly accurate when compared with results obtained using posterior simulation.

The Tierney–Kadane approach does suffer from some drawbacks (beyond being an approximation). If the number of parameters is large, then the researcher may wish to calculate $E[g(\theta)|y]$ for many different choices of $g(\theta)$. Each one will require a new function maximization. For instance, if the researcher wishes to calculate posterior means and standard deviations for each of the k parameters, then $2k + 1$ maximizations (and Hessian calculations) must be performed. If k is large, then this can become computationally burdensome. In contrast, adding a new $g(\theta)$ to a posterior simulation program will usually be virtually costless. Furthermore, the Laplace approximation requires $g(\theta)$ to be positive. For many things (e.g. regression coefficients) this may be unreasonable. One simple trick to get around this problem is to redefine the function of interest as $g(\theta) + c$ where c is a large positive constant. Tierney, Kass and Kadane (1989) discuss alternative ways of getting around this problem. A few relevant references for the reader interested in learning more about Laplace approximations and their use in Bayesian inference are Carlin and Louis (2000, pp. 121–129), Poirier (1995, pp. 306–309) and Tierney and Kadane (1986). Kass and Raftery (1995, pp. 777–779) discuss the use of Laplace and other approximations for calculating Bayes factors.

12.3 OTHER ISSUES

Prior elicitation remains one of the most controversial areas of Bayesian econometrics. In this book, we have focused mainly on subjective elicitation of priors. Furthermore, prior sensitivity analyses can be used to investigate the influence of prior on posterior results. However, we have stressed that, if the researcher does not have subjective prior information, noninformative priors or empirical Bayesian methods can be used. Noninformative priors have the drawback that they tend to be improper and, as we have seen, this makes it difficult to calculate meaningful Bayes factors (e.g. see the discussion in Chapter 3, Section 3.6.2). Empirical Bayesian methods are quite popular when working with hierarchical priors (see Chapter 8, Section 8.2.3) and seem to work well in practice. However, from a theoretical perspective they can be criticized on the grounds that they allow for data information to be used in selecting the prior. Accordingly, much research has been and is being done in the Bayesian statistical literature on the issue of prior elicitation. In this section, we briefly point the reader in the direction of some of this research. A good general discussion of the issues surrounding prior elicitation is provided in Poirier (1995, Chapter 6, Section 6.8).

There are many papers which aim to make it easier to elicit informative priors. As discussed in Chapters 2 and 3, natural conjugate priors, since they imply that prior and posterior have the same distributional form, imply that the prior can be interpreted as arising from a hypothetical sample of data. Different models, of course, have different natural conjugate priors (if they exist). Hence, there are many papers which derive natural conjugate priors for particular models. For instance, Koop and Poirier (1993) derives the natural conjugate prior for the multinomial logit model. Many other papers discuss prior elicitation about functions of the original model parameters. For instance, in the linear regression model the researcher may not have prior information about individual regression coefficients. However, functions of the regression coefficients may be easy to interpret and, hence, priors for such functions can be easier to elicit (e.g. see Poirier, 1996a). As an example, consider a microeconomic application where a production function is being estimated. The researcher may want to elicit prior information about returns to scale (which is a function of regression coefficients). A related line of research is epitomized by Kadane *et al.* (1980). Roughly speaking, this paper sets out a method where a prior over the model parameters is inferred based on a researcher's answers to questions about predictives (e.g. "if the explanatory variables were set with values $X = x$, what would you expect y to be?"). In general, many researchers recommend investigating what the prior says about hypothetical data through simulation. For instance, in the linear regression model, the researcher would take random draws, $\beta^{(s)}$ and $h^{(s)}$ from the prior, $p(\beta, h)$. For each of these draws, hypothetical data, $y^{(s)}$, can be drawn from $p(y|\beta, h)$. If these hypothetical data sets cover a reasonable range, then the prior is a sensible one.

Another strand of the literature seeks to develop noninformative priors for various models. In this paper, we have been quite informal about what we mean by a 'noninformative' prior. We have argued that if one simply takes a Normal prior with large (or infinite) variance or a Gamma prior with small (or zero) degrees of freedom, then data information will predominate and the prior can be referred to as noninformative. In practice, this strategy works well in most applications. However, there are some important theoretical issues relating to 'noninformativeness' that have been discussed in the literature. Poirier (1995, Chapter 6, Section 6.8) provides a good introduction to many of these issues (see also Zellner, 1971). Intuitively, Uniform priors sound as though they should be noninformative. However, a prior which is Uniform over a model's parameters will not be Uniform over nonlinear reparameterizations of these parameters (see Chapter 2, Exercise 1). Thus, this definition of 'noninformativeness' depends upon how precisely the researcher chooses to parameterize the model. In response to this, and other, issues, some Bayesians prefer to avoid the term 'noninformative'. Instead, they seek to develop priors which allow the data to be dominant and can be used in an automatic fashion (i.e. the researcher does not have to subjectively elicit prior hyperparameters). These priors are called *reference priors*. We do not discuss the principles underlying reference prior analysis in any detail here. The textbook *Bayesian Theory* (Bernardo and Smith, 1994) contains a statistically rigorous discussion of these principles. Other influential recent papers include Bernardo and Ramon (1998) and Kass and Wasserman (1996).

One popular reference prior is referred to as *Jeffreys' prior* (see Jeffreys, 1946). This is motivated by worries over the lack of invariance to transformation of some noninformative priors. Using our usual general notation, where θ denotes the model's parameters and $p(y|\theta)$ the likelihood function, the *information matrix* is defined as

$$I(\theta) = E\left[-\frac{\partial^2 p(y|\theta)}{\partial\theta\partial\theta'}\right]$$

where the expected value is taken with respect to y (i.e. y is treated as the random variable with p.d.f. $p(y|\theta)$ in the expectation calculation).[2] Jeffreys' prior is defined as

$$p(\theta) \propto |I(\theta)|^{\frac{1}{2}} \tag{12.7}$$

This prior has the desirable property that it is invariant to reparameterization. That is, consider the model reparameterized in terms of $\alpha = h(\theta)$, where $h()$ is a vector of k functions. If a Jeffreys prior is used for α

$$p(\alpha) \propto \left|E\left[-\frac{\partial^2 p(y|\alpha)}{\partial\alpha\partial\alpha'}\right]\right|^{\frac{1}{2}}$$

then the same posterior inference would result as if $p(y|\theta)$ were combined with (12.7). The exact form of the Jeffreys' prior, since it depends upon the information matrix, is model specific (see Poirier (1994) for logit models). However, Jeffreys' prior seems to work well as a reference prior. One drawback is that it is often improper, which precludes calculation of meaningful Bayes factors in many cases.

For researchers who do not want to subjectively elicit informative priors, but do want to have a proper prior (e.g. so as to calculate meaningful Bayes factors), *training sample priors* are a possible choice. The basic idea underlying training sample priors is easy to explain, although there are many subtleties of relevance for the practitioner. The idea is to begin with a reference prior. This reference prior is then combined with a subset of the data (referred to as a *training sample*) to create a proper posterior. This 'posterior' is then used as a prior in a standard Bayesian analysis involving the remainder of the data. A key question is what training sample to use. One common suggestion (with independent data) is to find

[2] For the reader unfamiliar with matrix calculus, $\frac{\partial^2 p(y|\theta)}{\partial\theta\partial\theta'}$ is the matrix of second derivatives with ijth element given by $\frac{\partial^2 p(y|\theta)}{\partial\theta_i\partial\theta_j}$.

out the minimum number of data points which, when combined with the reference prior, yields a proper 'posterior'. A training sample involving this minimum number of data points is referred to as a *minimal training sample*. Bayesian analysis of any feature of interest (e.g. a Bayes factor) can be done using every possible minimal training sample, and final results presented which average across all training samples. Berger and Pericchi (1996) is an example of an influential recent paper in this literature that the interested reader may wish to look at.

Regardless of which approach to prior elicitation is used, it is sensible to carry out a prior sensitivity analysis. In this book, we have been informal about this procedure, offering little advice other than to try various sensible priors and see how they affect posterior results. However, there are some more rigorous methods for carrying out prior sensitivity analysis. In the context of the Normal linear regression model, we have briefly mentioned extreme bounds analysis (see the discussion in Chapter 2, Section 2.4, and Chapter 3, Exercise 6). This literature provides various bounds or intervals in which the posterior means of regression coefficients must lie. The reader interested in learning more about extreme bounds analysis and related issues is referred to Leamer (1978, 1982), Poirier (1995, pp. 529–537) and Iwata (1996). More broadly, there is a *Bayesian robustness* literature, which (among other things) considers how sensitive posterior results are to prior assumptions for many models. A few references for the interested reader are Berger (1985, 1990) and Kadane (1984).

Prior elicitation becomes especially important when a model suffers from an *identification problem*. We have come across a model with an identification problem in Chapter 9 (Section 9.6) with the multinomial probit model. In that model, all values for β and σ_{11} which satisfy $\frac{\beta}{\sqrt{\sigma_{11}}} = a$ (for any constant a) yield the same value for the likelihood function. There we defined identification as implying that many parameter values yielded the same value for the likelihood function. Many other models used by econometricians potentially involve identification problems, including simultaneous equations models (to be discussed shortly), some sorts of regime switching models (see Koop and Poirier, 1997) and nonlinear time series models, and several structural specifications based on theoretical models of individual behavior (e.g. job search models; see Koop and Poirier, 2001).

The following provides a formal definition.

Definition 12.1: Identification

Let $p(y|\theta)$ be a likelihood function which depends upon a k-vector of parameters, θ, defined on a set Θ. Then θ *is identified at the point* $\theta = \theta_0 \in \Theta$ if and only if, for all $\theta \in \Theta$, $\theta \neq \theta_0$ implies $p(y|\theta) \neq p(y|\theta_0)$ for all possible values of y. If θ is identified at all points $\theta_0 \in \Theta$, then we say θ *is identified.*

In one sense, lack of identification causes no problems for Bayesian inference. After all, provided $p(\theta)$ and $p(y|\theta)$ are valid p.d.f.s, then the posterior $p(\theta|y) \propto p(\theta)p(y|\theta)$ will be a valid p.d.f., and Bayesian inference can be carried out in the standard way. However, there are several issues which arise that are worth mentioning. Consider the case where the model is not identified, since $p(y|\theta) = c$ for all $\theta \in \Theta_0$ where $\Theta_0 \subseteq \Theta$. In this case, Bayes rule implies $p(\theta|y) \propto p(\theta)$ for all $\theta \in \Theta_0$. In words, the prior is equal to the posterior over the unidentified region, and we have not learned anything about the parameters in this region. If the unidentified region is unbounded, then the use of improper, noninformative, priors is precluded, as then the posterior is improper.

However, in some cases, prior information can allow us to learn about unidentified parameters in a manner that is impossible using frequentist econometrics. To illustrate this point, let us consider a concrete example where $\theta = (\theta_1, \theta_2)'$, and we have an extreme identification problem where $p(y|\theta_1, \theta_2) = p(y|\theta_1)$ for all values of θ_2. Hence, θ_2 is not identified. Bayesian inference can still be carried out through the posterior:

$$\begin{aligned} p(\theta_1, \theta_2|y) &\propto p(\theta_1, \theta_2)p(y|\theta_1, \theta_2) \\ &= p(\theta_1, \theta_2)p(y|\theta_1) \end{aligned} \tag{12.8}$$

If the prior exhibits independence between the two parameters, then $p(\theta_1, \theta_2) = p(\theta_1)p(\theta_2)$ and the posterior also exhibits independence, and we have

$$p(\theta_1|y) \propto p(\theta_1)p(y|\theta_1) \tag{12.9}$$

and

$$p(\theta_2|y) \propto p(\theta_2) \tag{12.10}$$

In words, we have a conventional posterior for the identified parameter, θ_1, but the posterior for θ_2 is simply its prior, and no data-based learning about θ_2 has occurred. However, if the prior does not exhibit independence, then (12.8) still holds, but we cannot break the posterior into two components as in (12.9) and (12.10). Technically, we have

$$p(\theta_2|y) = \int p(\theta_1, \theta_2)p(y|\theta_1)d\theta_1 \neq p(\theta_2)$$

and, thus, data-based learning does occur about θ_2. In words, the data allows us to learn about θ_1. However, the prior says θ_1 and θ_2 are correlated. Hence, the information we have learned about θ_1 spills over into learning about θ_2 through this correlation. Thus, if informative priors are available, learning about unidentified parameters can occur. These points about Bayesian inference in unidentified models (and many more) are made in Poirier (1998), and the reader is referred to this paper for further details.

12.4 OTHER MODELS

In this section, we briefly motivate several different classes of model and provide a few citations for the reader interested in learning more about them. We divide this section into subsections on time series, endogeneity, models with non-standard dependent variables, structural models and nonparametrics.

12.4.1 Time Series Models

Chapter 8 offered an introduction to time series methods via a discussion of state space models. State space models are a general and powerful tool for dealing with a wide variety of time series problems. However, there are many applications where other methods are commonly used or extensions of the linear state space models discussed in Chapter 8 are required. In this section, we briefly discuss some of these other methods and extensions. Bauwens, Lubrano and Richard (1999) is an excellent source for information about Bayesian time series methods. We provide additional citations for further readings on specific topics below.

Time Series Models for Finance

Financial economists often work with time series data (e.g. daily stock returns). However, since stock returns are difficult to predict,[3] interest often centers on the error variance. An empirical regularity seems to be that stock markets tend to move through periods of high/low volatility. Such volatility clustering suggests that the error variance might be predictable. This consideration motivates two commonly-used time series models: the *Autoregressive Conditionally Heteroskedastic* (or ARCH) model, and the *stochastic volatility* model.

If $y = (y_1, \ldots, y_T)'$ is a vector of time series data (e.g. y_t is the return of a stock on day t), then the ARCH model of order p (or ARCH(p)) is written as

$$y_t = \sqrt{\sigma_t^2}\varepsilon_t \tag{12.11}$$

where

$$\sigma_t^2 = \beta_0 + \beta_1 y_{t-1}^2 + \cdots + \beta_p y_{t-p}^2 \tag{12.12}$$

and ε_t is i.i.d. $N(0, 1)$.[4] It can be verified that the variance of the error is σ_t^2. Note that, since we need $\sigma_t^2 > 0$, restrictions need to be placed on β_j for $j = 0, \ldots, p$. This can be done through the prior. The error variance varies over time (i.e. is heteroskedastic) in a manner which depends upon (i.e. conditions on) past values of the dependent variable (i.e. in a similar fashion to an AR model). This motivates the name Autoregressive Conditional Heteroskedastic model. It is easy to extend (12.11) to allow for explanatory variables to create a regression model with ARCH errors. In some macroeconomic applications, such regression models with ARCH errors are popular. Another simple extension is to allow for ε_t to have Student-t (see Chapter 6, Section 6.4) or mixture of Normals (see Chapter 10, Section 10.3) errors. Another popular extension of the

[3]Financial theory suggests that predictable movements in stock market returns should be arbitraged away.

[4]The unit variance for ε_t is an identifying restriction.

ARCH model is the *Generalized Autoregressive Conditional Heteroskedastic* (or GARCH) model. This extends (12.12) to

$$\sigma_t^2 = \beta_0 + \beta_1 y_{t-1}^2 + \cdots + \beta_p y_{t-p}^2 + \gamma_1 \sigma_{t-1}^2 + \cdots + \gamma_q \sigma_{t-q}^3 \tag{12.13}$$

Chapter 7 of Bauwens, Lubrano and Richard (1999) discusses the properties of ARCH and GARCH models, and describes how Bayesian inference can be carried out. One problem with this class of models is that Bayesian computation is somewhat difficult. Early work such as Geweke (1988, 1989a) and Kleibergen and van Dijk (1993) used importance sampling. Bauwens and Lubrano (1998) develops a Griddy–Gibbs sampler, which seems to work well and avoids problems associated with fine-tuning importance functions.

The stochastic volatility model was mentioned briefly in Chapter 8 (Section 8.4). However, it is worth providing a few additional details here as it is similar to the ARCH and GARCH models. The difference lies in the fact that the variation in volatility is modeled through a latent variable. That is, a simple variant of the stochastic volatility model involves replacing (12.12) by

$$\log(\sigma_t^2) = \log(\sigma^2) + \beta \log(\sigma_{t-1}^2) + u_t \tag{12.14}$$

where u_t is i.i.d. $N(0, 1)$. The fact that (12.14) is specified in terms of logarithms of variances means that we do not have to worry about imposing restrictions to ensure $\sigma_t^2 > 0$. If we define $\alpha_t \equiv \log(\sigma_t^2)$, then it can be seen that (12.14) can be interpreted as a state equation (e.g. see Chapter 8 (8.39)), and the assumption that u_t is i.i.d. $N(0, 1)$ implies a hierarchical prior for α_t.[5] The stochastic volatility model is, thus, very similar to the models we discussed in Chapter 8. The fact that the measurement equation given in (12.11) is not linear in α_t causes some complications. Early Bayesian work with this model (Jacquier, Polson and Rossi, 1994) developed a posterior simulation algorithm which overcame these complications. However, the posterior simulator described in Chapter 8 (Section 8.3.1) can easily be adapted for the stochastic volatility model (DeJong and Shephard, 1995). In practice, this latter algorithm seems to be quite efficient (see also Chib, Nardari and Shephard, 2002). Other papers which discuss Bayesian inference in nonlinear and/or non-Normal state space models in general (i.e. including the stochastic volatility model, but also covering many others) are Carlin, Polson and Stoffer (1992) and Geweke and Tanizaki (1999, 2001).

Stock trades occur in continuous time, and a popular area of current research is to extend stochastic volatility models to allow for this. Such models are suitable for use with high frequency financial data sets (e.g. data sets where every trade is recorded). This literature is quite technically sophisticated, being based on concepts beyond the scope of this book.[6] Hence, we do not even try to motivate

[5] As with any hierarchical model, the Normality assumption can be relaxed through the use of mixtures-of-Normals (see Chapter 10, Section 10.3), or other methods such as that proposed by Steel (1998).

[6] For the reader with a background in this area, most of the work in this area involves the volatility behaving according to an Ornstein–Uhlenbeck process.

this class of model, but simply refer the interested reader to Elerian, Chib and Shephard (2001), Eraker (2001) and Griffin and Steel (2001).

There are many other models commonly used by Bayesians doing empirical work in finance. For instance, Vector Autoregressions (VARs), which we discuss shortly, are popular in many contexts. A few Bayesian VAR applications in finance include Barberis (2000), Kandel, McCulloch and Stambaugh (1995), Lamoureux and Zhou (1996) and Pastor (2000).

Nonlinear Time Series Models

In Chapter 8, all of the models considered were linear. Some of the models mentioned in the preceding section on financial time series were nonlinear. All of these were state space models, and papers such as Carlin, Polson and Stoffer (1992) describe methods for carrying out Bayesian inference in nonlinear state space models. However, especially in macroeconomics, other types of nonlinear time series models are becoming quite popular. These models are typically not written as state space models (although they can be as described in Kim and Nelson, 1999). Hence, it is worthwhile briefly introducing them here, and offering a few citations for the reader interested in more detail. A good overview of Bayesian methods for nonlinear time series models is provided in Bauwens, Lubrano and Richard (1999, Chapter 8).

The nonlinear time series models we discuss in this section are based on AR(p) models (see Chapter 8, Section 8.1). To refresh your memory, we say y_t follows an AR(p) process if it can be written as

$$y_t = \rho_1 y_{t-1} + \cdots + \rho_p y_{t-p} + \varepsilon_t \tag{12.15}$$

Alternative notation for the AR(p) model is

$$\begin{aligned} y_t &= (\rho_1 L + \cdots + \rho_p L^p) y_t + \varepsilon_t \\ &= \rho(L) y_t + \varepsilon_t \end{aligned} \tag{12.16}$$

where $\rho(L)$ is a polynomial in the lag operator, L and ε_t is i.i.d. $N(0, h^{-1})$.

The idea underlying several popular nonlinear time series models is that the dependent variable follows a different AR(p) in different regimes. We let S_t denote these regimes. That is, S_t is a random variable taking values in the set $\{1, 2, \ldots, S\}$. A wide class of nonlinear time series models can be written as:

$$y_t = \rho^{\{S_t\}}(L) y_t + \sigma^{\{S_t\}} v_t \tag{12.17}$$

where $\rho^{\{S_t\}}(L)$ is a polynomial in the lag operator like that defined in (12.16). Note, however, that there are now S different polynomials and, thus, S different regimes. We have written the error as $\sigma^{\{S_t\}} v_t$, with v_t being i.i.d. $N(0, 1)$, to make clear that we are allowing for a different error variance in each regime.

One of the most popular models which can be put in the form of (12.17) is the *Threshold Autoregressive model* (TAR). This defines the index variable, S_t, by choosing a value for a lag of the dependent variable, d, known as the delay parameter, and a collection of thresholds $\{r_s, s = 1, \ldots, S-1\}$ to construct S_t. An example would be

$$y_t = \begin{cases} \rho^{\{1\}}(L)y_t + \sigma^{\{1\}}v_t \text{ if } y_{t-d} < r_1 \\ \rho^{\{2\}}(L)y_t + \sigma^{\{2\}}v_t \text{ if } r_1 \leq y_{t-d} < r_2 \\ \vdots \qquad\qquad \vdots \\ \rho^{\{J\}}(L)y_t + \sigma^{\{J\}}v_t \text{ if } y_{t-d} \geq r_{S-1} \end{cases} \tag{12.18}$$

Thus, y_t follows one of S different AR(p) models, depending on what y_{t-d} was. This sort of specification (or extensions of it) are popular with macroeconomic applications involving things like GDP growth. It allows for GDP growth to exhibit different behavior, depending on whether the economy was in recession, expansion, etc. d periods ago. $\gamma = (r_1, \ldots, r_{S-1}, d)'$ is treated as a vector of unknown parameters. A simple extension is to allow the threshold trigger to be X_t rather than y_{t-d}, where X_t is a function of lags of the dependent variable and, possibly, unknown parameters. Examples include $X_t = y_{t-d}$ (i.e. the model given in (12.18)), $X_t = \Delta y_{t-d}$ (i.e. regime switches are triggered by the growth rate d periods ago), and $X_t = \frac{y_{t-1} - y_{t-d-1}}{d}$ (i.e. regime switches are triggered by average growth over the last d periods). Another popular extension is the *Smooth Transition Autoregressive*, or STAR model, which generalizes the TAR to allow for gradual switching between regimes (as opposed to the abrupt switches apparent in (12.18)).

Another popular nonlinear time series model is the so-called *Markov Switching model*. As in (12.18), this allows for different AR dynamics to hold in different regimes (called *states*). Markov switching models usually have two states which can hold at any point in time (e.g. the economy can be in a recessionary state or an expansionary state). These are labeled $S_t = 1$ or 2. S_t is assumed to obey

$$\begin{aligned} p(S_t = 1|S_{t-1} = 1) &= p \\ p(S_t = 2|S_{t-1} = 1) &= 1 - p \\ p(S_t = 2|S_{t-1} = 2) &= q \\ p(S_t = 1|S_{t-1} = 2) &= 1 - q \end{aligned} \tag{12.19}$$

where $0 \leq p \leq 1$ and $0 \leq q \leq 1$. From a Bayesian perspective, (12.19) can be interpreted as a hierarchical prior for the latent data S_t for $t = 1, \ldots, T$, which depends on unknown parameters p and q. By choosing a prior for p and q the researcher can completely specify a Bayesian model. Many extensions (e.g. to allow for the probabilities in (12.19) to depend upon explanatory variables) are possible.

Bayesian analysis of any of these regime-based time series models can be done in a straightforward fashion, since they all can be written as Normal linear regression models conditional on some parameter vector, γ, and possibly some latent data. Thus, a Gibbs sampler (possibly with data augmentation) can be set up. For instance, in the TAR model $\gamma = (r_1, \ldots, r_{J-1}, d)'$. Conditional upon knowing γ, we know which regime is relevant at each time period (i.e.

we know S_t for $t = 1, \ldots, T$). All observations with $S_t = s$ can be grouped together, and standard results for the Normal linear regression model used to derive the posterior for the AR(p) coefficients in the sth regime. This process can be repeated for $s = 1, \ldots, S$. This provides us with a posterior for $p(\rho^1, \ldots, \rho^S, h^1, \ldots, h^S|y, \gamma)$, where ρ^s is the vector containing all the AR coefficients in the sth regime and h^s is the error precision in the sth regime. Thus, we only need $p(\gamma|y, \rho^1, \ldots, \rho^S, h^1, \ldots, h^S)$ to complete specification of a Gibbs sampler. For the Markov switching model, we define $\gamma = (p, q)'$ and create a Gibbs sampler with data augmentation which sequentially draws from $p(\rho^1, \ldots, \rho^S, h^1, \ldots, h^S|y, \gamma, S_1, \ldots, S_T)$, $p(\gamma|y, \rho^1, \ldots, \rho^S, h^1, \ldots, h^S, S_1, \ldots, S_T)$ and $p(S_1, \ldots, S_T|y, \gamma, \rho^1, \ldots, \rho^S, h^1, \ldots, h^S)$.

The reader interested in more details about Bayesian inference in these models (as well as extensions and applications) is referred to Forbes, Kalb and Kofman (1999), Geweke and Terui (1993) and Koop and Potter (1999, 2003) for TAR models, and Albert and Chib (1993a), Filardo and Gordon (1998) and McCulloch and Tsay (1993, 1994) for Markov switching models.

Structural break models are another popular class of nonlinear time series model. These models can be put in the form of (12.17) with S_t switching at a particular point in time. For instance,

$$S_t = \begin{cases} 1 \text{ if } t \leq \gamma \\ 2 \text{ if } t > \gamma \end{cases}$$

This allows the behavior of the dependent variable to change at time γ. The breakpoint, γ, can be treated as an unknown parameter and estimated from the data. The model can be extended to allow for many breakpoints in a straightforward fashion. Bayesian estimation proceeds in a similar fashion as for the TAR model, or a hierarchical prior akin to that used in the Markov switching model can be used. A few representative citations include Carlin, Gelfand and Smith (1992), Chib (1998), DeJong (1996), Koop and Potter (2000) and Martin (2000).

In this section, we have offered only a very brief and incomplete introduction to Bayesian inference in nonlinear time series models, focusing on regime-based models which can be written as (12.18). It is worth stressing that there is a wide variety of other models and methods not discussed here that might be of interest when the researcher leaves the world of linear time series models.

Multivariate Time Series Models

In many applications, especially in the field of macroeconomics, interest centers on the interactions between many variables. For instance, many macroeconomic models involve working with the RMPY variables where R = the interest rate, M = the money supply, P = the price level and Y = real output. Multivariate models are typically used for prediction and to understand the relationships between many variables. Important concepts, such as cointegration, are inherently multivariate in nature. Other functions of the model parameters, such as *impulse response functions* or *forecast error decompositions*, can also be used to help understand the relationships between variables. It would be too much of a digression to describe precisely what these functions are. Hence, we note only that they can all be written as $g(\theta)$ for some appropriately chosen function. Thus, once a posterior simulator is developed it is easy to calculate features such as $E[g(\theta)|y]$ in the standard way. Two influential (non-Bayesian) econometrics textbooks which discuss multivariate models, impulse response functions, etc. in detail are Hamilton (1994) and Lutkephohl (1993). Other papers which discuss impulse response functions include Sims and Zha (1999) and Koop (1992, 1996).

In Chapter 8 (Section 8.4) we described multivariate state space models. These can be used for a wide variety of applications. However, the most popular multivariate time series model is the *Vector Autoregression* (or VAR). We have already discussed, in the preceding section on Time Series Models for Finance, how these are commonly used in financial applications. However, they are even more popular with macroeconomists. To briefly describe VARs, let y_{jt} denote the jth time series variable of interest for $j = 1, \ldots, J$ and $t = 1, \ldots, T$. Then a VAR of order p, written VAR(p), is a model with J equations of the form

$$\begin{aligned}
y_{1t} &= \beta_{01}^1 + \beta_{11}^1 y_{1,t-1} + \cdots + \beta_{p1}^1 y_{1,t-p} + \cdots + \beta_{J1}^1 y_{J,t-1} \\
&\quad + \cdots + \beta_{Jp}^1 y_{J,t-p} + \varepsilon_{1t} \\
y_{2t} &= \beta_{01}^2 + \beta_{11}^2 y_{1,t-1} + \cdots + \beta_{p1}^2 y_{1,t-p} + \cdots + \beta_{J1}^2 y_{J,t-1} \\
&\quad + \cdots + \beta_{Jp}^2 y_{J,t-p} + \varepsilon_{2t} \\
&\quad \vdots \\
y_{Jt} &= \beta_{01}^J + \beta_{11}^J y_{1,t-1} + \cdots + \beta_{p1}^J y_{1,t-p} + \cdots + \beta_{J1}^J y_{J,t-1} \\
&\quad + \cdots + \beta_{Jp}^J y_{J,t-p} + \varepsilon_{Jt} \qquad (12.20)
\end{aligned}$$

where, for example, β_{p1}^J denotes the coefficient on the pth lag of the 1st variable in the Jth equation. Thus, each variable depends upon p lags of itself and p lags of all of other variables in the model. This model is very similar to a SUR model and Bayesian inference is done using methods similar to those described in Chapter 6 (Section 6.6). Much of this literature grew out of Sims (1980) and Doan, Litterman and Sims (1984) was an influential early Bayesian reference. Kadiyala and Karlsson (1997) is a more recent reference which discusses prior elicitation and computation in detail.

A sampling of other Bayesian papers which use VARs or related techniques in macroeconomic applications (including forecasting) include DeJong, Ingram and Whiteman (1996, 2000), Geweke (1999a), Litterman (1986), Min and Zellner (1993), Otrok and Whiteman (1998), Poirier (1991), Sims and Zha (2002), Zellner and Hong (1989) and Zellner, Hong and Min (1991).

Unit Roots and Cointegration

The concepts of unit roots and cointegration were briefly introduced in Chapter 8 (see Sections 8.2 and 8.4, respectively). These concepts have played a large role in recent frequentist time series econometrics. There is also a large Bayesian literature, although some of this is skeptical about the importance of unit roots and cointegration for Bayesian analysis (e.g. see Sims (1998) and Sims and Uhlig (1991)). Rather than attempt to summarize this rather arcane debate, we will just offer a few relevant citations for the interested reader.

Early Bayesian unit root work included DeJong and Whiteman (1991), Koop (1991), Lubrano (1995) and Schotman and van Dijk (1991). Some controversies are presented in Phillips (1991, with discussion). An entire special issue of *Econometric Theory* (volume 10, 1994) is devoted to the topic and many of the papers in this volume are of interest. Bayesian unit root inference, including a discussion of the related controversies and additional citations, is covered in Bauwens, Lubrano and Richard (1999, Chapter 6).

With regards to cointegration, DeJong (1992) and Dorfman (1994) were early papers. Other influential papers include Bauwens and Lubrano (1996), Chao and Phillips (1999), Geweke (1996), Kleibergen and van Dijk (1994), Kleibergen and Paap (2002) and Strachan (2002). Bauwens, Lubrano and Richard (1999, Chapter 9) offers an excellent summary of this topic.

12.4.2 Endogeneity, Sample Selection and Related Issues

The earlier chapters in this book contained no discussion of models involving endogeneity. That is, all of our models are interested in investigating the effect of explanatory variables on a dependent variable (or dependent variables). The models assume that the conditional distribution of y given X is the appropriate one for investigating this effect. In the jargon of economics, we implicitly assumed X contained exogenous variables and y an endogenous variable. In many applications, y given X is the relevant distribution of interest. However, there are several important cases where y and X are determined jointly (i.e. endogenously) and, if so, an investigation using the distribution of y given X may not provide answers to the questions the researcher is interested in. Some other distribution (e.g. the joint distribution of y and X) may be the relevant one.

Similar issues hold for sample selection problems. An example of a sample selection problem would be a case where a training program for unemployed individuals exists. If some unemployed individuals are randomly chosen to participate in the training program (and the remainder left untrained), then regression-based methods with independent errors are suitable for analyzing the effect of the training program on future job success. However, if the unemployed are allowed to choose whether to enter the program or not, then regression-based methods are not appropriate. The decision to participate is made on the basis of the likely effect on future job success of the training program. Only the individuals who are most likely to benefit from the training program will choose to participate.

Hence, allocation to the training program is not random, and we have what is referred to as a sample selection problem. Of course, sample selection and endogeneity are closely related concepts. In our sample selection example, the explanatory variable (participation in the training program) and the dependent variable (future job success) are jointly (i.e. endogenously) determined.

As a more detailed example, let us suppose that the researcher is interested in the returns to schooling (i.e. the increase in wages which result from an extra year of schooling) and has data for many individuals. One thing the researcher could do is run a regression where the wage is the dependent variable and years of schooling is the explanatory variable (in practice, other explanatory variables should be included, but we ignore this here). Using the methods of Chapters 2, 3 or 4 we could obtain a posterior for β. But remember that these regression methods used the distribution of y given X to create the likelihood (see the discussion in Chapter 2, Section 2.2). Thus, β will simply look at the wages for given levels of schooling. It will simply measure the fact that individuals with more years of schooling tend to earn more. But the relevant policy question might be "The government is interested in a policy which will encourage more individuals to go to university. What will be the returns to schooling for those influenced by this policy?" The fact that individuals with university educations tend to earn β extra per year of schooling may not answer the policy question. For instance, individuals may vary in the benefits they get from university, and those most able to benefit may already be going to university. In this case, the benefits of the government policy to those lured into university will be less than β. Alternatively, β may not actually be measuring the returns to schooling. Suppose, for instance, that there is some quality called 'talent' which determines both success in the job market and success in education. Then the fact that β is positive may simply reflect that talented people tend to stay longer in education. Loosely speaking, the fact that university students will make more money in the future may not be due to the fine education provided by universities. It may simply be that these students are talented individuals to begin with who would experience job market success regardless of their education. If this story is true, then β will not answer the policy question of interest. In short, the schooling decision may be a complicated one involving many factors where schooling and wage are determined jointly. In such cases, the distribution of y given X may not be very informative, and more complicated models may be appropriate. The general Bayesian advice in such cases is to decide on the distribution of interest and choose an appropriate form for it (i.e. a suitable likelihood and prior).

In this section, we describe in some detail the most common case where endogeneity problems arise and the most common model used to deal with it: the *simultaneous equations model*. We then briefly describe *treatment models*. Specific citations are provided below. Another Bayesian textbook, which provides a discussion of a wide variety of endogeneity issues, is Lancaster (2003). We should also mention that there are other classes of models where related problems occur. For instance, similar issues arise in some sorts of errors-in-variables problems (i.e. regression models where the explanatory variables are observed with error). Zellner (1971) has a good chapter on this topic. Erickson (1989) is another important errors-in-variables citation.

The Simultaneous Equations Model

The simultaneous equations model is usually motivated using an economic model of supply and demand. The quantity supplied of a good in market i, q_i^S, depends upon the price, p_i, and other explanatory variables, x_{iS}. The quantity demanded of the good, q_i^D, also depends upon the price of the good and other explanatory variables x_{iD}. In equilibrium, quantity supplied equals quantity demanded and we can write $q_i^S = q_i^D \equiv q_i$. If we assume supply and demand curves follow linear regression models we have

$$q_i = \gamma_S p_i + x_{iS}' \beta_S + \varepsilon_{iS} \quad \text{(Supply)}$$

and

$$q_i = \gamma_D p_i + x_{iD}' \beta_D + \varepsilon_{iD} \quad \text{(Demand)}$$

The problems associated with simply running a regression of price on quantity (and other explanatory variables) or quantity on price (and other explanatory variables) can be seen in several ways. First, it can be seen that, if x_{iS} and x_{iD} were not in the equations, then the supply and demand equations would both involve only quantity and price. Given a data set, one could run a regression of quantity on price, but it is unclear what the resulting estimate would tell you. Would it be an estimate of the slope of the demand curve? supply curve? neither? Without more information it would be impossible to know. Secondly, we can use the fact that regressions relate to the distribution of the dependent variable conditional on the explanatory variables to work out the interpretation of the coefficient on price in a regression of quantity on price (and other explanatory variables). Assuming ε_{iS} and ε_{iD} are Normally distributed, we leave the reader to use Appendix B, Theorem B.9, to show that the conditional distribution of q given p (and other explanatory variables) is Normal. However, the mean of this conditional distribution (which is the regression line) implies that the coefficient on price is neither γ_S nor γ_D, but rather a more complicated function of the model parameters. In short, simply running a regression of q on p (and other explanatory variables) will not provide an estimate of the slope of the demand curve, nor the slope of the supply curve. More sophisticated methods are necessary to shed light on these economically important parameters.

The general form of the Normal linear simultaneous equations model with M endogenous variables (and, thus, M equations) contained in the vector y_i and K exogenous variables contained in the vector x_i is

$$\Gamma y_i = B x_i + \varepsilon_i \tag{12.21}$$

where $i = 1, \ldots, N$, Γ is an $M \times M$ matrix of coefficients, B a $M \times K$ matrix of coefficients and ε_i is i.i.d. $N(0, H^{-1})$. This is referred to as the *structural form* of the model. The parameters in the structural form of the model are usually those of interest to the researcher (e.g. the Supply and Demand equations above are in structural form). This model can also be written as a *reduced form* model, with one equation for each endogenous variable and only exogenous variables on the right-hand side by multiplying both sides of (12.21) by Γ^{-1}. This yields

$$\begin{aligned} y_i &= \Gamma^{-1} B x_i + \Gamma^{-1} \varepsilon_i \\ &= \Pi x_i + v_i \end{aligned} \tag{12.22}$$

where $\Pi = \Gamma^{-1}B$, v_i is i.i.d. $N(0, \Sigma)$ and $\Sigma = \Gamma^{-1}H^{-1}\Gamma^{-1}$. A consideration of (12.21) and (12.22) allows us to formalize the intuition of the supply-demand example. Bayesian inference in the reduced form in (12.22) can be implemented using results for the SUR model (or simple extensions thereof) in Chapter 6 (Section 6.6). This allows us to learn about the parameters Π and Σ. However, interest usually centers on the parameters of the structural form: Γ, B and H. Thus, one method of carrying out Bayesian inference in the simultaneous equations model is to work with the reduced form and then use the facts that $\Pi = \Gamma^{-1}B$ and $\Sigma = \Gamma^{-1}H^{-1}\Gamma^{-1}$ to work out the posteriors of the structural form parameters. However, this is not as easy as it sounds, since Π and Σ together contain $MK + \frac{M(M+1)}{2}$ free parameters, whereas Γ, B and H contain $M^2 + MK + \frac{M(M+1)}{2}$ free parameters. Without placing restrictions on the structural form parameters, there are simply not enough reduced form parameters available. It is easy to find M restrictions to impose, since it is standard to normalize the coefficient on the jth endogenous variable to equal one in the jth equation (i.e. the diagonal elements of Γ are all set to one). However, additional restrictions depend on economic theory in the application under study. In terms of language introduced earlier in this chapter, identification issues are quite important in the simultaneous equations model.

There are also new computational issues that must be addressed if the researcher wishes to do anything other than work directly with the reduced form, (12.22), using a natural conjugate Normal-Wishart or independent Normal-Wishart prior. However, the researcher may not wish to elicit a prior directly using the reduced form. After all, the structural form parameters are usually related to economic theory, and the researcher is likely to be able to use such theory to guide prior elicitation. Furthermore, prior elicitation and identification issues can be related (e.g. a Normal prior over Π typically does not rule regions of the parameters space which imply a lack of identification). Dreze and Richard (1983) and Bauwens (1984) are classic references which admirably survey most of the Bayesian work in this area before the mid-1980s. Zellner (1971) also contains an excellent chapter on the topic. Kleibergen (1997), Kleibergen and van Dijk (1998) and Kleibergen and Zivot (2002) are recent references which (among many other things) discuss identification and prior elicitation (including the development of noninformative priors). Some of the issues raised in the cointegration literature are also relevant to the simultaneous equations literature (see Geweke (1996) and Kleibergen and Paap (2002)). Li (1998) develops methods of Bayesian inference for a simultaneous equations model with limited dependent variables.

Treatment Effect Models

In medical statistics, there are many cases where interest centers on the effect of a treatment on a health outcome. Models have been developed to estimate such *treatment effects*. In economics, there are also many cases where interest centers on the effect of 'treatments' (e.g. participating in a training program) on 'outcomes' (e.g. the salary achieved). In many economic and medical applications, statistical analysis is complicated by the fact that individuals are not randomly assigned to treatments. A related problem is non-compliance, where an individual allocated to a certain treatment may not take up the treatment. Thus, this is an area where the medical statistics and econometrics literatures have a lot in common. In this section, we briefly describe some of the relevant Bayesian econometrics literature. But it is worth stressing that there is a vast Bayesian medical statistics literature on similar issues. An example of a recent medical paper in this area is Hirano, Imbens, Rubin and Zhou (2000). Influential papers in the Bayesian statistics literature (which are relevant for both econometrics and medical statistics) include Rubin (1978) and Imbens and Rubin (1997).

A common model for investigating treatment effects is the *dummy endogenous variable model* (see Angrist, Imbens and Rubin, 1996). A simplified variant can be written as

$$\begin{aligned} y_i &= \beta_0 + \beta_1 s_i + \varepsilon_i \\ s_i^* &= \alpha_0 + \alpha_1 z_i + v_i \end{aligned} \tag{12.23}$$

where s_i^* is unobserved, but we do observe

$$s_i = \begin{cases} 1 \text{ if } s_i^* \geq 0 \\ 0 \text{ if } s_i^* < 0 \end{cases} \tag{12.24}$$

In this model, we observe whether the individual received the treatment or not (i.e. whether $s_i = 0$ or 1) and the outcome, y_i. We are interested in the effect of this treatment, which is β_1. The individual decides whether to take the treatment (or not) based on whether a latent variable, s_i^*, is positive (or negative). The latent variable depends upon an explanatory variable z_i. The econometric methods necessary to estimate β_1 depend upon the assumptions made about the errors and z_i. However, in many cases endogeneity can be a problem. For instance, the decision to take the treatment would often depend upon the likely outcome, and thus could reflect y_i. An extreme case would set $z_i = y_i$, and then (12.23) would be an unidentified simultaneous equations model. A related endogeneity problem occurs if unobservable individual characteristics affect both the outcome and the decision to take treatment (i.e. ε_i and v_i are correlated). We shall not discuss this model in any more detail, other than to note that simply running a regression of y on s may not provide a reasonable estimate of β_1. Bayesian inference can, however, be done by drawing on ideas from the simultaneous equations literature (for (12.23)) and the probit literature (for (12.24)).

A closely related model is the *Roy model.* The Normal variant of this model can be written as

$$\begin{pmatrix} s_i^* \\ y_{i0} \\ y_{i1} \end{pmatrix} \sim N\left(\begin{bmatrix} z_i'\gamma \\ x_{i0}'\beta_0 \\ x_{i1}'\beta_1 \end{bmatrix}, \begin{bmatrix} 1 & \sigma_{12} & \sigma_{13} \\ \sigma_{12} & \sigma_{22} & \sigma_{23} \\ \sigma_{13} & \sigma_{23} & \sigma_{22} \end{bmatrix} \right) \tag{12.25}$$

where the definitions are as in (12.23) and (12.24), except that z_i is a vector, x_{i0} and x_{i1} are vectors of explanatory variables, and the outcome, y_i, is defined as

$$y_i = \begin{cases} y_{i0} \text{ if } s_i = 0 \\ y_{i1} \text{ if } s_i = 1 \end{cases}$$

In words, the latent utility relevant for deciding whether to take the treatment or not (s_i^*), the outcome if treatment is received (y_{i1}) and the outcome if treatment is not received (y_{i0}) all follow Normal linear regression models. An interesting property of this model arises, since we only observe y_{i0} or y_{i1}, but not both. That is, for individuals who took the treatment, we do not observe the outcome which would have occurred had they not taken the treatment (and vice versa for individuals who did not take the treatment). This raises an interesting identification problem discussed in Koop and Poirier (1997) and Poirier and Tobias (2002).[7] Nevertheless, Bayesian inference in this model can be implemented using a Metropolis-within-Gibbs sampler with data augmentation as described in Chib and Hamilton (2000). In many applications, the Normality assumption in (12.25) has been found to be too restrictive, and Chib and Hamilton (2002) use a mixture of Normals model (see Chapter 10, Section 10.3) to allow for a more flexible distribution.

12.4.3 Models with Non-standard Dependent Variables

In Chapter 9, we presented methods for Bayesian inference in tobit and probit models. We also informally discussed logit models. The tobit model is suitable for use when the dependent variable is censored. The probit and logit models can be used when the dependent variable is qualitative. However, there are several other types of non-standard dependent variable which the econometrician might run into. We briefly mention some of these in this chapter.

There is a large literature on methods which can be used when the dependent variable is a count. Much of this literature is in the field of medical statistics, since count data often arises in medical applications (e.g. number of deaths, number of health events, etc.). However, there are some economic and business applications involving count data (e.g. number of patents received by a firm or the number of times a consumer purchased a particular product). A common model for count

[7]Note that another identification problem, of the sort noted in the multinomial probit model, has been solved by imposing the restriction $\sigma_{11} = 1$ in the error covariance matrix in (12.25).

data uses the Poisson distribution (see Appendix B, Definition B.2)

$$y_i|\beta \sim Po(\mu_i) \tag{12.26}$$

where $\mu_i \geq 0$ is the mean of the Poisson distribution (see Appendix B, Theorem B.5) and

$$\mu_i = \exp(x_i'\beta) \tag{12.27}$$

Note that (12.27) allows the mean of the dependent variable to depend upon explanatory variables, x_i, and fact that $\exp(x_i'\beta) \geq 0$ ensures a non-negative mean. Chib, Greenberg and Winkelmann (1998) is an example of a paper which carries out Bayesian inference in this model using an MCMC algorithm. In fact, this paper is much more general, since it considers the case where panel data is available and develops a model with individual effects and a hierarchical prior.

The Poisson model given in (12.26) and (12.27) is just one example of a so-called *generalized linear model.* This model assumes a likelihood based on the assumption that y_i lies in a broad family of distributions known as the *exponential family of distributions.* If, as in (12.26) we denote the mean of y_i by μ_i, then the generalized linear model assumes that μ_i is related to explanatory variables, x_i according to

$$g(\mu_i) = x_i'\beta$$

where $g(\mu_i)$ is a known function known as the *link function.* Many models, including several discussed in this book (e.g. the Normal linear regression model, logit and probit models and Poisson model of (12.26) and (12.27)), can be put into this form. We do not discuss generalized mixed models any more at this stage, other than to note that general MCMC algorithms have been developed for working with this class of models. A few key references include Clayton (1996), Dellaportas and Smith (1993) and Zeger and Karim (1991).

Another class of model, closely related to generalized linear models, can be used when the dependent variable is a duration (e.g. number of weeks in unemployment). The statistical literature for working with data of this sort largely developed in medicine and, hence, the terminology used can seem unfamiliar to econometricians. For instance, in a medical application, the researcher might administer different doses of a potentially toxic substance to different animals and observe how long it took for them to die. In this case, the dependent variable is time to death of animal i, which we denote by t_i, which is assumed to depend upon explanatory variables x_i (e.g. the dose administered to animal i). The likelihood function is based on the *hazard function*, $\lambda(t_i, x_i)$, a p.d.f. which describes the probability of death at a given time (or, in an economic application, the time that an individual finds a job). A common example is the *proportional hazards model*, which assumes the hazard function depends upon explanatory variables

$$\lambda(t_i, x_i) = \lambda_0(t_i)\exp(x_i'\beta)$$

and $\lambda_0(t_i)$ is called the *baseline hazard*. Different choices of the baseline hazard define various popular models. For instance, $\lambda_0(t_i) = \lambda$ yields the *exponential regression model*, and $\lambda_0(t_i) = \lambda t_i^{\lambda-1}$ yields the *Weibull regression model*. Dellaportas and Smith (1993) develop a posterior simulator which allows the researcher to carry out Bayesian inference in proportional hazard models. Ibrahim, Chen and Sinha (2001) is an excellent textbook reference in the medical statistics literature which covers a wider range of models of potential use with duration data (see also Volinsky and Raftery, 2000). Bayesian econometric applications involving duration data include Campolieti (2001) and Li (1999a).

12.4.4 Structural Models

The ideal way of carrying out empirical economic research is for an economic theorist to develop a model (e.g. based on an optimization problem for consumers or firms), which implies a particular likelihood function. The empirical economist is then left to collect the data and use Bayesian or frequentist methods to analyze the parameters of this likelihood function. Unfortunately, economic theory rarely provides the precise form for the likelihood function. For instance, microeconomic theories of production imply that the output of a firm should depend upon inputs and prices. Economic theory thus tells us what explanatory variables are relevant. Economic theory may even provide us with some additional restrictions (e.g. that output should be non-decreasing in inputs). However, in this case it does not provide us with a precise likelihood function. Thus, in most cases the empirical economist chooses a specification that may be consistent with economic theory (e.g. chooses a regression model where the dependent variable is firm output and the explanatory variables are inputs and prices) and then uses statistical methods to present evidence that the model is a sensible one (e.g. posterior odds ratios or posterior predictive p-values). You can think of most of the models outlined in this book as being relevant for the empirical economist who proceeds in this manner. However, there are a few cases where economic theorists have provided empirical economists with precise likelihoods. In this section, we briefly mention two such structural models: *job search models* and *auction models*.

When unemployed individuals search for jobs, their decisions depend upon various factors. For instance, if on-the-job search is impossible, then a worker who receives a job offer must decide whether to accept the offer (but then receive no future offers), or reject it in the hope of getting a better offer in the future. Economic theorists have developed various ways of modeling individual behavior in such situations. Many of these job search models imply likelihood functions. Empirical economists can use data on the unemployment durations and accepted wages of many individuals to analyze these likelihood functions. Bayesian work with job search models include Kiefer and Steel (1998) and Lancaster (1997). Koop and Poirier (2001) discusses identification issues in job search models, and shows how Bayesian methods can be used to test whether the implications of economic theory hold. These papers are based on simple job search models which only consider the decision problem of the unemployed individual. Recent economic theory papers model the decision problem of the individual and the wage setting problem of the firm jointly. This strand of the literature yields so-called *equilibrium search models*. Bayesian work with equilibrium search models includes Koop (2001). Although most of the empirical work in this literature relates to job search, it is worth mentioning that similar models arise in marketing when consumers are searching for products and stores are deciding on optimal price setting behavior.

In modern economies, auctions are a common way of buying and selling products. In addition to areas where auctions have traditionally been used (e.g. fine art and some commodities), auctions of virtually any product are now popular on the internet. Governments, too, are increasingly turning to auctions when public contracts are tendered or government assets sold. There is a large and growing theoretical literature on the properties of such auctions. Many of the theories developed in this literature imply specific likelihood functions. This has stimulated a growing empirical literature which attempts to test the implications of various theories relating to auctions. Most of this literature is non-Bayesian. However, there are some Bayesian researchers working in this area. The reader interested in more detail is referred to Albano and Jouneau (1998), Bajari (1997, 1998) and Sareen (1999).

Currently, many of the models that theorists have derived are too abstract for serious empirical work. Nevertheless, as both theory and data collection improve, structural modeling is becoming increasingly popular. Since these models provide a likelihood function and prior elicitation of easy-to-interpret parameters of structural models is typically simple, Bayesian methods have a promising future in this area of the literature.

12.4.5 Bayesian Nonparametric Methods

In Chapter 10, we described some simple Bayesian methods for carrying out flexible econometric modeling. Although the models described in that chapter all involved parametric likelihood functions, we referred to them as 'nonparametric', since they had many properties similar to non-Bayesian procedures which went by this name. Bayesian nonparametrics is currently a very hot field, and there are too many other approaches, with new ones being proposed each year. Dey, Muller and Sinha (1998) provide an introduction to many such approaches in this rapidly developing field. In this section, we offer a very brief list of citations for several popular approaches.

Spline models have been used to do nonparametric regression and, in a time series context, model trends of an unknown form. As an example, let us suppose we have a time series problem

$$y_t = f(t) + \varepsilon_t$$

where $f(t)$ is an unknown trend and $t = 1, \ldots, T$. A spline model would choose several time points, referred to as *knots*, and fit a function of known form between the knots. Spline models differ in their choice of this known function. We illustrate this for the thin plate spline.[8] If we let n_j denote the knot at time j and N the number of knots, then the unknown trend implied by the thin plate spline is given by

$$f(t) = \alpha_0 + \sum_{j=1}^{N} \alpha_j b_j(t)$$

where

$$b_j(t) = (t - n_j)^2 \log(|t - n_j|)$$

From a statistical point of view, the key point to note is that $b_j(t)$ can be interpreted as an explanatory variable and α_j as a regression coefficient. Thus spline models can be analyzed using regression methods. Green and Silverman (1994), Silverman (1985) or Wahba (1983) are good citations with at least some Bayesian content. Note that spline methods depend upon the choice of knots and theory rarely tells us precisely where the knots should be located. To get around this problem, it is common to put in many knots and then use Bayesian model averaging or Bayesian model selection methods (see Chapter 11) to deal with problems caused by having many explanatory variables. Smith and Kohn (1996) is a good example of a paper which adopts this approach.

In Chapter 10, we described how mixtures of Normals can be used to approximate virtually any distribution. In that chapter, we only discussed finite mixtures of Normals. These are extremely flexible but, in the terminology of the literature, are not fully nonparametric. Infinite mixtures of Normals (or other distributions) are more worthy of the name nonparametric. The reader interested in more detail about carrying out Bayesian nonparametric inference using infinite mixtures is referred to Robert (1996). The most popular such mixture model involves *Dirichlet process priors* (a concept which is not explained here). Escobar and West (1995) and West, Muller and Escobar (1994) provide thorough discussions of this model. Campolieti (2001) and Ruggiero (1994) are empirical Bayesian applications using this model.

Other popular Bayesian nonparametric methods use wavelets (e.g. see Müller and Vidakovic, 1998) and Polya tree processes (e.g. see Berger and Guglielmi, 2001).

12.5 SUMMARY

The purpose of this chapter was to briefly describe several topics which were not covered in previous chapters. References were given so that interested readers could follow up these topics on their own. The chapter was organized into sections on other methods, issues and models.

[8]Do not worry about this terminology. It is introduced simply to illustrate the practical implications of spline modeling.

The Other Methods section included discussion of additional ways of doing posterior simulation (e.g. acceptance sampling and the Griddy–Gibbs sampler). Various approximations were discussed which allow the researcher to avoid posterior simulation. The Tierney–Kadane approximation was emphasized as being particularly useful for applied work.

The Other Issues section mainly discussed prior elicitation issues. The concepts of a reference prior and a training sample prior were introduced. In addition, the concept of identification was defined and discussed.

The Other Models section discussed additional time series models (i.e. time series models for finance, nonlinear time series models, multivariate time series models, unit roots and cointegration), endogeneity and sample selection issues (i.e. the simultaneous equations model and two treatment effects models), additional models for non-standard dependent variables (i.e. the Poisson regression model, generalized linear models and duration models), structural models relating to auctions and job search and additional nonparametric models.

We have stressed throughout this chapter that there are many other models which are not discussed in this book. However, Bayesian methods can be used with any model. Accordingly, a researcher should be encouraged to develop the model appropriate for her own application rather than simply choosing a model someone else has developed. This book has shown how any new model should be approached and described methods of computation, which can be used in virtually any model the researcher may care to develop.

Appendix A: Introduction to Matrix Algebra

This appendix presents only the elements of matrix algebra which are used in this book. Advanced econometrics or statistics textbooks such as Poirier (1995, Appendices A and B) or Greene (2000, Chapter 2) have more extensive discussions of matrix algebra and many references for further reading. Some notes have been added below to help the reader by providing intuition or motivation, or highlighting issues of importance. Proofs of theorems are not provided.

Definition A.1: Matrix and vector

An $N \times K$ matrix, A, is an arrangement of NK elements (e.g. numbers or random variables) into N rows and K columns:

$$A = \begin{bmatrix} a_{11} & a_{12} & \cdot & \cdot & a_{1K} \\ a_{21} & a_{22} & \cdot & \cdot & \cdot \\ \cdot & \cdot & \cdot & \cdot & \cdot \\ \cdot & \cdot & \cdot & \cdot & \cdot \\ a_{N1} & a_{N2} & \cdot & \cdot & a_{NK} \end{bmatrix}$$

where a_{nk} is the element in the nth row and kth column. If $K = 1$ then A is a column vector and if $N = 1$ then A is a row vector. A matrix with $N = 1$ and $K = 1$ is called a scalar.

Note: a matrix is a convenient way of organizing large numbers of data points, parameters, etc.

Definition A.2: Matrix Addition/Subtraction

If A and B are two $N \times K$ matrices, then $A + B$ is an $N \times K$ matrix with the element in the nth row and kth column given by $a_{nk} + b_{nk}$. $A - B$ is an $N \times K$ matrix with the element in the nth row and kth column given by $a_{nk} - b_{nk}$.

Note: matrix addition/subtraction is a straightforward extension of regular addition/subtraction. You merely add/subtract each element of one matrix to/from the corresponding element of the other. Note that the matrices being added must have the same number of rows and columns.

Theorem A.1: Properties of Matrix Addition

Matrix addition satisfies the commutative and associative properties. That is, if A, B and C are all $N \times K$ matrices, then

- $A + B = B + A$, and
- $A + (B + C) = (A + B) + C$.

Note: these properties are the same as in regular arithmetic.

Definition A.3: Scalar Multiplication

If c is a scalar and A is an $N \times K$ matrix then cA is an $N \times K$ matrix with the element in the nth row and kth column given by ca_{nk}.

Note: with scalar multiplication you simply multiply every element of the matrix by the scalar.

Theorem A.2: Properties of Scalar Multiplication

If c and d are scalars and A and B are $N \times K$ matrices, then

- $(c + d)A = cA + dA$, and
- $c(A + B) = cA + cB$.

Note: these properties are the same as in regular arithmetic.

Definition A.4: Matrix Multiplication

Let A be an $N \times K$ matrix and B be a $K \times J$ matrix, then $C = AB$ is a $N \times J$ matrix with the element in the nth row and jth column given by c_{nj}, where

$$c_{nj} = \sum_{k=1}^{K} a_{nk} b_{kj}$$

Notes: matrix multiplication is not an obvious extension of standard arithmetic. Working through some examples (e.g. set $N = 3$, $K = 2$ and $J = 4$, etc.) is a good way of developing a feeling for what matrix multiplication does. A and B do not have to have exactly the same dimensions, but matrix multiplication is only defined if the number of columns in A is the same as the number of rows in B.

Theorem A.3: Properties of Matrix Multiplication

Matrix multiplication is not commutative, but does satisfy associative and distributive properties. That is, in general, $AB \neq BA$ (indeed BA is not defined unless $N = J$). However, $A(BC) = (AB)C$ and $A(B + C) = AB + AC$, provided A, B and C are matrices with dimensions such that all of the previous operations are defined.

Definition A.5: Transpose of a Matrix

Let A be an $N \times K$ matrix with the element in the nth row and kth column given by a_{nk}. Then the transpose of A, denoted A', is a $K \times N$ matrix with the element in the kth row and nth column given by a_{nk}.

Note: transposing interchanges rows and columns so that the nth row of A is the nth column of A'.

Theorem A.4: Properties of the Transpose Operator

Let A be an $N \times K$ matrix and B be a $K \times J$ matrix, then $(AB)' = B'A'$.

Definition A.6: Special Matrices

A *square matrix* has an equal number of rows and columns. A *diagonal matrix* is a square matrix with all non-diagonal elements equal to zero (i.e. $a_{nk} = 0$ if $n \neq k$). An *upper triangular matrix* has all elements below the diagonal equal to zero (i.e. $a_{nk} = 0$ if $n > k$). A *lower triangular matrix* has all elements above the diagonal equal to zero (i.e. $a_{nk} = 0$ if $n < k$). A *symmetric matrix* is a square matrix with $a_{nk} = a_{kn}$.

Note: a symmetric matrix is characterized by the property that $A = A'$.

Definition A.7: Useful Matrices

The *null matrix* is an $N \times K$ matrix with all elements equal to zero, denoted $0_{N \times K}$. If $K = 1$ it is written as 0_N. The *unitary matrix* is an $N \times K$ matrix with all elements equal to one, denoted $\iota_{N \times K}$. If $K = 1$ it is written as ι_N. The *identity matrix* is an $N \times N$ diagonal matrix with all diagonal elements equal to one (i.e. $a_{nk} = 1$ if $n = k$ and $a_{nk} = 0$ if $n \neq k$), denoted by I_N. If the dimension of any of these matrices is clear from the context, it is common to drop subscripts and write 0, ι or I.

Theorem A.5: Some Properties of Useful Matrices

Let A be an $N \times K$ matrix, then

- $A + 0_{N \times K} = 0_{N \times K} + A = A$,
- $A0_{K \times J} = 0_{N \times J}$, $0_{J \times N}A = 0_{J \times N}$, and
- $AI_N = I_K A = A$.

Definition A.8: Linear Independence

Let $A_1, \ldots, A_K$ denote the K columns of the $N \times K$ matrix, A. These columns are *linearly dependent* if scalars $c_1, \ldots, c_K$ exist (not all zero) such that

$$c_1 A_1 + c_2 A_2 + \cdots + c_K A_k = 0_N$$

If no such scalars exist, then the columns of A are *linearly independent*.

Definition A.9: Rank of a Matrix

The *rank* of a matrix, A, denoted by rank(A), is the maximum number of linearly independent columns of A.

Definition A.10: The Determinant of a Matrix

The precise definition of a determinant is quite complicated and will not be given here (see Poirier (1995) or Greene (2000), or any advanced econometrics text or matrix algebra book). For the purposes of applied Bayesian analysis, you need only know that the determinant of an $N \times N$ matrix, A, is denoted by $|A|$ and is a scalar with an intuitive interpretation relating to the size of the matrix. The scalar analogue is absolute value. In Bayesian econometrics, it usually arises in a context where it must be calculated using computer software. Every relevant computer package (e.g. MATLAB or Gauss) will calculate it automatically for you. Additional useful properties which provide some aid in understanding determinants will be given below.

Definition A.11: The Trace of a Matrix

The *trace* of a matrix, A, denoted by $tr(A)$, is the sum of its diagonal elements.

Theorem A.6: Some Properties of Determinants

Let A and B be $N \times N$ square matrices and c be a scalar, then

- $|AB| = |A||B|$,
- $|cA| = c^N|A|$ and
- $|A'| = |A|$.

However, in general, $|A + B| \neq |A| + |B|$.

Note: many more useful properties of determinants are given in Poirier (1995, pp. 624–626).

Definition A.12: Inverse of a Matrix

The *inverse* of an $N \times N$ matrix A, denoted by A^{-1}, is an $N \times N$ matrix with the property that $AA^{-1} = I_N$. If A^{-1} exists, then A is said to be *nonsingular*. If A^{-1} does not exist, then A is said to be *singular*.

Note: the scalar analogue of the statement $AA^{-1} = I$ is $a\frac{1}{a} = 1$. Thus matrix inversion allows for something which might be called matrix division, although this term is not used.

Theorem A.7: Determining Singularity of a Matrix

Let A be an $N \times N$ matrix, then the columns of A are linearly dependent if and only if $|A| = 0$. Equivalently, if $|A| = 0$, then rank $(A) < N$. Furthermore, A will be singular if and only if $|A| = 0$ (or equivalently rank $(A) < N$).

Note: the scalar analogue of the fact that A^{-1} does not exist if $|A| = 0$ is that $\frac{1}{a}$ does not exist if $a = 0$ where a is a scalar.

Theorem A.8: Some Properties Involving Inverses

Let A and B be nonsingular $N \times N$ matrices, then

- $(A^{-1})^{-1} = A$,
- $(AB)^{-1} = B^{-1}A^{-1}$,
- $(A')^{-1} = (A^{-1})'$ and
- $|A^{-1}| = |A|^{-1}$.

Notes: aside from a few special cases, calculating the inverse or determinant of a matrix is usually a hard thing to do analytically. However, the computer packages relevant for Bayesian econometrics (e.g. MATLAB or Gauss) will all have functions for numerically calculating determinants or inverses. The following theorem can sometimes be used to simplify the task of calculating inverses or determinants.

Theorem A.9: Inverse and Determinant of a Partitioned Matrix

Let A be an $N \times N$ nonsingular matrix which is partitioned as:

$$A = \begin{bmatrix} A_{11} & A_{12} \\ A_{21} & A_{22} \end{bmatrix}$$

where A_{11} and A_{22} are, respectively, $N_1 \times N_1$ and $N_2 \times N_2$ nonsingular matrices where $N_1 + N_2 = N$. A_{21} and A_{12} are, respectively, $N_2 \times N_1$ and $N_1 \times N_2$ matrices. A^{-1} is partitioned in the same manner as:

$$A^{-1} = \begin{bmatrix} A^{11} & A^{12} \\ A^{21} & A^{22} \end{bmatrix}$$

With these definitions,

- $|A| = |A_{22}||A_{11} - A_{12}A_{22}^{-1}A_{21}| = |A_{11}||A_{22} - A_{21}A_{11}^{-1}A_{12}|$, and
- the components of A^{-1} can be calculated as: $A^{11} = (A_{11} - A_{12}A_{22}^{-1}A_{21})^{-1}$, $A^{22} = (A_{22} - A_{21}A_{11}^{-1}A_{12})^{-1}$, $A^{12} = -A_{11}^{-1}A_{12}A^{22}$ and $A^{21} = -A_{22}^{-1}A_{21}A^{11}$.

Note: Theorem A.44 of Poirier (1995) provides some additional expressions for the inverse of a partitioned matrix.

Definition A.13: Quadratic Form

Let x be an $N \times 1$ vector and A be a symmetric $N \times N$ matrix, then the scalar $x'Ax$ is referred to as a *quadratic form*.

Notes: let x_i denote the elements of x and a_{ij} denote the elements of a, where $a_{ij} = a_{ji}$, $i = 1, \ldots, N$ and $j = 1, \ldots, N$. Then $x'Ax = \sum_{i=1}^{N}\sum_{j=1}^{N} a_{ij}x_i x_j$ is a quadratic function in the squares and cross-products of the elements of x. Loosely speaking, quadratic forms are the matrix generalization of sums of squares.

Definition A.14: Definiteness of a Matrix

An $N \times N$ symmetric matrix A is said to be

- positive definite if and only if $x'Ax > 0$ for all non-zero x,
- negative definite if and only if $-A$ is positive definite,
- positive semidefinite if and only if $x'Ax \geq 0$ for all x, and $x'Ax = 0$ for some non-zero x, and
- negative semidefinite if and only if $-A$ is positive semidefinite.

Notes: the covariance matrix of a random vector is positive definite (or positive semidefinite). The scalar analogue of this is that the variance of a random variable is positive (or non-negative). A useful property to know is that a positive definite matrix is nonsingular. Appendix B provides definitions of the terms random variable, variance and covariance.

Theorem A.10: Diagonalization of a Symmetric Matrix

Let A be an $N \times N$ matrix, then there exist $N \times N$ matrices X and D such that $X'X = I_N$ and D is a diagonal matrix and $X'AX = D$. If A is positive definite, then X and D are nonsingular.

Note: this result is most commonly used to transform a model with general error covariance matrices to one with an error covariance matrix equal to cI where c is a scalar.

Theorem A.11: Cholesky Decomposition

Let A be an $N \times N$ positive definite matrix, then there exists a nonsingular $N \times N$ lower triangular matrix X with the property that $A = XX'$.

Notes: X is not unique. A common way of making it unique is to specify that all its diagonal elements are positive. The scalar analogue of the Cholesky decomposition is the square root operator. If A is a covariance matrix, then the scalar analogue of calculating the Cholesky decomposition is the calculation of the standard deviation. A property of the Cholesky decomposition is that $A^{-1} = (X')^{-1}X^{-1}$. This property is often used in computer programs for calculating the inverse of a positive definite matrix. Cholesky decompositions also come in useful if you are using a computer package which only provides random draws from the standard Normal distribution. That is, the Cholesky decomposition of a covariance matrix can be used to transform standard Normal draws to those from the multivariate Normal distribution with arbitrary covariance matrix. Appendix B provides a definition of the Normal distribution.

Appendix B: Introduction to Probability and Statistics

This appendix presents only the elements of probability which are used in this book. There is a myriad of probability and statistics books which will cover this material in more detail (e.g. Wonnacott and Wonnacott, 1990). In addition, Chapters 2–5 of Poirier (1995) are especially useful. Most frequentist econometrics textbooks will also have a chapter on basic probability (e.g. Greene, 2000, Chapter 3). Some notes have been added below to help the reader by providing intuition, motivation or highlighting issues of importance. Proofs of theorems are not provided.

B.1 BASIC CONCEPTS OF PROBABILITY

Some of the definitions and concepts relating to probability are surprisingly subtle. The first part of this appendix is fairly intuitive and lacks mathematical rigor. The reader interested in a more fundamental and mathematically rigorous discussion of probability is referred to Poirier (1995), especially pages 9–35.

Definition B.1: Experiments and Events

An *experiment* is a process whose outcome is not known in advance. The possible outcomes of an experiment are called *events*. The set of all possible outcomes is called the *sample space*.

Definition B.2: Discrete and Continuous Variables

A variable is *discrete* if the number of values it can take on is finite or countable. A variable is *continuous* if it can take on any value on the real line or in a specified real interval.

Definition B.3: Random Variables and Probability (informal definition)

Usually (and everywhere in this book), issues relating to probability, experiments and events are represented by a variable (either continuous or discrete). Since the outcome of an experiment is not known in advance, such a variable is known as a *random variable*. The precise definition and interpretation of *probability* is a source of dispute. In this book, it is enough to have an intuitive grasp of probability (perhaps as reflecting the likelihood that each event will occur) and know its properties as described below. The probability of event A occurring will be denoted by $\Pr(A)$. The following example will serve to clarify these basic concepts.

Suppose an experiment involves tossing a single fair die (i.e. each of the six faces of the die is equally likely to come up). Then the sample space is $\{1, 2, 3, 4, 5, 6\}$ and the discrete random variable, X, takes on values $1, 2, 3, 4, 5, 6$ with probabilities given by $\Pr(X = 1) = \Pr(X = 2) = \cdots = \Pr(X = 6) = \frac{1}{6}$. Alternatively, the random variable X is a function defined at the points $1, 2, 3, 4, 5, 6$. The function is implicitly defined through the probabilities $\Pr(X = 1) = \Pr(X = 2) = \cdots = \Pr(X = 6) = \frac{1}{6}$.

Note: it is important to distinguish between the random variable, X, which can take on values $1, 2, 3, 4, 5, 6$ and the *realization* of the random variable which is the value which actually arises when the experiment is run (e.g. if the die is tossed, a 4 might appear which is the realization of the random variable in a particular experiment). It is common to denote random variables by upper-case letters (e.g. X) with associated realizations denoted by lower-case letters (e.g. x). We adopt such a convention below.

Definition B.4: Independence

Two events, A and B, are *independent* if $\Pr(A, B) = \Pr(A)\Pr(B)$, where $\Pr(A, B)$ is the probability of A and B occurring.

Definition B.5: Conditional Probability

The conditional probability of A given B, denoted by $\Pr(A|B)$, is the probability of event A occurring given event B has occurred.

Theorem B.1: Rules of Conditional Probability Including Bayes Theorem

Let A and B denote two events, then

- $\Pr(A|B) = \frac{\Pr(A,B)}{\Pr(B)}$, and
- $\Pr(B|A) = \frac{\Pr(A,B)}{\Pr(A)}$.

These two rules can be combined to yield *Bayes Theorem*:

$$\Pr(A|B) = \frac{\Pr(B|A)\ \Pr(A)}{\Pr(B)}$$

Note: Theorem B.1 and Definitions B.4 and B.5 are expressed in terms of two events, A and B. However, they also can be interpreted as holding for two random variables, A and B with probability or probability density functions (see below) replacing the Pr()s in the previous formulae.

Definition B.6: Probability and Distribution Functions

The *probability function* associated with a discrete random variable X with sample space $\{x_1, x_2, x_3, \ldots, x_N\}$ is denoted by $p(x)$, where

$$p(x) = \begin{cases} \Pr(X = x_i) & \text{if } x = x_i \\ 0, & \text{otherwise} \end{cases}$$

for $i = 1, 2, \ldots, N$. The *distribution function*, denoted by $P(x)$, is defined as

$$P(x) = \Pr(X \leq x) = \sum_{j \in J} \Pr(x_j)$$

where J is the set of j's with the property that $x_j \leq x$.

Probability and distribution functions satisfy the following conditions:

- $p(x_i) > 0$ for $i = 1, 2, \ldots, N$, and
- $\sum_{i=1}^{N} p(x_i) = P(x_N) = 1$.

Notes: N can be infinite in the previous definition. In words, a probability function simply gives the probability of each event occurring and the distribution function gives the cumulative probability of all events up to some point occurring.

Definition B.7: Probability Density and Distribution Functions

The distribution function of a continuous random variable, X, is $P(x) = \Pr(X \leq x) = \int_{-\infty}^{x} p(t)dt$, where $p()$ is the *probability density function* or *p.d.f.* Probability density and distribution functions satisfy the following conditions:

- $p(x) \geq 0$ for all x,
- $\int_{-\infty}^{\infty} p(t)dt = P(\infty) = 1$, and
- $p(x) = \frac{dP(x)}{dx}$.

Notes: with discrete random variables the definition of a probability function is clear, it is simply the probability of each event in the sample space occurring. With continuous random variables, such a definition is not possible since there is an uncountably infinite number of events. For instance, there is an uncountably infinite number of real numbers in the interval [0, 1] and, hence, if this is the sample space in an experiment we cannot simply attach a probability to each point in the interval. With continuous random variables, probabilities are thus defined only for intervals, and represented as areas under (i.e. integrals of) p.d.f.s. For instance, Definition B.7 implies that $\Pr(a \leq x \leq b) = P(b) - P(a) = \int_a^b p(x)dx$.

Definition B.8: Expected Value

Let $g()$ be a function, then the *expected value* of $g(X)$, denoted $E[g(X)]$, is defined by:

$$E[g(X)] = \sum_{i=1}^{N} g(x_i)p(x_i)$$

if X is a discrete random variable with sample space $\{x_1, x_2, x_3, \ldots, x_N\}$, and

$$E[g(X)] = \int_{-\infty}^{\infty} g(x)p(x)dx$$

if X is a continuous random variable (provided $E[g(X)] < \infty$).

Important special cases of this general definition include:

- the mean, $\mu \equiv E(X)$,
- the variance, $\sigma^2 \equiv var(X) = E[(X - \mu)^2] = E(X^2) - \mu^2$,
- the rth moment, $E(X^r)$, and
- the rth moment about the mean, $E[(X - \mu)^r]$.

Notes: the mean and variance are common measures of the location (central tendency or average) and dispersion, respectively, of a random variable. The third and fourth moments about the mean are common measures of skewness and kurtosis (i.e. fatness of tails of the p.d.f.), respectively, of a random variable. The *standard deviation* is the square root of the variance.

Theorem B.2: Properties of Expected Value Operator

Let X and Y by two random variables, $g()$ and $h()$ be two functions and a and b be constants, then

- $E[ag(X) + bh(Y)] = aE[g(X)] + bE[g(Y)]$, and
- $var[ag(X) + bh(Y)] = a^2 var[g(X)] + b^2 var[g(Y)]$ if X and Y are independent.

Note: the first result implies that the expected value operator is additive. However, it is not multiplicative. For instance, in general, $E[XY] \neq E(X)E(Y)$.

Definition B.9: Mode, Median and Interquartile Range

The mean is the most common measure of central tendency of a p.d.f. or probability function; alternatives include the *median* and *mode*. The median, x_{med} is characterized by $P(x_{med}) = \frac{1}{2}$. The mode, x_{mod}, is characterized by $x_{\text{mod}} = \arg\max[p(x)]$. The *quartiles* of a p.d.f. or probability function, $x_{.25}, x_{.50}, x_{.75}$, are characterized by $P(x_{.25}) = 0.25$, $P(x_{.50}) = 0.50$ and $P(x_{.75}) = 0.75$ (of course, $x_{med} = x_{.50}$). The variance is the most common measure of the dispersion of a p.d.f. or probability function, the chief alternative is the *interquartile range*, defined as $x_{.75} - x_{.25}$.

Definition B.10: Joint Probability and Distribution Functions

Let $X = (X_1, \ldots, X_N)'$ be a vector of N discrete random variables with the sample space of X_i given by $\{x_{i1}, \ldots, x_{iN_i}\}$, then the probability function of X, denoted by $p(x)$ is given by

$$p(x) = \Pr(X_1 = x_1, \ldots, X_N = x_N)$$

where x is an N-vector given by $x = (x_1, \ldots, x_N)'$. If x is not in the sample space for X, then $p(x) = 0$. The distribution function, denoted by $P(x)$, is defined as

$$P(x) = \Pr(X_1 \leq x_1, \ldots, X_N \leq x_N)$$

Note: this is the multivariate extension of Definition B.6 and similar intuition holds, but joint probability relates events for X_1 and X_2 and ... and X_N all holding.

Definition B.11: Joint Probability Density and Distribution Functions

The distribution function of a continuous random vector, $X = (X_1, \ldots, X_N)'$, is denoted by $P(x)$ where $x = (x_1, \ldots, x_N)'$ and

$$P(x) = \Pr(X_1 \leq x_1, \ldots, X_N \leq x_N) = \int_{-\infty}^{x_1} \cdots \int_{-\infty}^{x_N} p(t) dt_1 \cdots dt_N$$

where $p(x)$ is the joint probability density function.

Notes: this is the multivariate extension of Definition B.7 and similar intuition and properties hold.

Definition B.12: Marginal Probability (Density) and Distribution Functions

Let $X = (X_1, \ldots, X_N)'$ be a vector of N random variables (either discrete or continuous) and $X^* = (X_1, \ldots, X_J)'$, where $J < N$, then the *joint marginal distribution function* of X^* is related to the distribution function of X by

$$P(x^*) = P(x_1, \ldots, x_J) = \lim_{x_i \to \infty} P(x_1, \ldots, x_J, \infty, \ldots, \infty)$$

If X is a continuous random vector, then the joint marginal probability density function, $p(x^*)$, based on the joint probability density function, $p(x)$, can be defined using

$$p(x^*) = \int_{-\infty}^{\infty} \cdots \int_{-\infty}^{\infty} p(x) dx_{J+1} \cdots dx_N$$

If X is a discrete random vector, then the definition of its joint marginal probability function is the obvious generalization of the previous expression (essentially just replace $\int$'s with $\sum$'s). When $J = 1$, the terminology *marginal distribution/probability density/probability function* is used. By suitably re-ordering the subscripts, the joint marginal distribution/probability density/probability function of any subset of the original N random variables is defined.

Notes: the joint marginal p.d.f. is used for calculating probabilities relating to only some elements of X, ignoring the other elements. For instance, if $N = 2$, then $p(x_1)$ will simply be the p.d.f. of x_1, regardless of what the value of the other random variable X_2, is.

Definition B.13: Joint Probability Density and Distribution Functions

The distribution function (also called the *cumulative distribution function* or c.d.f.) of a continuous random vector, $X = (X_1, \ldots, X_N)'$, is denoted by $P(x)$ where $x = (x_1, \ldots, x_N)'$ and

$$P(x) = \Pr(X_1 \leq x_1, \ldots, X_N \leq x_N) = \int_{-\infty}^{x_1} \cdots \int_{-\infty}^{x_N} p(t)dt_1 \cdots dt_N$$

where $p(x)$ is the joint probability density function.

Notes: this is the multivariate extension of Definition B.7 and similar intuition and properties hold.

Definition B.14: Marginal Probability (Density) and Distribution Functions

Let $X = (X_1, \ldots, X_N)'$ be a vector of N random variables (either discrete or continuous) and $X^* = (X_1, \ldots, X_J)'$ where $J < N$, then the *joint marginal distribution function* of X^* is related to the distribution function of X by

$$P(x^*) = P(x_1, \ldots, x_J) = \lim_{x_i \to \infty} P(x_1, \ldots, x_J, \infty, \ldots, \infty)$$

If X is a continuous random vector, then the joint marginal probability density function, $p(x^*)$, based on the joint probability density function, $p(x)$, can be defined using

$$p(x^*) = \int_{-\infty}^{\infty} \cdots \int_{-\infty}^{\infty} p(x)dx_{J+1} \cdots dx_N$$

If X is a discrete random vector, then the definition of its joint marginal probability function is the obvious generalization of the previous expression (essentially just replace $\int$'s with $\sum$'s). When $J = 1$, the terminology *marginal distribution/probability density/probability function* is used. By suitably re-ordering the subscripts, the joint marginal distribution/probability density/probability function of any subset of the original N random variables is defined.

Notes: the joint marginal p.d.f. is used for calculating probabilities relating to only some elements of X, ignoring the other elements. For instance, if $N = 2$, then $p(x_1)$ will simply be the p.d.f. of x_1, regardless of what the value of the other random variable X_2, is.

Definition B.15: Conditional Probability Density and Distribution Functions

Let $X = (X_1, \ldots, X_N)'$ be a vector of N continuous random variables, define $X^* = (X_1, \ldots, X_J)'$, $X^{**} = (X_{J+1}, \ldots, X_N)'$, and let x, x^* and x^{**} be the accompanying realizations, then the *conditional p.d.f.* of X^* given X^{**} is defined by

$$p(x_1, \dots, x_J | x_{J+1}, \dots, x_N) = p(x^* | x^{**}) = \frac{p(x)}{p(x^{**})} = \frac{p(x^*, x^{**})}{p(x^{**})}$$

The *conditional distribution function*, denoted by $P(x^*|x^{**})$, is defined by

$$\begin{aligned} P(x^*|x^{**}) &= \Pr(X_1 \le x_1, \dots, X_J \le x_j | X_{J+1} = x_{J+1}, \dots, X_N = x_N) \\ &= \int_{-\infty}^{x_1} \cdots \int_{-\infty}^{x_J} p(x_1, \dots, x_J | x_{J+1}, \dots, x_N) dx_1 \dots dx_J \end{aligned}$$

Definitions for the case where X is a discrete random vector are the obvious generalizations of the above.

Notes: the conditional p.d.f. is used for calculating probabilities relating to some elements of X given that the remaining elements of X take on specific values. Definition B.15 is the extension of Definition B.5 to the case of random variables and has a similar intuition. Theorem B.1 is expressed in terms of events, A and B, but can be extended to any number of random variables by replacing probabilities by p.d.f.s (or probability functions). Bayes theorem and the rules of conditional probability for random variables are of particular importance for Bayesian econometrics. The definition of independence also extends to random variables if we replace probability statements by p.d.f.s (or probability functions).

Definition B.16: Multivariate Expected Values

Let $X = (X_1, \dots, X_N)'$ be a vector of N continuous random variables with p.d.f. $p(x)$ where $x = (x_1, \dots, x_N)'$, then the *expected value* of a scalar function $g(X)$ is denoted by $E[g(X)]$ and given by

$$E[g(X)] \equiv \int_{-\infty}^{\infty} \cdots \int_{-\infty}^{\infty} g(x)p(x)dx_1 \cdots dx_N$$

Notes: this is the multivariate generalization of Definition B.8 and similar intuition holds. Extensions such as conditional expected value, denoted by $E[g(X^*)| x^{**}]$, can be done by replacing $p(x)$ by $p(x^*|x^{**})$ in the previous integral. Multivariate expectation for a discrete random vector involves the straightforward generalization of the above material.

Definition B.17: Covariance and Correlation

Let X_1 and X_2 be random variables with $E(X_1) = \mu_1$ and $E(X_2) = \mu_2$, then the *covariance* between X_1 and X_2 is denoted by $cov(X_1, X_2)$ and defined by

$$\begin{aligned} cov(X_1, X_2) &= E[(X_1 - \mu_1)(X_2 - \mu_2)] \\ &= E(X_1 X_2) - \mu_1 \mu_2 \end{aligned}$$

the *correlation* between X_1 and X_2 is denoted by $corr(X_1, X_2)$ and defined by

$$corr(X_1, X_2) = \frac{cov(X_1, X_2)}{\sqrt{var(X_1)var(X_2)}}$$

Notes: if $X_1 = X_2$ then covariance is simply variance. Correlation can be interpreted as the degree of association between two random variables. It satisfies $-1 \leq corr(X_1, X_2) \leq 1$ with larger positive/negative values indicating stronger positive/negative relationships between X_1 and X_2. If X_1 and X_2 are independent, then $corr(X_1, X_2) = 0$ (although the converse does not necessarily hold).

Definition B.18: The Covariance Matrix

Let $X = (X_1, \ldots, X_N)'$ be a vector of N random variables and define the N-vector $\mu \equiv E(X) \equiv [E(X_1), \cdots, E(X_N)]' \equiv [\mu_1, \cdots, \mu_N]'$. Then the *covariance matrix* is denoted by $var(X)$ and is the $N \times N$ matrix containing the variances and covariances of all the elements of X arranged as

$$\begin{aligned} var(X) &= E[(X-\mu)(X-\mu)'] \\ &= \begin{bmatrix} var(X_1) & cov(X_1, X_2) & . & . & cov(X_1, X_N) \\ cov(X_1, X_2) & var(X_2) & . & . & . \\ . & . & . & . & . \\ . & . & . & . & cov(X_{N-1}, X_N) \\ cov(X_1, X_N) & . & . & cov(X_{N-1}, X_N) & var(X_N) \end{bmatrix} \end{aligned}$$

Theorem B.3: Properties Involving Multivariate Expectations and Covariance

Let A be a fixed (non-random) $M \times N$ matrix, $Y = AX$ and other elements be as given in Definition B.18. Then

- $E(Y) = AE(X) = A\mu$, and
- $var(Y) = Avar(X)A'$.

Note: this is the multivariate extension of Theorem B.2. An implication is that $var(aX_1 + bX_2) = a^2var(X_1) + b^2var(X_2) + 2abcov(X_1, X_2)$ where a and b are scalars.

B.2 COMMON PROBABILITY DISTRIBUTIONS

In this section, the p.d.f.s used in the book are defined and some of their properties described. There are, of course, many other p.d.f.s which arise in other models. Poirier (1995, Chapter 3) discusses many of these. Probably the most complete source of information on probability distributions is the *Distributions in Statistics* series (see Johnson, Kotz and Balakrishnan (1994, 1995, 2000) and Johnston, Kotz and Kemp (1993)).

Following standard practice, we introduce two sorts of notation. The first is for the p.d.f. or probability function itself. The second is the symbol '$\sim$' meaning 'is distributed as'. For instance, we use the notation $f_N(Y|\mu, \Sigma)$ for the Normal p.d.f. meaning "the random variable y has a p.d.f. given by $f_N(Y|\mu, \Sigma)$". An equivalent way of saying this is "Y is Normally distributed with mean μ and variance Σ" or "$Y \sim N(\mu, \Sigma)$". Hence, we have two types of notation, $f_N(Y|\mu, \Sigma)$ indicating the p.d.f. itself and $N(\mu, \Sigma)$.

It is also worth noting that in this appendix we fully specify the p.d.f.s. In practice, it is often enough to know the *kernel* of the p.d.f. The kernel is the p.d.f., ignoring integrating constants (i.e. multiplicative constants which do not involve the random variable).

Definition B.19: The Binomial Distribution

A discrete random variable, Y, has a *Binomial distribution* with parameters T and p, denoted $Y \sim B(T, p)$, if its probability function is given by

$$f_B(y|T, p) = \begin{cases} \frac{T!}{(T-y)!y!} p^y (1-p)^{T-y} & \text{if } y = 0, 1, \ldots, T \\ 0 & \text{otherwise} \end{cases}$$

where $0 \leq p \leq 1$ and T is a positive integer.

Theorem B.4: Mean and Variance of the Binomial Distribution

If $Y \sim B(T, p)$ then $E(Y) = Tp$ and $var(Y) = Tp(1-p)$.

Note: this distribution is used in cases where an experiment, the outcome of which is either 'success' or 'failure', is repeated independently T times. The probability of success in an experiment is p. The distribution of the random variable Y, which counts the number of successes, is $B(T, p)$.

Definition B.20: The Poisson Distribution

A discrete random variable, Y, has a *Poisson distribution* with parameter λ, denoted $Y \sim Po(\lambda)$, if its probability function is given by

$$f_{Po}(y|\lambda) = \begin{cases} \frac{\lambda^y \exp(-\lambda)}{y!} & \text{if } y = 0, 1, 2, \ldots \\ 0 & \text{otherwise} \end{cases}$$

where λ is a positive real number.

Theorem B.5: Mean and Variance of the Poisson Distribution

If $Y \sim Po(\lambda)$ then $E(Y) = \lambda$ and $var(Y) = \lambda$.

Definition B.21: The Uniform Distribution

A continuous random variable, Y, has a *Uniform distribution* over the interval $[a, b]$, denoted $Y \sim U(a, b)$, if its p.d.f. is given by

$$f_U(y|a, b) = \begin{cases} \frac{1}{b-a} & \text{if } a \leq y \leq b \\ 0 & \text{otherwise} \end{cases}$$

where $-\infty < a < b < \infty$.

Theorem B.6: Mean and Variance of the Uniform Distribution

If $Y \sim U(a, b)$ then $E(Y) = \frac{a+b}{2}$ and $var(Y) = \frac{(b-a)^2}{12}$.

Definition B.22: The Gamma Distribution

A continuous random variable Y has a *Gamma* distribution with mean $\mu > 0$ and degrees of freedom $\nu > 0$, denoted by $Y \sim G(\mu, \nu)$, if its p.d.f. is:

$$f_G(y|\mu, \nu) \equiv \begin{cases} c_G^{-1} y^{\frac{\nu-2}{2}} \exp\left(-\frac{y\nu}{2\mu}\right) & \text{if } 0 < y < \infty \\ 0 & \text{otherwise} \end{cases}$$

where the integrating constant is given by $c_G = \left(\frac{2\mu}{\nu}\right)^{\frac{\nu}{2}} \Gamma\left(\frac{\nu}{2}\right)$ where $\Gamma(a)$ is the Gamma function (see Poirier, 1995, p. 98).

Theorem B.7: Mean and Variance of the Gamma Distribution

If $Y \sim G(\mu, \nu)$ then $E(Y) = \mu$ and $var(Y) = \frac{2\mu^2}{\nu}$.

Notes: the Gamma distribution is a very important one in Bayesian econometrics as it usually relates to the error precision. Further properties of the Gamma distribution are given in Poirier (1995, pp. 98–102). Distributions related to the Gamma include the *Chi-squared distribution* which is a Gamma distribution with $\nu = \mu$. It is denoted by $Y \sim \chi^2(\nu)$. The *exponential distribution* is a Gamma distribution with $\nu = 2$. The *inverted Gamma distribution* has the property that, if Y has an inverted Gamma distribution, then $\frac{1}{Y}$ has a Gamma distribution. In some other Bayesian books, the authors work with error variances (instead of error precisions) and the inverted Gamma is used extensively.

Definition B.23: The Multinomial Distribution

A discrete N-dimensional random vector, $Y = (Y_1, \ldots, Y_N)'$, has a *Multinomial distribution* with parameters T and p, denoted $Y \sim M(T, p)$, if its probability function is given by

$$f_M(y|T, p) = \begin{cases} \frac{T!}{y_1! \cdots y_N!} p_1^{y_1} \cdots p_N^{y_N} & \text{if } y_i = 0, 1, \ldots, T \text{ and } \sum_{i=1}^{N} y_i = T \\ 0 & \text{otherwise} \end{cases}$$

where $p = (p_1, \ldots, p_N)'$, $0 \leq p_i \leq 1$ for $i = 1, \ldots, N$, $\sum_{i=1}^{N} p_i = 1$ and T is a positive integer.

Theorem B.8: Mean and Variance of the Multinomial Distribution
If $Y \sim M(T, p)$ then $E(Y_i) = Tp_i$ and $var(Y_i) = Tp_i(1-p_i)$ for $i = 1, \ldots, N$.

Note: this distribution is generalization of the Binomial to the case where an experiment involving N outcomes is repeated T times. The random vector, Y, counts the number of times each outcome occurs. Since $\sum_{i=1}^{N} Y_i = T$, one of the elements of Y can be eliminated and the Multinomial distribution can be written in terms of an $(N-1)$ random vector.

Definition B.24: The Multivariate Normal Distribution
A continuous k-dimensional random vector, $Y = (Y_1, \ldots, Y_k)'$, has a *Normal distribution* with mean μ (a k-vector) and covariance matrix Σ (a $k \times k$ positive definite matrix), denoted $Y \sim N(\mu, \Sigma)$, if its p.d.f. is given by

$$f_N(y|\mu, \Sigma) = \frac{1}{2\pi^{\frac{k}{2}}}|\Sigma|^{-\frac{1}{2}} \exp\left[-\frac{1}{2}(y-\mu)'\Sigma^{-1}(y-\mu)\right]$$

Note: the special case where $k = 1$, $\mu = 0$ and $\Sigma = 1$ is referred to as the *standard Normal* distribution. Tables providing percentiles of the standard Normal are available in many econometrics and statistics textbooks.

Theorem B.9: Marginals and Conditionals of the Multivariate Normal Distribution
Suppose the k-vector $Y \sim N(\mu, \Sigma)$ is partitioned as

$$Y = \begin{pmatrix} Y_{(1)} \\ Y_{(2)} \end{pmatrix}$$

where $Y_{(i)}$ is a k_i-vector for $i = 1, 2$ with $k_1 + k_2 = k$ and μ and Σ have been partitioned conformably as

$$\mu = \begin{pmatrix} \mu_{(1)} \\ \mu_{(2)} \end{pmatrix}$$

and

$$\Sigma = \begin{pmatrix} \Sigma_{(11)} & \Sigma_{(12)} \\ \Sigma'_{(12)} & \Sigma_{(22)} \end{pmatrix}$$

Then the following results hold:

- The marginal distribution of $Y_{(i)}$ is $N(\mu_{(i)}, \Sigma_{(ii)})$ for $i = 1, 2$.
- The conditional distribution of $Y_{(1)}$ given $Y_{(2)} = y_{(2)}$ is $N(\mu_{(1|2)}, \Sigma_{(1|2)})$ where

$$\mu_{(1|2)} = \mu_{(1)} + \Sigma_{(12)}\Sigma_{(22)}^{-1}(y_{(2)} - \mu_{(2)})$$

and

$$\Sigma_{(1|2)} = \Sigma_{(11)} - \Sigma_{(12)}\Sigma_{(22)}^{-1}\Sigma_{(12)}'$$

The conditional distribution of $Y_{(2)}$ given $Y_{(1)} = y_{(1)}$ can be obtained by reversing subscripts 1 and 2 in the previous formulae.

Theorem B.10: Linear Combinations of Normals are Normal

Let $Y \sim N(\mu, \Sigma)$ be a k-dimensional random vector and A be a fixed (non-random) $m \times k$ matrix with $\text{rank}(A) = m$, then $AY \sim N(A\mu, A\Sigma A')$.

Theorem B.11: Relationship between Normal and Chi-squared Distributions

Suppose the k-dimensional random vector $Y \sim N(\mu, \Sigma)$, then the random variable $Q = (Y - \mu)'\Sigma^{-1}(Y - \mu)$ has a Chi-squared distribution with k degrees of freedom (i.e. $Q \sim \chi^2(k)$ or, equivalently, $Q \sim G(k, k)$).

Definition B.25: The Multivariate t Distribution

A continuous k-dimensional random vector, $Y = (Y_1, \ldots, Y_k)'$, has a *t distribution* with parameters μ (a k-vector), Σ (a $k \times k$ positive definite matrix) and ν (a positive scalar referred to as a *degrees of freedom* parameter), denoted $Y \sim t(\mu, \Sigma, \nu)$, if its p.d.f. is given by

$$f_t(y|\mu, \Sigma, \nu) = \frac{1}{c_t}|\Sigma|^{-\frac{1}{2}}[\nu + (y - \mu)'\Sigma^{-1}(y - \mu)]^{-\frac{\nu+k}{2}}$$

where

$$c_t = \frac{\pi^{\frac{k}{2}}\Gamma\left(\frac{\nu}{2}\right)}{\nu^{\frac{\nu}{2}}\Gamma\left(\frac{\nu+k}{2}\right)}$$

Notes: the univariate case with $k = 1$ is sometimes referred to as the *Student-t* distribution. Tables providing percentiles of the Student-t with $\mu = 0$ and $\Sigma = 1$ are available in many econometrics and statistics textbooks. The case where $\nu = 1$ is referred to as the *Cauchy distribution.*

Theorem B.12: Mean and Variance of the t Distribution

If $Y \sim t(\mu, \Sigma, \nu)$ then $E(Y) = \mu$ if $\nu > 1$ and $var(Y) = \frac{\nu}{\nu-2}\Sigma$ if $\nu > 2$.

Notes: the mean and variance only exist if $\nu > 1$ and $\nu > 2$, respectively. This implies, for instance, that the mean of the Cauchy does not exist even though it is a valid p.d.f. and, hence, its median and other quantiles exist. Σ is not exactly the same as the covariance matrix and, hence, is given another name: the *scale matrix.*

Theorem B.13: Marginals and Conditionals of the Multivariate t Distribution

Suppose the k-vector $Y \sim t(\mu, \Sigma, \nu)$ is partitioned as in Theorem B.9 as are μ and Σ. Then the following results hold:

- The marginal distribution of $Y_{(i)}$ is $t(\mu_{(i)}, \Sigma_{(ii)}, \nu)$ for $i = 1, 2$.
- The conditional distribution of $Y_{(1)}$ given $Y_{(2)} = y_{(2)}$ is $t(\mu_{(1|2)}, \Sigma_{(1|2)}, \nu+k_1)$ where

$$\mu_{(1|2)} = \mu_{(1)} + \Sigma_{(12)}\Sigma_{(22)}^{-1}(y_{(2)} - \mu_{(2)})$$

$$\Sigma_{(1|2)} = h_{(1|2)}[\Sigma_{(11)} - \Sigma_{(12)}\Sigma_{(22)}^{-1}\Sigma_{(12)}']$$

and

$$h_{(1|2)} = \frac{1}{\nu + k_2}[\nu + (y_{(2)} - \mu_{(2)})'\Sigma_{(22)}^{-1}(y_{(2)} - \mu_{(2)})]$$

The conditional distribution of $Y_{(2)}$ given $Y_{(1)} = y_{(1)}$ can be obtained by reversing subscripts 1 and 2 in the previous formulae.

Theorem B.14: Linear Combinations of t's are t

Let $Y \sim t(\mu, \Sigma, \nu)$ be a k-dimensional random vector and A be a fixed (non-random) $m \times k$ matrix with rank equal to m, then $AY \sim t(A\mu, A\Sigma A', \nu)$.

Definition B.26: The Normal-Gamma Distribution

Let Y be a k-dimensional random vector and H a scalar random variable. If the conditional distribution of Y given H is Normal and the marginal distribution for H is Gamma then (Y, H) is said to have a *Normal-Gamma distribution*. Formally, if $Y|H \sim N(\mu, \Sigma)$ and $H \sim G(m, \nu)$ then $\theta = (Y', H)'$ has a Normal-Gamma distribution denoted $\theta \sim NG(\mu, \Sigma, m, \nu)$. The corresponding p.d.f. is denoted $f_{NG}(\theta|\mu, \Sigma, m, \nu)$.

Theorem B.15: Marginal Distributions Involving the Normal-Gamma

If $\theta = (Y', H)' \sim NG(\mu, \Sigma, m, \nu)$, then the marginal for Y is given by $Y \sim t(\mu, m^{-1}\Sigma, \nu)$. Of course, by definition the marginal for H is given by $H \sim G(m, \nu)$.

Definition B.27: The Wishart Distribution

Let H be an $N \times N$ positive definite (symmetric) random matrix, A be a fixed (non-random) $N \times N$ positive definite matrix and $\nu > 0$ a scalar degrees of freedom parameter. Then H has a Wishart distribution, denoted $H \sim W(\nu, A)$, if its p.d.f. is given by

$$f_W(H|\nu, A) = \frac{1}{c_W}|H|^{\frac{\nu-N-1}{2}}|A|^{-\frac{\nu}{2}}\exp\left[-\frac{1}{2}tr(A^{-1}H)\right]$$

where

$$c_W = 2^{\frac{\nu N}{2}}\pi^{\frac{N(N-1)}{4}}\prod_{i=1}^{N}\Gamma\left(\frac{\nu+1-i}{2}\right)$$

Note: if $N = 1$, then the Wishart reduces to a Gamma distribution (i.e. $f_W(H|\nu, A) = f_G(H|\nu A, \nu)$ if $N = 1$).

Theorem B.16: Means, Variances and Covariances Relating to the Wishart Distribution

If $H \sim W(\nu, A)$ then $E(H_{ij}) = \nu A_{ij}$, $var(H_{ij}) = \nu(A_{ij}^2 + A_{ii}A_{jj})$ for $i, j = 1, \ldots, N$ and $cov(H_{ij}, H_{km}) = \nu(A_{ik}A_{jm} + A_{im}A_{jk})$ for $i, j, k, m = 1, \ldots, N$, where subscripts i, j, k, m refer to elements of matrices.

Definition B.28: The Dirichlet and Beta Distribution

Let $Y = (Y_1, \ldots, Y_N)'$ be a vector of continuous random variables with the property that $Y_1 + \cdots + Y_N = 1$. Then Y has a *Dirichlet distribution*, denoted $Y \sim D(\alpha)$, if its p.d.f. is given by

$$f_D(Y|\alpha) = \left[\frac{\Gamma(a)}{\prod_{i=1}^N \Gamma(\alpha_i)}\right] \prod_{i=1}^N y_i^{\alpha_i - 1}$$

where $\alpha = (\alpha_1, \ldots, \alpha_N)'$, $\alpha_i > 0$ for $i = 1, \ldots, N$ and $a = \sum_{i=1}^N \alpha_i$. The *Beta distribution*, denoted by $Y \sim B(\alpha_1, \alpha_2)$, is the Dirichlet distribution for the case $N = 2$. Its p.d.f. is denoted by $f_B(Y|\alpha_1, \alpha_2)$.

Note: in the case $N = 2$, the restriction $Y_1 + Y_2 = 1$ can be used to remove one of the random variables. Thus, the Beta distribution is a univariate distribution.

Theorem B.17: Means and Variances of the Dirichlet Distribution

Suppose $Y \sim D(\alpha)$ where α and a are as given in Definition B.28, then for $i, j = 1, \ldots, N$,

- $E(Y_i) = \frac{\alpha_i}{a}$,
- $var(Y_i) = \frac{\alpha_i(a - \alpha_i)}{a^2(a+1)}$, and
- $cov(Y_i, Y_j) = -\frac{\alpha_i \alpha_j}{a^2(a+1)}$.

B.3 INTRODUCTION TO SOME CONCEPTS IN SAMPLING THEORY

In this section, we introduce a few concepts in sampling theory which are used in this book. Poirier (1995, Chapter 5), provides a much more detailed discussion of this topic. Asymptotic theory is an important tool for frequentist econometrics and there are a huge number of relevant theorems. The interested reader is referred to White (1984) for further details.

Definition B.29: Random Sample

Suppose Y_i for $i = 1, \ldots, T$ are random variables which are independent of one another and have identical distributions (denoted as *i.i.d.* for *independent and identically distributed*) with p.d.f. depending on a vector of parameters, θ, denoted by $p(y_i|\theta)$. Then $Y = (Y_1, \ldots, Y_T)'$ has p.d.f.

$$p(y|\theta) = \prod_{i=1}^{T} p(y_i|\theta)$$

and Y_i for $i = 1, \ldots, T$ is referred to as a *random sample.*

Theorem B.18: Properties of Random Sample from the Normal Distribution

Suppose Y_i is a random sample from the $N(\mu, \sigma^2)$ and define the *sample mean* $\overline{Y} = \frac{\sum_{i=1}^{T}}{T}$, then $\overline{Y} \sim N\left(\mu, \frac{\sigma^2}{T}\right)$.

Definition B.30: Convergence in Probability

Let $\{Y_T\}$ be a sequence of random variables. Then $\{Y_T\}$ *converges in probability* to a constant Y, denoted $\text{plim}Y_T = Y$ or $Y_T \to^p Y$, if

$$\lim_{T\to\infty} \Pr(|Y_T - Y| > \varepsilon) = 0$$

for every $\varepsilon > 0$.

Notes: for some intuition about convergence in probability consider what happens to $\overline{Y}$ in Theorem B.18 as $T \to \infty$. The variance of $\overline{Y}$ goes to zero and its distribution collapses to a point at μ. This is an example of convergence in probability. Convergence in probability has the property that, if $Y_T \to^p Y$ then $g(Y_T) \to^p g(Y)$, where $g()$ is a continuous function.

Definition B.31: Weak Laws of Large Numbers

Let $\{Y_T\}$ be a sequence of random variables with corresponding sequence of finite means $\{\mu_T\}$, and denote the sample mean based on a sample of size T as

$$\overline{Y}_T = \frac{\sum_{t=1}^{T} Y_t}{T}$$

and define

$$\overline{\mu}_T = \frac{\sum_{t=1}^{T} \mu_t}{T}$$

Then $\overline{Y}_T$ satisfies a *weak law of large numbers (W.L.L.N.)* if $\overline{Y}_T \to^p \overline{\mu}_T$.

Notes: Bayesian computation often proceeds by taking a sequence of draws from the posterior distribution. In such cases, a W.L.L.N. is invoked to show that the average of this sequence converges to the desired expectation. W.L.L.N.s are, thus, crucial to Bayesian computation. There is a myriad of different W.L.L.N.s which can be invoked, depending on the properties of the sequence (e.g. whether the sequence contains independent random variables or whether they are dependent, whether the sequence is drawn from the same distribution, or different distributions, etc.). The reader is referred to White (1984) for more detail. As an example, the following theorem gives a W.L.L.N. which is relevant when the sequence is a random sample. Although some of the posterior simulators used in this book (e.g. the Gibbs sampler) provide a sequence which is not a random sample, we reassure the reader that they all satisfy W.L.L.N.s necessary to ensure convergence in probability.

Theorem B.19: A Weak Law of Large Numbers for a Random Sample

Let $\{Y_T\}$ be a sequence of i.i.d. random variables drawn from some distribution with mean μ and variance σ^2, then $\overline{Y}_T \rightarrow^p \mu$.

Note: this theorem holds for *any* distribution, and works even if σ^2 is infinite.

Definition B.32: Convergence in Distribution

Let $\{Y_T\}$ be a sequence of random variables, $\{P_T(.)\}$ be the corresponding sequence of distribution functions, and let Y be a random variable with distribution function $P(y)$. Then $\{Y_T\}$ *converges in distribution* to a random variable Y, denoted $Y_T \rightarrow^d Y$, if

$$\lim_{T\rightarrow\infty} P_T(y) = P(y)$$

$P(y)$ is referred to as the *limiting distribution.* Convergence in distribution has the property that, if $Y_T \rightarrow^d Y$, then $g(Y_T) \rightarrow^d g(Y)$ where $g()$ is a continuous function.

Definition B.33: Central Limit Theorems

Let $\{Y_T\}$ be a sequence of random variables, Y be a random variable and denote the sample mean based on a sample of size T as

$$\overline{Y}_T = \frac{\sum_{t=1}^{T} Y_t}{T}$$

Then $\overline{Y}_T$ satisfies a *Central Limit Theorem (C.L.T.)* if $\overline{Y}_T \rightarrow^d Y$.

Notes: in econometric problems, the limiting distribution is virtually always Normal. In Bayesian computation, C.L.T.s are used to calculate numerical standard errors for estimates obtained using a sequence of random draws from a posterior simulator. Just as with W.L.L.N.s, there is a myriad of different C.L.T.s which can be invoked, depending on the properties of the sequence (e.g. whether the sequence contains independent random variables or whether they are dependent, whether the sequence is drawn from the same distribution or different distributions, etc.). The reader is referred to White (1984) for more details. As an example, the following theorem gives a C.L.T. which is relevant when the sequence is a random sample. Although some of the posterior simulators used in this book (e.g. the Gibbs sampler) provide a sequence which is not a random sample, we reassure the reader that they all satisfy C.L.T.s necessary to ensure convergence in distribution.

Theorem B.20: A Central Limit Theorem for a Random Sample

Let $\{Y_T\}$ be a sequence of i.i.d. random variables drawn from some distribution with mean μ and variance σ^2. Define a new sequence of i.i.d. random variables $\{Z_T\}$, where

$$Z_T = \frac{\sqrt{T}(\overline{Y}_T - \mu)}{\sigma}$$

Then $Z_T \to^d Z$ where $Z \sim N(0, 1)$.

Notes: this is often referred to as the *Lindeberg–Levy Central Limit Theorem.* There are many extensions to the basic C.L.T.s and W.L.L.N.s provided here. For instance, multivariate extensions to the univariate theorems provided here are available.

B.4 OTHER USEFUL THEOREMS

Theorem B.21: Change of Variable Theorem (Univariate version)

Let X be a continuous random variable with p.d.f. $p_X(x)$ defined over the interval A and let $Y = g(X)$, where $g()$ is a one-to-one function that maps the interval A into the interval B. Let $g^{-1}()$ denote the inverse of $g()$ such that $X = g^{-1}(Y)$. Assume that

$$\frac{dg^{-1}(y)}{dy} = \frac{dx}{dy}$$

is continuous, and does not equal zero for any value of y in B. Then the p.d.f. of Y is given by

$$p_Y(y) = \begin{cases} \left|\frac{dx}{dy}\right| p_X[g^{-1}(y)] & \text{if } y \in B \\ 0 & \text{otherwise} \end{cases}$$

Note: the change of variable theorem for discrete random variables is the same, except that the term $\left|\frac{dx}{dy}\right|$ is omitted.

Theorem B.22: Change of Variable Theorem (Multivariate version)

Let $X = (X_1, \ldots, X_N)'$ be a continuous random vector with joint p.d.f. $p_X(x)$ where $x = (x_1, \ldots, x_N)'$ is defined over the set A. Let $Y_i = g_i(X)$ for $i =$

$1, \ldots, N$ where $g_i()$ are one-to-one functions that map the set A into the set B. Let $f_i()$ denote the inverse transformation such that $X_i = f_i(Y)$ where $Y = (Y_1, \ldots, Y_N)'$. Assume that the $N \times N$ determinant

$$J = \left|\begin{bmatrix} \frac{\partial x_1}{\partial y_1} & \frac{\partial x_1}{\partial y_2} & \cdot & \cdot & \frac{\partial x_1}{\partial y_N} \\ \frac{\partial x_2}{\partial y_1} & \frac{\partial x_2}{\partial y_2} & \cdot & \cdot & \cdot \\ \cdot & \cdot & \cdot & \cdot & \cdot \\ \cdot & \cdot & \cdot & \cdot & \frac{\partial x_{N-1}}{\partial y_N} \\ \frac{\partial x_N}{\partial y_1} & \cdot & \cdot & \frac{\partial x_N}{\partial y_{N-1}} & \frac{\partial x_N}{\partial y_N} \end{bmatrix}\right|$$

does not equal zero for any value of y in B. Then the joint p.d.f. of Y is given by

$$p_y(y) = \begin{cases} |J| p_x[f_1(y), \ldots, f_N(y)] & \text{if } y \in B \\ 0 & \text{otherwise} \end{cases}$$

Note: J is referred to as the *Jacobian*.

Bibliography

Albano, G. and Jouneau, F. (1998) A Bayesian Approach to the Econometrics of First Price Auctions, Center for Operations Research and Econometrics, Universite Catholique de Louvain, Discussion Paper 9831.

Albert, J. and Chib, S. (1993) Bayesian Analysis of Binary and Polychotomous Response Data, *Journal of the American Statistical Association*, **88**, 669–679.

Albert, J. and Chib, S. (1993a) Bayesian Analysis via Gibbs Sampling of Autoregressive Time Series Subject to Markov Mean and Variance Shifts, *Journal of Business and Economic Statistics*, **11**, 1–15.

Allenby, G. and Rossi, P. (1999) Marketing Models of Consumer Heterogeneity, *Journal of Econometrics*, **89**, 57–78.

Anglin, P. and Gencay, R. (1996) Semiparametric Estimation of a Hedonic Price Function, *Journal of Applied Econometrics*, **11**, 633–648.

Angrist, J., Imbens, G. and Rubin, D. (1996) Identification of Causal Effects Using Instrumental Variables, *Journal of the American Statistical Association*, **91**, 444–455.

Bajari, P. (1997) Econometrics of the First Price Auction with Asymmetric Bidders, available at http://www.stanford.edu/~bajari/.

Bajari, P. (1998). Econometrics of Sealed Bid Auction, available at http://www.stanford.edu/~bajari/.

Barberis, N. (2000) Investing for the Long Run When Returns are Predictable, *Journal of Finance*, **55**, 225–264.

Bauwens, L. (1984) *Bayesian Full Information Analysis of Simultaneous Equations Models Using Integration by Monte Carlo*. Berlin: Springer-Verlag.

Bauwens, L. and Lubrano, M. (1996) Identification Restrictions and Posterior Densities in Cointegrated Gaussian VAR Systems, in Fomby, T. (ed.), *Advances in Econometrics: Bayesian Methods Applied to Time Series Data*, vol. 11, pp. 3–28, part B. Greenwich, CT: JAI Press.

Bauwens, L. and Lubrano, M. (1998) Bayesian Inference in GARCH Models Using the Gibbs Sampler, *The Econometrics Journal*, **1**, C23–C46.

Bauwens, L., Lubrano, M. and Richard, J.-F. (1999) *Bayesian Inference in Dynamic Econometric Models*. Oxford: Oxford University Press.

Bayarri, M. and Berger, J. (2000) P-Values for Composite Null Models, *Journal of the American Statistical Association*, **95**, 1127–1142.

Bayarri, M., DeGroot, M. and Kadane, J. (1988) What is the Likelihood Function?, in Gupta, S. and Berger, J. (eds.), *Statistical Decision Theory and Related Topics IV*, vol. 1, pp. 1–27. New York: Springer-Verlag.

Berger, J. (1985) *Statistical Decision Theory and Bayesian Analysis*, second edition. New York: Springer-Verlag.

Berger, J. (1990) Robust Bayesian Analysis: Sensitivity to the Prior, *Journal of Statistical Planning and Inference*, **25**, 303–328.

Berger, J. and Guglielmi, A. (2001) Bayesian and Conditional Frequentist Testing of a Parametric Model versus Nonparametric Alternatives, *Journal of the American Statistical Association*, **96**, 174–184.

Berger, J. and Pericchi, L. (1996) The Intrinsic Bayes Factor for Model Selection and Prediction, *Journal of the American Statistical Association*, **91**, 109–122.

Bernardo, J.M. and Ramón, J.M. (1998) An Introduction to Bayesian Reference Analysis, *The Statistician*, **47**, 101–135.

Bernardo, J. and Smith, A.F.M. (1994) *Bayesian Theory*. Chichester: John Wiley & Sons.

Best, N., Cowles, M. and Vines, S. (1995) *CODA: Manual version 0.30*. Available at http://www.mrc-bsu.cam.ac.uk/bugs/.

van den Broeck, J., Koop, G., Osiewalski, J. and Steel, M.F.J. (1994) Stochastic Frontier Models: A Bayesian Perspective, *Journal of Econometrics*, **61**, 273–303.

Brown, P., Fearn, T. and Vannucci, M. (1999) The Choice of Variables in Multivariate Regression: A Non-Conjugate Bayesian Decision Theory Framework, *Biometrika*, **86**, 635–648.

Campolieti, M. (2001) Bayesian Semiparametric Estimation of Discrete Duration Models: An Application of the Dirichlet Process Prior, *Journal of Applied Econometrics*, **16**, 1–22.

Carlin, B. and Chib, S. (1995) Bayesian Model Choice via Markov Chain Monte Carlo Methods, *Journal of the Royal Statistical Society, Series B*, **57**, 473–484.

Carlin, B., Gelfand, A. and Smith, A.F.M. (1992) Hierarchical Bayesian analysis of changepoint problems, *Applied Statistics*, **41**, 389–405.

Carlin, B. and Louis, T. (2000) *Bayes and Empirical Bayes Methods for Data Analysis*, second edition. Boca Raton: Chapman & Hall.

Carlin, B., Polson, N. and Stoffer, D. (1992) A Monte Carlo Approach to Nonnormal and Nonlinear State Space Modeling, *Journal of the American Statistical Association*, **87**, 493–500.

Carter, C. and Kohn, R. (1994) On Gibbs Sampling for State Space Models, *Biometrika*, **81**, 541–553.

Chao J. and Phillips P.C.B. (1999) Model Selection in Partially Nonstationary Vector Autoregressive Processes with Reduced Rank Structure, *Journal of Econometrics*, **91**, 227–271.

Chen, M.H., Shao, Q.-M. and Ibrahim, J. (2000) *Monte Carlo Methods in Bayesian Computation*. New York: Springer-Verlag.

Chib, S. (1992) Bayes Inference in the Tobit Censored Regression Model, *Journal of Econometrics*, **51**, 79–99.

Chib, S. (1993) Bayes Regression with Autoregressive Errors, *Journal of Econometrics*, **58**, 275–294.

Chib, S. (1995) Marginal Likelihood from the Gibbs Sampler, *Journal of the American Statistical Association*, **90**, 1313–1321.

Chib, S. (1998) Estimation and Comparison of Multiple Change Point Models, *Journal of Econometrics*, **86**, 221–241.

Chib, S. and Jeliazkov, I. (2001) Marginal Likelihood from the Metropolis–Hastings Output, *Journal of the American Statistical Association*, **96**, 270–281.

Chib, S. and Greenberg, E. (1995) Understanding the Metropolis–Hastings Algorithm, *The American Statistician*, **49**, 327–335.

Chib, S., Greenberg, E. and Winkelmann, R. (1998) Posterior Simulation and Bayes Factors in Panel Count Data Models, *Journal of Econometrics*, **86**, 33–54.

Chib, S. and Hamilton, B. (2000) Bayesian Analysis of Cross-Section and Clustered Data Treatment Models, *Journal of Econometrics*, **97**, 25–50.

Chib, S. and Hamilton, B. (2002) Semiparametric Bayes Analysis of Longitudinal Data Treatment Models, *Journal of Econometrics*, **110**, 67–89.

Chib, S., Nardari, F. and Shephard, N. (2002) Markov Chain Monte Carlo Methods for Stochastic Volatility Models, *Journal of Econometrics*, **108**, 281–316.

Chib, S. and Winkelmann. R. (2001) Markov Chain Monte Carlo Analysis of Correlated Count Data, *Journal of Business and Economic Statistics*, **19**, 428–435.

Chipman, H., George, E. and McCulloch, R. (1998) Bayesian CART Model Search, *Journal of the American Statistical Association*, **93**, 935–960.

Clayton, D. (1996) Generalized Linear Mixed Models, In Gilks, Richardson and Speigelhalter (1996).

Clyde, M. (1999) Bayesian Model Averaging and Model Search Strategies (with discussion), in Bernardo, J., Dawid, A.P., Berger, J.O. and Smith, A.F.M. (eds.), *Bayesian Statistics 6*, pp. 157–185 Oxford: Oxford University Press.

Clyde, M., Desimone, H. and Parmigiani, G. (1996) Prediction via Orthogonalized Model Mixing, *Journal of the American Statistical Association*, **91**, 1197–1208.

Damien, P., Wakefield, J. and Walker, S. (1999) Gibbs Sampling for Bayesian Nonconjugate and Hierarchical Models by Using Auxiliary Variables, *Journal of the Royal Statistical Society, Series B*, **61**, 331–344.

DeJong D. (1992) Co-integration and trend-stationarity in macroeconomic time series, *Journal of Econometrics*, **52**, 347–370.

DeJong, D. (1996) A Bayesian Search for Structural Breaks in US GNP, in Fomby, T. (ed.), *Advances in Econometrics: Bayesian Methods Applied to Time Series Data*, vol. 11, pp. 109–146, part B. Greenwich, CT: JAI Press.

DeJong, D., Ingram, B. and Whiteman, C. (1996) A Bayesian Approach to Calibration, *Journal of Business and Economic Statistics*, **14**, 1–10.

DeJong, D., Ingram, B. and Whiteman, C. (2000) A Bayesian Approach to Dynamic Macroeconomics, *Journal of Econometrics*, **15**, 311–320.

DeJong, D. and Whiteman, C. (1991) The Temporal Stability of Dividends and Stock Prices: Evidence from the Likelihood Function, *American Economic Review*, **81**, 600–617.

DeJong, P. and Shephard, N. (1995) The Simulation Smoother for Time Series Models, *Biometrika*, **82**, 339–350.

Dellaportas, P. and Smith, A.F.M (1993) Bayesian Inference for Generalized Linear and Proportional Hazards Models via Gibbs Sampling, *Applied Statistics*, **42**, 443–459.

Devroye, L. (1986) *Non-Uniform Random Number Generation*. New York: Springer-Verlag.

Dey, D., Muller, P. and Sinha, D. (eds.) (1998) *Practical Nonparametric and Semiparametric Bayesian Statistics*. New York: Springer-Verlag.

Doan, T., Litterman, R. and Sims, C. (1984) Forecasting and Conditional Projection Using Realistic Prior Distributions, *Econometric Reviews*, **3**, 1–100.

Dorfman, J. (1994) A Numerical Bayesian Test for Cointegration of AR Processes, *Journal of Econometrics*, **66**, 289–324.

Dorfman, J. (1997) *Bayesian Economics through Numerical Methods*. New York: Springer-Verlag.

Draper, D. (1995) Assessment and Propagation of Model Uncertainty (with discussion), *Journal of the Royal Statistical Society, Series B*, **56**, 45–98.

Dreze, J. and Richard, J.-F. (1983) Bayesian Analysis of Simultaneous Equation Systems, in Griliches, Z. and Intriligator M. (eds.), *Handbook of Econometrics*, vol. 1, Amsterdam: North-Holland.

Durbin, J. and Koopman, S. (2001) *Time Series Analysis by State Space Methods*. Oxford: Oxford University Press.

Elerian, O., Chib, S. and Shephard, N. (2001) Likelihood Inference for Discretely Observed Nonlinear Diffusions, *Econometrica*, **69**, 959–993.

Enders, W. (1995) *Applied Econometric Time Series*. New York: John Wiley & Sons.

Eraker, B. (2001) MCMC Analysis of Diffusion Models with Applications to Finance, *Journal of Business and Economic Statistics*, **19**, 177–191.

Erickson, T. (1989) Proper Posteriors from Improper Priors for an Unidentified Errors-in-Variables Model, *Econometrica*, **57**, 1299–1316.

Escobar, M. and West, M. (1995) Bayesian Density Estimation Using Mixtures, *Journal of the American Statistical Association*, **90**, 577–588.

Fernandez, C., Ley, E. and Steel, M. (2001a) Model uncertainty in cross-country growth regressions, *Journal of Applied Econometrics*, **16**, 563–576.

Fernandez, C., Ley, E. and Steel, M. (2001b) Benchmark priors for Bayesian model averaging, *Journal of Econometrics*, **100**, 381–427.

Fernandez, C., Osiewalski, J. and Steel, M.F.J. (1997) On the Use of Panel Data in Stochastic Frontier Models with Improper Priors, *Journal of Econometrics*, **79**, 169–193.

Filardo, A. and Gordon. S. (1998) Business Cycle Durations, *Journal of Econometrics*, **85**, 99–123.

Forbes, C., Kalb, G. and Kofman, P. (1999) Bayesian Arbitrage Threshold Analysis, *Journal of Business and Economic Statistics*, **17**, 364–372.

Fruhwirth-Schnatter, S. (1995) Bayesian model discrimination and Bayes factors for linear Gaussian state space models, *Journal of the Royal Statistical Society, Series B*, **56**, 237–246.

Gelfand, A. and Dey, D. (1994) Bayesian Model Choice: Asymptotics and Exact Calculations, *Journal of the Royal Statistical Society Series B*, **56**, 501–514.

Gelman, A. (1996) Inference and Monitoring Convergence, in Gilks, Richardson and Speigelhalter (1996).

Gelman, A. and Meng. X. (1996) Model Checking and Model Improvement, in Gilks, Richardson and Speigelhalter (1996).

Gelman, A. and Rubin, D. (1992) Inference from Iterative Simulation Using Multiple Sequences, *Statistical Science*, **7**, 457–511.

George, E. and McCulloch, R. (1993) Variable Selection via Gibbs Sampling, *Journal of the American Statistical Association*, **88**, 881–889.

George, E. and McCulloch, R. (1997) Approaches for Bayesian Variable Selection, *Statistica Sinica*, **7**, 339–373.

Geweke, J. (1988) Exact Inference in Models with Autoregressive Conditional Heteroscedasticity, in Barnett, W., Berndt, E. and White, H. (eds.), *Dynamic Econometric Modeling*, pp. 73–104. Cambridge: Cambridge University Press.

Geweke, J. (1989) Bayesian Inference in Econometric Models using Monte Carlo Integration, *Econometrica*, **57**, 1317–1340.

Geweke, J. (1989a) Exact Predictive Densities in Linear Models with ARCH Disturbances, *Journal of Econometrics*, **40**, 63–86.

Geweke, J. (1991) Efficient Simulation from the Multivariate Normal and Student-t Distributions Subject to Linear Constraints, in Keramidas E. (ed.), *Computer Science and Statistics: Proceedings of the Twenty-Third Symposium on the Interface*, p. 571–578. Fairfax: Interface Foundation of North America, Inc.

Geweke, J. (1992) Evaluating the Accuracy of Sampling-Based Approaches to the Calculation of Posterior Moments, in Bernardo, J., Berger, J., Dawid, A. and Smith, A. (eds.), *Bayesian Statistics 4*, pp. 641–649. Oxford: Clarendon Press.

Geweke, J. (1993) Bayesian Treatment of the Independent Student-t Linear Model, *Journal of Applied Econometrics*, **8**, S19–S40.

Geweke, J. (1996) Bayesian Reduced Rank Regression in Econometrics, *Journal of Econometrics*,**75**, 121–146.

Geweke, J. (1999) Using Simulation Methods for Bayesian Econometric Models: Inference, Development, and Communication (with discussion and rejoinder), *Econometric Reviews*, **18**, 1–126.

Geweke, J. (1999a) Computational Experiments and Reality, Department of Economics, University of Iowa working paper available at www.biz.uiowa.edu/faculty/jgeweke/papers.html.

Geweke, J. and Keane, M. (1999) Mixture of Normals Probit Models, in Hsiao, C., Lahiri, K., Lee, L.-F. and Pesaran, M. H. (eds.), *Analysis of Panels and Limited Dependent Variables: A Volume in Honor of G. S. Maddala*. Cambridge: Cambridge University Press.

Geweke, J., Keane, M. and Runkle, D. (1994) Alternative Computational Approaches to Statistical Inference in the Multinomial Probit Model, *Review of Economics and Statistics*, **76**, 609–632.

Geweke, J., Keane, M. and Runkle, D. (1997) Statistical Inference in the Multinomial Multiperiod Probit Model, *Journal of Econometrics*, **80**, 125–165.

Geweke, J. and Tanizaki, H. (1999) On Markov Chain Monte Carlo Methods for Nonlinear and Non-Gaussian State-Space Models, *Communications in Statistics*, **28**, 867–894.

Geweke, J. and Tanizaki, H. (2001) Bayesian Estimation of Nonlinear State-Space Models Using Metropolis–Hastings Algorithm with Gibbs Sampling, *Computational Statistics and Data Analysis*, **37**, 151–170.

Geweke, J. and Terui, N. (1993) Bayesian Threshold Autoregressive Models for Nonlinear Time Series, *Journal of Times Series Analysis*, **14**, 441–454.

Gilks, W. (1996) Full Conditional Distributions, in Gilks, Richardson and Speigelhalter (1996).

Gilks, W. and Roberts, G. (1996) Strategies for Improving MCMC, in Gilks, Richardson and Speigelhalter (1996).

Gilks, W. and Wild, P. (1992) Adaptive Rejection Sampling for Gibbs Sampling, *Applied Statistics*, **41**, 337–348.

Gilks, W., Richardson, S. and Speigelhalter, D. (1996) *Markov Chain Monte Carlo in Practice*. New York: Chapman & Hall.

Gilks, W., Richardson, S. and Speigelhalter, D. (1996a) Introducing Markov Chain Monte Carlo, in Gilks, Richardson and Speigelhalter (1996).

Greasley, D. and Oxley, L. (1994) Rehabilitation Sustained: The Industrial Revolution as a Macroeconomic Epoch, *Economic History Review, 2nd Series*, **47**, 760–768.

Green, P. and Silverman, B. (1994) *Nonparametric Regression and Generalized Linear Models*. London: Chapman & Hall.

Greene, W. (2000) *Econometric Analysis*, fourth edition. New Jersey: Prentice-Hall.

Griffin, J. and Steel, M.F.J. (2001) Inference with Non-Gaussian Ornstein–Uhlenbeck Processes for Stochastic Volatility, Institute for Mathematics and Statistics, University of Kent at Canterbury working paper available at http://www.ukc.ac.uk/IMS/statistics/people/M.F.Steel/.

Griffiths, W. (2001) Heteroskedasticity, in Baltagi, B. (ed.), *A Companion to Theoretical Econometrics*. Oxford: Blackwell.

Gujarati, D. (1995) *Basic Econometrics*, third edition. New York: McGraw-Hill.

Hamilton, J. (1994). *Time Series Analysis*. Princeton: Princeton University Press.

Hill, C., Griffiths, W. and Judge, G. (1997) *Undergraduate Econometrics*. New York: John Wiley & Sons.

Hirano, K., Imbens, G., Rubin, D. and Zhou, A. (2000) Estimating the Effect of Flu Shots in a Randomized Encouragement Design, *Biostatistics*, **1**, 69–88.

Hobert, J. and Casella, G. (1996) The Effect of Improper Priors on Gibbs Sampling in Hierarchical Linear Mixed Models, *Journal of American Statistical Association*, **91**, 1461–1473.

Hodges, J. (1987) Uncertainty, Policy Analysis and Statistics, *Statistical Science*, **2**, 259–291.

Hoeting, J., Madigan, D., Raftery, A. and Volinsky, C. (1999) Bayesian Model Averaging: A Tutorial, *Statistical Science*, **14**, 382–417.

Horowitz, J. (1998) *Semiparametric Methods in Econometrics*. New York: Springer-Verlag.

Ibrahim, J., Chen, M. and Sinha, D. (2001) *Bayesian Survival Analysis*. New York: Springer-Verlag.

Imbens, G. and Rubin, D. (1997) Bayesian Inference for Causal Effects in Randomized Experiments with Noncompliance, *Annals of Statistics*, **25**, 305–327.

Iwata, S. (1996) Bounding Posterior Means by Model Criticism, *Journal of Econometrics*, **75**, 239–261.

Jacquier, E., Polson, N. and Rossi, P. (1994) Bayesian Analysis of Stochastic Volatility, *Journal of Business and Economic Statistics*, **12**, 371–417.

Jain, D.C., Vilcassim, N. and Chintagunta, P. (1994) A Random-Coefficient Logit Brand-Choice Model Applied to Panel Data, *Journal of Business & Economic Statistics*, **12**, 317–328.

Jeffreys, H. (1946) An Invariant Form for the Prior Probability in Estimation Problems, *Proceedings of the Royal Statistical Society of London, Series A*, **186**, 453–461.

Johnson, N., Kotz, S. and Balakrishnan, N. (1994) *Continuous Univariate Distributions*, vol. 1, second edition. New York: John Wiley & Sons.

Johnson, N., Kotz, S. and Balakrishnan, N. (1995) *Continuous Univariate Distributions*, vol. 2, second edition. New York: John Wiley & Sons.

Johnson, N., Kotz, S. and Balakrishnan, N. (2000) *Continuous Multivariate Distributions*, vol. 1, second edition. New York: John Wiley & Sons.

Johnson, N., Kotz, S. and Kemp, A. (1993) *Univariate Discrete Distributions*, second edition. New York: John Wiley & Sons.

Judge, G., Griffiths, W., Hill, R., Lutkepohl, H. and Lee, T. (1985) *The Theory and Practice of Econometrics*. New York: John Wiley & Sons.

Kadane, J. (ed.) (1984) *Robustness of Bayesian Analysis*. Amsterdam: Elsevier.

Kadane, J., Dickey, J., Winkler, R., Smith, W. and Peters, S. (1980) Interactive Elicitation of Opinion for a Normal Linear Model, *Journal of the American Statistical Association*, **75**, 845–854.

Kadiyala, K. and Karlsson, S. (1997) Numerical Methods for Estimation and Inference in Bayesian VAR-Models, *Journal of Applied Econometrics*, **12**, 99–132.

Kandel, S., McCulloch, R. and Stambaugh, R. (1995) Bayesian Inference and Portfolio Allocation, *Review of Financial Studies*, **8**, 1–53.

Kass, R. and Raftery, A. (1995) Bayes Factors, *Journal of the American Statistical Association*, **90**, 773–795.

Kass, R. and Wasserman, L. (1996) The Selection of Prior Distributions by Formal Rules, *Journal of the American Statistical Association*, **91**, 1343–1370.

Kiefer, N. and Steel, M.F.J. (1998) Bayesian Analysis of the Prototypical Search Model, *Journal of Business and Economic Statistics*, **16**, 178–186.

Kim, C. and Nelson, C. (1999) *State Space Models with Regime Switching*. Cambridge: MIT Press.

Kleibergen, F. (1997) Bayesian Simultaneous Equations Analysis Using Equality Restricted Random Variables, *American Statistical Association*, *Proceedings of the Section on Bayesian Statistical Science*, 141–147.

Kleibergen, F. and Paap, R. (2002) Priors, Posteriors and Bayes factors for a Bayesian Analysis of Cointegration, *Journal of Econometrics*, **111**, 223–249.

Kleibergen, F. and van Dijk, H. (1993) Non-stationarity in GARCH Models: A Bayesian Analysis, *Journal of Applied Econometrics*, **S8**, 41–61.

Kleibergen, F. and van Dijk, H. (1994) On the Shape of the Likelihood/Posterior in Cointegration Models, *Econometric Theory*, **10**, 514–551.

Kleibergen, F. and van Dijk, H. (1998) Bayesian Simultaneous Equations Analysis Using Reduced Rank Structures, *Econometric Theory*, **14**, 699–744.

Kleibergen, F. and Zivot, E. (2002) Bayesian and Classical Approaches to Instrumental Variable Regression, *Journal of Econometrics*, forthcoming.

Kloek, T. and van Dijk, H. (1978) Bayesian Estimates of Equation System Parameters, An Application of Integration by Monte Carlo, *Econometrica*, **46**, 1–19.

Koop, G. (1991) Intertemporal Properties of Real Output: A Bayesian Analysis, *Journal of Business and Economic Statistics*, **9**, 253–266.

Koop, G. (1992) Aggregate Shocks and Macroeconomic Fluctuations: A Bayesian Approach, *Journal of Applied Econometrics*, **7**, 395–411.

Koop, G. (1996) Parameter Uncertainty and Impulse Response Analysis, *Journal of Econometrics*, **72**, 135–149.

Koop, G. (2000) *Analysis of Economic Data*. New York: John Wiley & Sons.

Koop, G. (2001) Bayesian Inference in Models Based on Equilibrium Search Theory, *Journal of Econometrics*, **102**, 311–338.

Koop, G. and Steel, M.F.J. (2001) Bayesian Analysis of Stochastic Frontier Models, in Baltagi, B. (ed.), *A Companion to Theoretical Econometrics*. Oxford: Blackwell Publishers.

Koop, G., Osiewalski, J. and Steel, M.F.J. (1994) Bayesian Efficiency Analysis with a Flexible Cost Function, *Journal of Business and Economic Statistics*, **12**, 93–106.

Koop, G., Osiewalski, J. and Steel, M.F.J. (1997) Bayesian Efficiency Analysis through Individual Effects: Hospital Cost Frontiers, *Journal of Econometrics*, **76**, 77–105.

Koop, G., Osiewalski, J. and Steel, M.F.J. (2000) Modeling the Sources of Output Growth in a Panel of Countries, *Journal of Business and Economic Statistics*, **18**, 284–299.

Koop, G. and Poirier, D. (1993) Bayesian Analysis of Logit Models Using Natural Conjugate Priors, *Journal of Econometrics*, **56**, 323–340.

Koop, G. and Poirier, D. (1994) Rank-Ordered Logit Models: An Empirical Analysis of Ontario Voter Preferences before the 1988 Canadian Federal Election, *Journal of Applied Econometrics*, **9**, 369–388.

Koop, G. and Poirier, D. (1997) Learning About the Cross-regime Correlation in Switching Regression Models, *Journal of Econometrics*, **78**, 217–227

Koop, G. and Poirier, D. (2001) Testing for Optimality in Job Search Models, *The Econometrics Journal*, **4**, 257–272.

Koop, G. and Poirier, D. (2002) Bayesian Variants on Some Classical Semiparametric Regression Techniques, *Journal of Econometrics*, forthcoming.

Koop, G. and Potter, S. (1999) Dynamic Asymmetries in US Unemployment, *Journal of Business and Economic Statistics*, **17**, 298–312.

Koop, G. and Potter, S. (2000) Nonlinearity, Structural Breaks or Outliers in Economic Time Series? in Barnett, W., Hendry, D., Hylleberg, S., Terasvirta, T., Tjostheim, D. and Wurtz, A. (eds.), *Nonlinear Econometric Modeling in Time Series Analysis*. Cambridge: Cambridge University Press.

Koop, G. and Potter, S. (2001) Are Apparent Findings of Nonlinearity Due to Structural Instability in Economic Time Series? *The Econometrics Journal*, **4**, 37–55.

Koop, G. and Potter, S. (2003) Bayesian Analysis of Endogenous Delay Threshold Models, *Journal of Business and Economic Statistics*, **21**, 93–103.

Koop, G., Steel, M.F.J. and Osiewalski, J. (1995) Posterior Analysis of Stochastic Frontier Models Using Gibbs Sampling, *Computational Statistics*, **10**, 353–373.

Koop, G. and van Dijk, H. (2000) Testing for Integration Using Evolving Trend and Seasonals Models: A Bayesian Approach, *Journal of Econometrics*, **97**, 261–291.

Lancaster, A. (1997) Exact Structural Inference in Job-search Models, *Journal of Business and Economic Statistics*, **15**, 165–179.

Lancaster, A. (2003) *An Introduction to Modern Bayesian Econometrics*, forthcoming.

Leamer, E. (1978) *Specification Searches*. New York: Wiley.

Leamer, E. (1982) Sets of Posterior Means with Bounded Variance Priors, *Econometrica*, **50**, 725–736.

LeSage, J. (1999) *Applied Econometrics Using MATLAB*. Available at http://www.spatial-econometrics.com/.

Li, K. (1998) Bayesian Inference in a Simultaneous Equation Model with Limited Dependent Variables, *Journal of Econometrics*, **85**, 387–400.

Li, K. (1999) Exchange Rate Target Zone Models: A Bayesian Evaluation, *Journal of Applied Econometrics*, **14**, 461–490.

Li, K. (1999a) Bayesian Analysis of Duration Models: An Application to Chapter 11 Bankruptcy, *Economics Letters*, **63**, 305–312.

Lindley, D. and Smith, A.F.M. (1972) Bayes Estimates for the Linear Model, *Journal of the Royal Statistical Society, Series B*, **34**, 1–41.

Litterman, R. (1986) Forecasting with Bayesian Vector Autoregressions: Five Years of Experience, *Journal of Business and Economic Statistics*, **4**, 25–38.

Liu, J. and Wu, Y. (1999) Parameter Expansion for Data Augmentation, *Journal of the American Statistical Association*, **94**, 1264–1274.

Lubrano, M. (1995) Testing for Unit Roots in a Bayesian Framework, *Journal of Econometrics*, **69**, 81–109.

Lutkepohl, H. (1993) *Introduction to Multiple Time Series* (second edition). New York: Springer-Verlag.

McCausland, B. and Stevens, J. (2001) *Bayesian Analysis, Computation and Communication: The PC MATLAB Version of the BACC Software*. Available at http://www.econ.umn.edu/~bacc/bacc2001.

McCulloch, R., Polson, N. and Rossi, P. (2000) A Bayesian Analysis of the Multinomial Probit Model with Fully Identified Parameters, *Journal of Econometrics*, **99**, 173–193.

McCulloch, R. and Rossi, P. (1994) An Exact Likelihood Analysis of the Multinomial Probit Model, *Journal of Econometrics*, **64**, 207–240.

McCulloch, R. and Rossi, P. (2000) Bayesian Analysis of the Multinomial Probit Model, Mariano, M., Schuermann T. and Weeks, M., (eds.), in *Simulation Based Inference in Econometrics*. Cambridge: Cambridge University Press.

McCulloch, R. and Tsay, R. (1993) Bayesian Inference and Prediction for Mean and Variance Shifts in Autoregressive Time Series, *Journal of the American Statistical Association*, **88**, 968–978.

McCulloch, R. and Tsay, R. (1994) Statistical Analysis of Economic Time Series via Markov Switching Models, *Journal of Times Series Analysis*, **15**, 523–539.

Madigan, D. and York, J. (1995) Bayesian Graphical Models for Discrete Data, *International Statistical Review*, **63**, 215–232.

Martin, G. (2000) US Deficit Sustainability: A New Approach Based on Multiple Endogenous Breaks, *Journal of Applied Econometrics*, **15**, 83–105.

Meng, X. and van Dyk, D. (1999) Seeking Efficient Data Augmentation Schemes via Conditional and Marginal Augmentation, *Biometrika*, **86**, 301–320.

Min, C. and Zellner, A. (1993) Bayesian and Non-Bayesian Methods for Combining Models and Forecasts with Applications to Forecasting International Growth Rates, *Journal of Econometrics*, **56**, 89–118.

Müller, P. and Vidakovic, V. (1998) Bayesian Inference with Wavelets: Density Estimation, *Journal of Computational and Graphical Statistics*, **7**, 456–468.

Nobile, A. (2000) Comment: Bayesian Multinomial Probit Models with a Normalization Constraint, *Journal of Econometrics*, **99**, 335–345.

Otrok, C. and Whiteman, C. (1998) Bayesian Leader Indicators: Measuring and Predicting Economic Conditions in Iowa, *International Economic Review*, **39**, 997–1014.

Paap, R. and Franses, P. (2000) A Dynamic Multinomial Probit Model for Brand Choice with Different Long-run and Short-run Effects of Marketing Mix Variables, *Journal of Applied Econometrics*, **15**, 717–744.

Pagan, A. and Ullah, A. (1999) *Nonparametric Econometrics*. Cambridge: Cambridge University Press.

Pastor, L. (2000) Portfolio Selection and Asset Pricing Models, *Journal of Finance*, **55**, 179–223.

Phillips, D. and Smith, A.F.M. (1996) Bayesian Model Comparison via Jump Diffusions, in Gilks, Richardson and Speigelhalter (1996).

Phillips, P.C.B. (1991) To Criticize the Critics: An Objective Bayesian Analysis of Stochastic Trends (with discussion), *Journal of Applied Econometrics*, **6**, 333–474.

Poirier, D. (1991) A Bayesian View of Nominal Money and Real Output through a New Classical Macroeconomic Window (with discussion), *Journal of Business & Economic Statistics*, **9**, 125–148.

Poirier, D. (1994) Jeffreys' Prior for Logit Models, *Journal of Econometrics*, **63**, 327–339.

Poirier, D. (1995) *Intermediate Statistics and Econometrics: A Comparative Approach*. Cambridge: The MIT Press.

Poirier, D. (1996) A Bayesian Analysis of Nested Logit Models, *Journal of Econometrics*, **75**, 163–181.

Poirier, D. (1996a) Prior Beliefs about Fit, in Bernardo, J.M., Berger, J.O., Dawid, A.P. and Smith, A.F.M. (eds.), *Bayesian Statistics 5*, pp. 731–738. Oxford: Oxford University Press.

Poirier, D. (1998) Revising Beliefs in Non-identified Models, *Econometric Theory*, **14**, 483–509.

Poirier, D. and Tobias, J. (2002) On the Predictive Distributions of Outcome Gains in the Presence of an Unidentified Parameter, *Journal of Business and Economic Statistics*, forthcoming.

Press, S. J. (1989) *Bayesian Statistics: Principles, Models and Applications*. New York: Wiley.

Raftery, A. (1996) Hypothesis Testing and Model Selection, in Gilks, Richardson and Speigelhalter (1996).

Raftery, A. and Lewis, S. (1996) Implementing MCMC, in Gilks, Richardson and Speigelhalter (1996).

Raftery, A., Madigan, D. and Hoeting, J. (1997) Bayesian Model Averaging for Linear Regression Models, *Journal of the American Statistical Association*, **92**, 179–191.

Richard, J.-F. and Steel, M.F.J. (1988) Bayesian Analysis of Systems of Seemingly Unrelated Regression Equations under a Recursive Extended Natural Conjugate Prior Density, *Journal of Econometrics*, **38**, 7–37.

Richard, J.-F. and Zhang, W. (2000) Accelerated Monte Carlo Integration: An Application to Dynamic Latent Variable Models, in Mariano, M., Schuermann T. and Weeks, M. (eds.), *Simulation Based Inference in Econometrics*. Cambridge: Cambridge University Press.

Ritter, C. and Tanner, M. (1992) Facilitating the Gibbs Sampler: The Gibbs Stopper and the Griddy–Gibbs Sampler, *Journal of the American Statistical Association*, **48**, 276–279.

Robert, C. (1996) Mixtures of Distributions: Inference and Estimation, in Gilks, Richardson and Speigelhalter (1996).

Rossi, P., McCulloch, R. and Allenby, G. (1996) The Value of Purchase History Data in Target Marketing, *Marketing Science*, **15**, 321–340.

Rubin, D. (1978) Bayesian Inference for Causal Effects, *Annals of Statistics*, **6**, 34–58.

Ruggiero, M. (1994) Bayesian Semiparametric Estimation of Proportional Hazards Models, *Journal of Econometrics*, **62**, 277–300.

Sala-i-Martin, X. (1997) I Just Ran Two Million Regressions, *American Economic Review*, **87**, 178–183.

Sareen, S. (1999) Posterior Odds Comparison of a Symmetric Low-Price, Sealed-Bid Auction within the Common-Value and the Independent-Private-Values Paradigms, *Journal of Applied Econometrics*, **14**, 651–676.

Schwarz, G. (1978) Estimating the Dimension of a Model, *Annals of Statistics*, **6**, 461–464.
Schotman, P. (1994) Priors for the AR(1) Model: Parameterisation Issues and Time Series Considerations, *Econometric Theory*, **10**, 579–595.
Schotman, P. and van Dijk, H. (1991) A Bayesian Analysis of the Unit Root Hypothesis in Real Exchange Rates, *Journal of Econometrics*, **49**, 195–238.
Shively, T. and Kohn, R. (1997) A Bayesian Approach to Model Selection in Stochastic Coefficient Regression Models and Structural Time Series Models, *Journal of Econometrics*, **76**, 39–52.
Silverman, B. (1985) Some Aspects of the Spline Smoothing Approach to Nonparametric Regression Curve Fitting (with discussion), *Journal of the Royal Statistical Society, Series B*, **47**, 1–52.
Sims, C. (1980) Macroeconomics and Reality, *Econometrica*, **48**, 1–48.
Sims, C. (1988) Bayesian Skepticism on Unit Root Econometrics, *Journal of Economic Dynamics and Control*, **12**, 463–474.
Sims, C. and Uhlig, H. (1991) Understanding Unit Rooters: A Helicopter Tour, *Econometrica*, **59**, 1591–1600.
Sims, C. and Zha, T. (1999) Error Bands for Impulse Responses, *Econometrica*, **67**, 1113–1155.
Sims, C. and Zha, T. (2002) Macroeconomic Switching, Princeton University, Department of Economics working paper available at http://www.princeton.edu/~sims/.
Smith, M. and Kohn, R. (1996) Nonparametric Regression using Bayesian Variable Selection, *Journal of Econometrics*, **75**, 317–343.
Steel, M.F.J. (1998) Posterior Analysis of Stochastic Volatility Models with Flexible Tails, *Econometric Reviews*, **17**, 109–143.
Strachan, R. (2002) Valid Bayesian Estimation of the Cointegrating Error Correction Model, *Journal of Business and Economic Statistics*, forthcoming.
Tierney, L. (1996) Introduction to General State Space Markov Chain Theory, in Gilks, Richardson and Speigelhalter (1996).
Tierney, L. and Kadane, J. (1986) Accurate Approximations for Posterior Moments and Marginal Densities, *Journal of the American Statistical Association*, **81**, 82–86.
Tierney, L., Kass, R. and Kadane, J. (1989) Fully Exponential Laplace Approximations to Expectations and Variances of Nonpositive Functions, *Journal of the American Statistical Association*, **84**, 710–716.
Tobias, J. and Zellner, A. (2001) Further Results on Bayesian Method of Moments Analysis of the Multiple Regression Model, *International Economic Review*, **42**, 121–139.
Tsionas, E. (2000) Full Likelihood Inference in Normal-Gamma Stochastic Frontier Models, *Journal of Productivity Analysis*, **13**, 183–205.
van Dyk, D. and Meng, X. (2001) The Art of Data Augmentation (with discussion), *Journal of Computational and Graphical Statistics*, **10**, 1–111.
Verdinelli, I. and Wasserman, L. (1995) Computing Bayes Factors Using a Generalization of the Savage–Dickey Density Ratio, *Journal of the American Statistical Association*, **90**, 614–618.
Volinsky, C. and Raftery, A. (2000) Bayesian information criterion for censored survival models, *Biometrics*, **56**, 256–262.
Wahba, G. (1983) Bayesian Confidence Intervals for the Cross-validated Smoothing Spline, *Journal of the Royal Statistical Society, Series B*, **45**, 133–150.
West, M. and Harrison, P. (1997) *Bayesian Forecasting and Dynamic Models*, second edition. Berlin: Springer.
West, M., Muller, P. and Escobar, M. (1994) Hierarchical Priors and Mixture Models, in Freeman, P. and Smith, A.F.M. (eds.) *Aspects of Uncertainty*. New York: John Wiley & Sons.
White, H. (1984) *Asymptotic Theory for Econometricians*. New York: Academic Press.

Wonnacott, T. and Wonnacott, R. (1990) *Introductory Statistics for Business and Economics*, fourth edition. New York: John Wiley & Sons.

Zeger, S. and Karim, M. (1991) Generalized Linear Models with Random Effects: A Gibbs Sampling Approach, *Journal of the American Statistical Association*, **86**, 79–86.

Zellner, A. (1971) *An Introduction to Bayesian Inference in Econometrics*. New York: John Wiley & Sons.

Zellner, A. (1976) Bayesian and Non-Bayesian Analysis of the Regression Model with Multivariate Student-t Errors, *Journal of the American Statistical Association*, **71**, 400–405.

Zellner, A. (1985) Bayesian Econometrics, *Econometrica*, **53**, 253–269.

Zellner, A. (1986) On Assessing Prior Distributions and Bayesian Regression Analysis with g-Prior Distributions, in Goel, P. K. and Zellner, A. (eds.), *Bayesian Inference and Decision Techniques: Essays in Honour of Bruno de Finetti*. Amsterdam: North-Holland.

Zellner, A. (1986a) Bayesian Estimation and Prediction Using Asymmetric Loss Functions, *Journal of the American Statistical Association*, **81**, 446–451.

Zellner, A. (1988) Bayesian Analysis in Econometrics, *Journal of Econometrics*, **37**, 27–50.

Zellner, A. (1996) Bayesian Method of Moments/Instrumental Variable (BMOM/IV) Analysis of Mean and Regression Problems, in Lee, J., Zellner, A. and Johnson, W. (eds.), *Modeling and Prediction: Honoring Seymour Geisser*. New York: Springer-Verlag.

Zellner, A. (1997) The Bayesian Method of Moments (BMOM): Theory and Applications, in Fomby, T. and Hill, R. (eds.), *Advances in Econometrics: Applying Maximum Entropy to Econometric Problems*.

Zellner, A. (1997a) *Bayesian Analysis in Econometrics and Statistics: The Zellner View and Papers*. Cheltenham: Edward Elgar.

Zellner, A. and Hong, C. (1989) Forecasting International Growth Rates Using Bayesian Shrinkage and Other Procedures, *Journal of Econometrics*, **40**, 183–202.

Zellner, A., Hong, C. and Min, C. (1991) Forecasting Turning Points in International Output Growth Rates Using Bayesian Exponentially Weighted Autoregression, Time-Varying Parameter and Pooling Techniques, *Journal of Econometrics*, **49**, 275–304.

Zellner, A. and Min, C. (1995) Gibbs Sampler Convergence Criteria, *Journal of the American Statistical Association*, **90**, 921–927.

Zellner, A. and Rossi. P. (1984) Bayesian Analysis of Dichotomous Quantal Response Models, *Journal of Econometrics*, **25**, 365–393.

Index

Note: Figures and Tables are indicated by *italic page numbers*, footnotes by suffix 'n' (e.g. 'ARMA models 181n(1)' means footnote (1), page 181).

9 780470 845677